John Mearns
65

UNIFIED ALGEBRA AND TRIGONOMETRY

ADDISON-WESLEY MATHEMATICS SERIES

ERIC REISSNER, *Consulting Editor*

Apostol—MATHEMATICAL ANALYSIS, A MODERN APPROACH TO ADVANCED CALCULUS
Bardell and Spitzbart—COLLEGE ALGEBRA
Bardell and Spitzbart—INTERMEDIATE ALGEBRA
Dadourian—PLANE TRIGONOMETRY
Daus and Whyburn—INTRODUCTION TO MATHEMATICAL ANALYSIS
Davis—MODERN COLLEGE GEOMETRY
Davis—THE TEACHING OF MATHEMATICS
Fuller—ANALYTIC GEOMETRY
Gnedenko and Kolmogorov—LIMIT DISTRIBUTIONS FOR SUMS OF INDEPENDENT RANDOM VARIABLES
Kaplan—ADVANCED CALCULUS
Kaplan—A FIRST COURSE IN FUNCTIONS OF A COMPLEX VARIABLE
LeVeque—TOPICS IN NUMBER THEORY, VOLS. I AND II
Martin and Reissner—ELEMENTARY DIFFERENTIAL EQUATIONS
May—ELEMENTS OF MODERN MATHEMATICS
Meserve—FUNDAMENTAL CONCEPTS OF ALGEBRA
Meserve—FUNDAMENTAL CONCEPTS OF GEOMETRY
Munroe—INTRODUCTION TO MEASURE AND INTEGRATION
Perlis—THEORY OF MATRICES
Richmond—INTRODUCTORY CALCULUS
Spitzbart and Bardell—COLLEGE ALGEBRA AND PLANE TRIGONOMETRY
Spitzbart and Bardell—PLANE TRIGONOMETRY
Springer—INTRODUCTION TO RIEMANN SURFACES
Stabler—AN INTRODUCTION TO MATHEMATICAL THOUGHT
Struik—DIFFERENTIAL GEOMETRY
Struik—ELEMENTARY ANALYTIC AND PROJECTIVE GEOMETRY
Thomas—CALCULUS
Thomas—CALCULUS AND ANALYTIC GEOMETRY
Vance—TRIGONOMETRY
Vance—UNIFIED ALGEBRA AND TRIGONOMETRY
Wade—THE ALGEBRA OF VECTORS AND MATRICES
Wilkes, Wheeler, and Gill—THE PREPARATION OF PROGRAMS FOR AN ELECTRONIC DIGITAL COMPUTER

UNIFIED ALGEBRA AND TRIGONOMETRY

by

ELBRIDGE P. VANCE
Oberlin College

ADDISON-WESLEY PUBLISHING COMPANY, INC.
READING, MASSACHUSETTS, U.S.A.
LONDON, ENGLAND

Copyright © *1955*
ADDISON-WESLEY PUBLISHING COMPANY, Inc.

Printed in the United States of America

ALL RIGHTS RESERVED. THIS BOOK, OR PARTS THERE-
OF, MAY NOT BE REPRODUCED IN ANY FORM WITHOUT
WRITTEN PERMISSION OF THE PUBLISHERS.

Library of Congress Catalog No. 55–6559

Third printing — October 1959

PREFACE

The entire approach to the study of mathematics has constantly been changing in recent years. The general direction of these changes has been toward a unification or integration of the subjects previously considered separately. Such a unification seems most efficient, for, in this way, the student acquires a real feeling for mathematical thought and processes common to all mathematical subjects, in addition to the true integration of the specific subjects considered. Since algebra and trigonometry together form the foundation for analytic geometry and the calculus, and in fact, for all more advanced mathematics, and since each supplements the other in many ways, these two subjects may efficiently be combined. It was with this in mind that the author wrote the present book.

The book is designed to serve two classes of students: (1) those students who wish to elect a small amount of mathematics and to obtain an insight into mathematics as it is applied in the world today, and (2) those who wish to secure a strong foundation for further study in mathematics and other physical sciences.

The choice of topics, with the integration of algebra and trigonometry where it seems natural, the proportionate time spent on each, and the logical treatment throughout are intended to emphasize the modern point of view. The problems which appear at the end of each article also help to unify the subjects. A brief discussion of the different number systems and the one-dimensional coordinate system is basic for the introduction of the rectangular coordinate system, fundamental to the study of number systems as well as all mathematical functions. The definitions of the circular functions in terms of this coordinate system, a feature which distinguishes this text from many, is the basic unifying link between trigonometry and analytic geometry, and enables the author to employ many simpler, more direct methods. The analytic rather than the computational part of trigonometry has been emphasized, although the latter has been included. Both degree and radian measure of angles are introduced at the beginning, and are treated together thereafter.

Certain topics and approaches not usually found in either algebra or trigonometry books are included. The distinctions between identities and conditional equations, and between absolute and conditional inequalities are emphasized. A discussion of the circular functions of $\pi/5$ is included. The approach, by means of ordered pairs, to the study of complex numbers, the emphasis on graphing, the insight into some of the applications of the circular functions in the description of many periodic phenomena, and the explicit and detailed treatment of determinants are other distinguishing features.

It was the aim of the author to write a versatile text, which could be used for any purpose from a one-semester course in the bare essentials, to a two-semester course. For example, most of the first two chapters could be omitted by the more advanced students, and Chapters 7, 10, 15, and 17 may be omitted without loss of continuity.

Frequently problems are used to introduce important concepts and theorems to which later reference is made. Such problems, which are necessary for the continuity of the discussions, are marked with an asterisk, and should always be assigned. The answers for the odd-numbered problems are given in the Answer Section. Included at the end of the book are tables of powers and roots, four-place tables of the circular functions, logarithms, and the logarithms of the circular functions, as well as the American Experience Table of Mortality.

The author expresses his appreciation to his colleagues who have given advice and suggestions, and to the staff of Addison-Wesley Publishing Company for their valuable assistance in the preparation of this book. We are indebted to Rinehart & Company, publishers, for permission to use Tables I, II, and III, which were taken from *Fundamentals of College Mathematics* by Johnson, McCoy, and O'Neill.

E. P. VANCE

January, 1955.

CONTENTS

CHAPTER 1. NUMBERS AND ELEMENTARY OPERATIONS. 1
  1-1 Numbers 1
  1-2 Equality 3
  1-3 Fundamental operations and laws 4
  1-4 Addition of algebraic expressions 6
  1-5 Multiplication of algebraic expressions 9
  1-6 Division of algebraic expressions 11
  1-7 Special products 17
  1-8 Prime numbers, factors, and factoring 19

CHAPTER 2. FRACTIONS, EXPONENTS, AND RADICALS 24
  2-1 Simplification of fractions 24
  2-2 Addition of fractions 26
  2-3 Multiplication and division of fractions. 28
  2-4 Integral and zero exponents 31
  2-5 Rational exponents 32
  2-6 Radicals 34
  2-7 Addition and subtraction of radicals 37
  2-8 Multiplication and division of radicals 37

CHAPTER 3. COORDINATE SYSTEMS, FUNCTIONS, AND GRAPHICAL REPRESENTATION 41
  3-1 A one-dimensional coordinate system 41
  3-2 A two-dimensional coordinate system 44
  3-3 The distance formula 47
  3-4 The circle and arc length 50
  3-5 Angles 53
  3-6 Functions 59
  3-7 Graphical representation of functions 62
  3-8 Graphical representation of empirical data 64

CHAPTER 4. THE CIRCULAR FUNCTIONS. 67
  4-1 Trigonometry 67
  4-2 Definitions of the circular functions 68
  4-3 The unit circle 73
  4-4 Values of special angle functions 79
  4-5 Exact values of the functions for $\theta = \pi/5$ 82

CHAPTER 5. CIRCULAR FUNCTIONS INVOLVING MORE THAN ONE ANGLE 85
  5-1 Proof of the formula for $\cos(\alpha - \beta)$ 85
  5-2 Special reduction formulas 86
  5-3 General addition formulas 88
  5-4 General reduction formulas 92
  5-5 Function values of any angle 95

CONTENTS

CHAPTER 6. LINEAR AND QUADRATIC FUNCTIONS 100
  6-1 The linear function 100
  6-2 The quadratic function 104
  6-3 Solution of the quadratic equation 107
  6-4 Conditional inequalities 112
  6-5 Relation between zeros and coefficients of the quadratic function 117
  6-6 Equations in quadratic form 122
  6-7 Equations involving radicals 123
  6-8 Variation 125
  6-9 Solution of two linear equations in two unknowns 129
  6-10 Algebraic solution of three linear equations in three unknowns 134
  6-11 Solution of one linear and one quadratic equation 137

CHAPTER 7. DETERMINANTS 141
  7-1 Determinants of order two and three 141
  7-2 Determinants of order n 147
  7-3 Expansion of a determinant by minors 152
  7-4 Solution of a system of linear equations by determinants . . . 157

CHAPTER 8. FUNCTIONS AND EQUATIONS OF HIGHER DEGREE . . . 160
  8-1 Certain theorems 160
  8-2 Graphing of polynomial functions 165
  8-3 General remarks concerning zeros and roots 167
  8-4 Rational roots 168
  8-5 Irrational roots 171

CHAPTER 9. INVERSE FUNCTIONS 174
  9-1 Inverse functions 174
  9-2 Inverse circular functions 177
  9-3 Principal values 180
  9-4 Operations involving inverse circular functions 182

CHAPTER 10. CIRCULAR FUNCTION GRAPHS WITH APPLICATIONS . . 185
  10-1 Graphs of the curves $y = a \sin kx$ 185
  10-2 Graphs of the curves $y = a \sin (kx + b)$ 187
  10-3 Graphing by addition of ordinates 189
  10-4 Simple harmonic motion 190
  10-5 Addition of two general sine functions 193
  10-6 Harmonic analysis and synthesis 197

CHAPTER 11. MATHEMATICAL INDUCTION AND THE BINOMIAL THEOREM 200
  11-1 Mathematical induction 200
  11-2 The binomial theorem 203
  11-3 The expansion of $(1 + x)^n$ 208

CHAPTER 12. EXPONENTIAL AND LOGARITHMIC FUNCTIONS 211
  12-1 The exponential function $y = a^x$ 211
  12-2 The logarithmic function 213

12-3 Common logarithms	217
12-4 Computation by the use of logarithms	221
12-5 Compound interest and its generalization	223
12-6 Applications of the exponential functions	227

CHAPTER 13. SOLUTION OF TRIANGLES 231

13-1 General discussion	231
13-2 The Law of Sines	231
13-3 Solution of right triangles	232
13-4 The Law of Cosines	240
13-5 Applications involving oblique triangles	240

CHAPTER 14. IDENTITIES AND RELATED SUBJECTS 249

14-1 The fundamental circular function identities	249
14-2 General identities	253
14-3 Conversions of sums and products	260
14-4 Absolute inequalities	262
14-5 Approximations of functional values for small angles	264

CHAPTER 15. COMPLEX NUMBERS 268

15-1 Algebra of ordered pairs	268
15-2 Complex numbers	270
15-3 Complex roots of an equation	273
15-4 Graphical representation of complex numbers	274
15-5 Powers and roots of complex numbers	278

CHAPTER 16. PROGRESSIONS 282

16-1 Arithmetic progressions	282
16-2 Geometric progressions	287
16-3 Geometric progressions with infinitely many terms	290

CHAPTER 17. PERMUTATIONS, COMBINATIONS, AND PROBABILITY . . 295

17-1 The fundamental principle	295
17-2 Permutations	297
17-3 Combinations	301
17-4 Probability	303
17-5 Empirical probability	306
17-6 Probability of more than one event	308

ANSWERS TO PROBLEMS	312
TABLE I. Values of trigonometric functions	335
TABLE II. Logarithms of numbers	340
TABLE III. Logarithms of trigonometric functions	342
TABLE IV. Powers and roots	347
TABLE V. American Experience Table of Mortality	348
INDEX	349

CHAPTER 1

NUMBERS AND ELEMENTARY OPERATIONS

The basic quantitative procedures of science involve counting and measurement. Counting is the act that leads to the characterization of a collection of objects by a number. Measurement is the act that leads to the assignment of a number to a property of an object. Both counting and measurement are far from simple concepts, and both have been the subjects of many studies in the field of scientific methodology. The important thing for us in the present study is the fact that both counting and measurement lead to numbers, and through the use of numbers, it is possible to obtain much insight into the workings of nature. Although a thorough study of our number system is beyond the scope of this book, we shall begin with the classification of the various types of numbers and then, using them, we shall introduce certain basic operations and notions which are fundamental in the presentation of an analytic study of trigonometry, as well as algebra.

1-1 Numbers. The first numbers encountered by everyone are the counting or natural numbers $1, 2, 3, \cdots$. We refer to these as the set of *positive integers*. These numbers, together with the *negative integers*, $-1, -2, -3, \cdots$, and the zero integer 0, are called the *set of integers*. When any two integers are added, subtracted, or multiplied, the result is an integer. That is, if a and b are integers, $a + b$, $a - b$, and $a \times b$ are also integers. The zero integer 0 has the special properties that for any integer a, $a + 0 = a$, $a - 0 = a$, and $a \times 0 = 0$.

Next, in forming quotients of integers, the rational numbers are introduced. If a is an integer and b is an integer different from zero, then a/b is defined as the number c which, when multiplied by b, gives a. That is, c is defined by the equation $c \times b = a$. Fractions of the form $a/0$ are not defined. We therefore say that $a/0$ does not exist, and division by zero is not permitted. Any number which we can express as the quotient or ratio of two integers (excluding division by zero) is called a *rational number*, and the entire set of all such quotients is called the *set of rational numbers*. The num-

1

bers 5/3, −13/2, 3, and 1.414 are rational numbers. Any integer n, such as 3, is a rational number, since $n = n/1$. The number 1.414, which is an approximation for $\sqrt{2}$, is a rational number, since

$$1.414 = \frac{1414}{1000}.$$

Any rational number may be written as a decimal and, in fact, one way of identifying such a number is by its decimal expansion, which either terminates or is periodic.*

There are many numbers expressible as unending decimals which are not rational. It has been shown that the numbers $\sqrt{2}$, π (the ratio of the circumference of any circle to its diameter), and e (the base for logarithms used in calculus) cannot be expressed as the quotient of two integers and yet can be written as unending decimals. The entire set of numbers expressible as unending decimals is called the *set of real numbers*, and those in this set which are not rational form the *set of irrational numbers*.† Any irrational number, however, can be approximated by rational numbers. For example, the rational number approximation for $\sqrt{2}$, correct to three decimal places, is 1.414. The rational number 22/7 is a common approximation for π, while a more accurate one is 3.1416. The sum, difference, product, or quotient (division by other than zero) of any two real numbers is a real number. The set of numbers so far discussed constitutes the set of all real numbers and may be classified as follows:

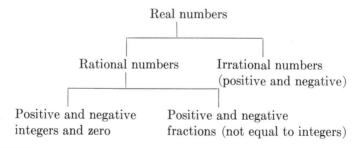

---

* Periodic, in this sense, means that the digits repeat either from the beginning (.161616···) or after a certain stage (3.2454545···).

† For a careful consideration of irrational numbers, see Garrett Birkhoff and Saunders MacLane, *A Survey of Modern Algebra*, Revised Edition. The Macmillan Company, 1953.

Although we shall deal primarily with real numbers in this book, there are still other types of numbers, which result from generalizations or extensions of real numbers. The *set of complex numbers* which includes imaginary numbers is one such type and will be discussed in Chapter 15. Throughout this book, unless otherwise stated, all numbers will be considered as belonging to the set of real numbers.

1–2 Equality. There are two main types of equations in mathematics. These two, the identity and the conditional equation, have many of the same properties but differ in meaning. In this book, we shall differentiate between the two, as they have an important place and use in algebra and trigonometry.

An *identity* is an equation that holds true for all permissible* values of the letters involved. Each such equation will be denoted by the symbol $\equiv$, in order to emphasize its nature. A simple example or two will help to clarify this concept.

ILLUSTRATION 1. $x + 2 + 5 - 4 - x \equiv 3$ holds for all values of x.

ILLUSTRATION 2. $x^2 - 4 \equiv (x + 2)(x - 2)$, ($x^2$ representing x times x) holds for all values of x. This would seem true by substituting any value for x in both sides of the identity, and is proved in Article 1–5.

The second type of equation, the *conditional equation*, does *not* hold true for all permissible values of the letters involved. One is usually required to solve such an equation, or to find all possible values (the solution) for which the equation is true. We shall discuss solving equations at the appropriate time, but consider the following examples.

ILLUSTRATION 1. $x + 2 = 4$ is a conditional equation for it is only true for $x = 2$.

ILLUSTRATION 2. $x^2 - 7x + 12 = 0$ is a conditional equation for it is only true for $x = 3$ or 4. Any other value substituted in this equation for x will not satisfy the condition.

---

* The permissible values are all those values for which each side of the equation has meaning.

The notation $\equiv$ for an identity, and $=$ for a conditional equation, will emphasize the distinction between the two types of equalities.

1–3 Fundamental operations and laws. As in arithmetic, there is a definite need for rules or laws in algebra. Before considering these, however, we must introduce the usual symbols for grouping two or more numbers together as a single quantity. Parentheses, ( ), brackets, [ ], or braces, { }, will be used for this purpose.

Certain fundamental laws or assumptions upon which algebra is based are familiar, and are stated here. These laws hold for all *real numbers*.

The *commutative law* for addition. The sum of two numbers is the same in whatever order the numbers are added. That is,

$$a + b \equiv b + a. \tag{1-1}$$

The *associative law* for addition. The sum of three or more numbers is the same in whatever manner the numbers are grouped for adding. That is,

$$a + b + c \equiv (a + b) + c \equiv a + (b + c). \tag{1-2}$$

The *commutative law* for multiplication. The product of two numbers is the same in whatever order the numbers are multiplied. That is,

$$a \times b \equiv b \times a. \tag{1-3}$$

Here, a and b are called *factors* of $a \times b$.

The *associative law* for multiplication. The product of three or more numbers is the same in whatever manner the numbers are grouped. That is,

$$a \times b \times c \equiv a \times (b \times c) \equiv (a \times b) \times c. \tag{1-4}$$

The *distributive law*. The product of one number and the sum of two or more numbers is equal to the sum of the products obtained by multiplying each of the other numbers by the first number. That is,

$$a(b + c + d) \equiv ab + ac + ad. \tag{1-5}$$

It seems wise at this time to state the *laws of signs*, which are considered as the rules for addition, subtraction, and multiplication (and therefore division) of positive and negative real numbers. If a and b denote two positive numbers, for addition we must have*

$$a + (-b) \equiv a - b, \quad -a + (+b) \equiv -a + b,$$
$$-a + (-b) \equiv -a - b. \quad (1\text{-}6)$$

From these equations, we have:

1. To add two numbers with like signs, add the numerical values and keep the common sign.

ILLUSTRATION 1. $2 + 3 = 5; -3 + (-4) = -7$.

2. To add two numbers with unlike signs, subtract the numerical values and prefix the sign of the larger number.

ILLUSTRATION 2. $-6 + 5 = -1; 4 + (-7) = -3$.

For subtraction, the defining equations are

$$a - (-b) \equiv a + b, \quad -a - (+b) \equiv -a - b,$$
$$-a - (-b) \equiv -a + b. \quad (1\text{-}7)$$

From these equations, we have:

3. To subtract one number from another, change the sign of the number to be taken away (subtrahend) and proceed as in addition.

ILLUSTRATION 3. $-8 - (3) = -11; 7 - (-4) = 11$.

For multiplication involving positive and negative numbers, our definitions are

$$a \times (-b) \equiv -ab, \quad -a \times (+b) \equiv -ab,$$
$$(-a) \times (-b) \equiv ab. \quad (1\text{-}8)$$

From these equations, we have:

4. To multiply (or divide) one number by another, multiply (or divide) their numerical values, and if the numbers have like signs, prefix the positive sign to the result; if the numbers have unlike signs, prefix the negative sign to the result.

ILLUSTRATION 4. $3 \times (-4) = -12; -3 \times (4) = -12; -3 \times (-4) = 12$.

---

* When no sign appears before the numbers, the $+$ sign is assumed.

Although these *laws of signs* are taken in this text as rules, in more advanced mathematics they *may be proved* from Eqs. (1–1,2,3,4,5) and certain other assumptions.

Problems

Find the sum, difference (first number minus the second), product and the quotient of the numbers in Problems 1–12.

1. 25 and 11
2. 7 and -5
3. -24 and 8
4. -12 and -8
5. 0 and -4
6. 6 and 0
7. $1/2$ and $1/3$
8. $4/7$ and $-2/3$
9. $-9/7$ and $-5/4$
10. $-9/8$ and $11/5$
11. 0 and $-5/4$
12. $-2/3$ and $2/3$

13. Using $(1-1,2,3,4,5)$, show that
    (a) $a(b + c) \equiv ba + ca$
    (b) $(b + c + d)a \equiv ab + ac + ad$
    (c) $(a + b)(c + d) \equiv ac + ad + bc + bd$

14. Show by a proper selection of numbers that (a) subtraction is not commutative, (b) division is not commutative.

15. Under what conditions will $-ab$ be (a) positive, (b) negative, (c) zero?

1–4 Addition of algebraic expressions. Any combination of ordinary and literal numbers by the fundamental operations of algebra is called an *algebraic expression*.

ILLUSTRATION 1. $5a + 6b$, $2ax^2$, and $(6x + 5y)2x$ are algebraic expressions.

Any algebraic expression consisting of distinct parts connected by plus or minus signs is called an *algebraic sum*. Each distinct part, together with its sign, is called a *term* of the expression.

ILLUSTRATION 2. In the algebraic sum $2x^2 - 3y^2 - 5$, $2x^2$ is one term, $-3y^2$ another, and -5 is a third term.

In a particular term consisting of two or more factors, any one of the factors or the product of any set of the factors may be called the *coefficient* of the product of the other factors. For example, in the term $2x^2y$, 2 is the coefficient of x^2y, $2x^2$ is the coefficient of y, or $2y$ is the coefficient of x^2. Frequently it is convenient to distinguish between the numerical coefficients and the other type. These coefficients are used in the process of determining the sum or difference of two algebraic expressions by combining similar terms.

ILLUSTRATION 3. The sum of $2x - 3y + 5$ and $x + 2y - 1$ is
$$(2x - 3y + 5) + (x + 2y - 1) \equiv 2x + x - 3y + 2y + 5 - 1$$
$$\equiv 3x - y + 4.$$

ILLUSTRATION 4. The difference between $2x - 3y + 5$ and $x + 2y - 1$ is
$$(2x - 3y + 5) - (x + 2y - 1) \equiv 2x - 3y + 5 - x - 2y + 1$$
$$\equiv x - 5y + 6.$$

Whether additions or subtractions are involved, we have stated that the algebraic expressions are called algebraic sums. If the expression consists of just one term, it is called a *monomial*. The algebraic sum or difference of two terms is called a *binomial* and in general, an algebraic expression consisting of a sum of any number of terms is called a *multinomial*. The expression $2x/3y^2$ is a monomial, while $3x^2 - 2y$ is a binomial.

EXAMPLE 1. Simplify the following expression by removing parentheses or brackets. Then combine similar terms.

Solution.
$$4x - [2x - 3y - (x + 4y)] + (x - 8) \equiv 4x - [2x - 3y - x - 4y] + x - 8$$
$$\equiv 4x - 2x + 3y + x + 4y + x - 8$$
$$\equiv 4x + 7y - 8.$$

The removal of parentheses (or any symbol of grouping) preceded by a minus sign requires changing the sign of each term within the parentheses, but parentheses preceded by a plus sign may be removed without changing the expression contained within the parentheses.

EXAMPLE 2. Find the value of $-2xy + 3x - 4y$ when $x = 3$ and $y = -2$.

Solution. Substitute 3 for x and -2 for y:
$$-2(3)(-2) + 3(3) - 4(-2) = 12 + 9 + 8 = 29.$$

Problems

In each of the following, (a) find the sum of the expressions and (b) subtract the second expression from the first.

1. $2a + 3b - 4$ and $a - 2b + 3$
2. $a - 2b + 3c$ and $2a + 4b - c$
3. $4x + 3y + z$ and $2x + 3y - 2z$
4. $2x + y + 5$ and $3y - 2z - 4$
5. $2(x - 3y)$ and $-5(2x + y)$
6. $-(x + 2y - z)$ and $3(x - y + 2z)$

In each of the following, remove all symbols of grouping and combine like terms.

7. $x - (2y + 3x) - 2y$
8. $3x - (2y - 4x) + 6y$
9. $(2x - 3y) - (8x + 6y + 4)$
10. $(2x - 3y) + (y - 4z) - (z - 3x)$
11. $8x + [(3x - 2y) + (6x - 9) - (x + y)]$
12. $3y - [2y + 3x - (2x - 3y)] + 4x$
13. $2x - \{3y - [5x - (7y - 6x)]\}$
14. $9x - (2y - 3x) - \{y - (2y - x)\} - [2y + (4x - 3y)]$

In each of the following expressions enclose the last three terms in parentheses preceded by a minus sign.

15. $a^2 - b^2 + 2bc - c^2$
16. $16 - x^2 + 2xy - y^2$
17. $4x^2 - 4y^2 - 4y - 1$
18. $9x^2 - 9y^2 - 6xy - x^2$

In each of the following expressions enclose the coefficients of x within parentheses preceded by (a) a plus sign, (b) a minus sign.

19. $cx + dy - ax - by$
20. $5x - ax + 3 - 4x$
21. $ax - by + bx - ay$
22. $6x^2 - 8x + 3y - 7y^2 + 6x - 4$

Find the value of each of the following expressions for the given values of the letters.

23. $3x + 4$ when $x = 1$
24. $x^2 - 7x + 10$ when $x = 2$
25. $3x^2 - 2x + 1$ when $x = -2$
26. $-3x^2 - 4x + 3$ when $x = -4$
27. $2x + 3y$ when $x = -2, y = 3$
28. $4x - 3y$ when $x = -1, y = 3$
29. $2x^2 - 3xy + y^2$ when $x = 2, y = -3$
30. $3x^2 + 2xy - 4y^2$ when $x = -3, y = 1$

1-5 Multiplication of algebraic expressions. When the factors of a product are equal the product is called a *power* of the repeated factor. We shall introduce a symbol to stand for such a product. The symbol a^2 represents the product $a \times a$. Similarly, $a \times a \times a$ may be represented by a^3 and, in general, we have:

If n is a positive integer, the symbol a^n, *the nth power of a*, is defined as the product of n factors each equal to a. Thus

$$a^n \equiv \underbrace{a \cdot a \cdot a \cdots a}_{n \text{ factors } a}. \tag{1-9}$$

In this symbol a^n, a is called the *base*, and n the *exponent* of the power.

ILLUSTRATION 1. $2^5 = 2 \cdot 2 \cdot 2 \cdot 2 \cdot 2 = 32.$*

From this definition, the following laws of exponents follow, where n and m are positive integers, and a and b are any real numbers.

$$a^n \cdot a^m \equiv \underbrace{a \cdot a \cdot a \cdots a}_{n \text{ factors } a} \underbrace{a \cdot a \cdot a \cdots a}_{m \text{ factors } a} \equiv \underbrace{a \cdot a \cdot a \cdots a}_{n+m \text{ factors } a} \equiv a^{n+m}. \tag{1-10}$$

The reader should also verify

$$(a^n)^m \equiv a^{nm} \tag{1-11}$$

and

$$(ab)^n \equiv a^n b^n. \tag{1-12}$$

ILLUSTRATION 2. $3^2 \cdot 3^4 = 3^6 = 729,$

$a^4 \cdot a^7 \equiv a^{11}.$

ILLUSTRATION 3. $(2^3)^2 = 2^6 = 64,$

$(a^5)^3 \equiv a^{15}.$

ILLUSTRATION 4. $(2 \cdot 3)^3 = 2^3 \cdot 3^3 = 8 \cdot 27 = 216.$

When no exponent is written, the exponent is understood to be 1.

---

* Since it is clear that equations of this type are identities and as such do not need to be emphasized, it is unnecessary to use the symbol $\equiv$.

We are now able, using these laws and recalling the distributive law, Eq. (1-5), to perform the multiplication of two algebraic expressions.

EXAMPLE 1. Multiply $(2x - 3y)$ by $(3x + 4y)$.

Solution. Each term of the second binomial must be multiplied by the first binomial. [See Problem 13(c), Article 1–3.]

$$(2x - 3y)(3x + 4y) \equiv (2x - 3y) \cdot 3x + (2x - 3y) \cdot 4y$$
$$\equiv 6x^2 - 9xy + 8xy - 12y^2,$$

and combining similar terms, we obtain

$$(2x - 3y)(3x + 4y) \equiv 6x^2 - xy - 12y^2.$$

In such problems it is sometimes convenient to arrange both multinomials in ascending (or descending) powers of one letter, write one below the other, carry out the multiplication, and add the products. Consider the following example.

EXAMPLE 2. Multiply $x^2 - 2y^2 + xy$ by $2x - y$.

Solution.
$$\begin{array}{l} x^2 + xy \quad\;\; - 2y^2 \\ 2x - y \\ \hline 2x^3 + 2x^2y - 4xy^2 \\ \quad\;\; - x^2y \;\; - xy^2 \;\; + 2y^3 \\ \hline 2x^3 + x^2y \;\; - 5xy^2 + 2y^3. \end{array}$$

PROBLEMS

Perform the indicated operations by using the laws of exponents.

1. $a^2 \cdot a^7$
2. $3x^4 \cdot 2x^3$
3. $3x^3 \cdot x^4 \cdot x^5$
4. $(-3)^4 \cdot (-3)^5$
5. $y^{13} \cdot y^{11}$
6. $(2^3)^4$
7. $(a^5)^4$
8. $(3b)^5$
9. $(5c)^3$
10. $(3a)^4$
11. $(2a^3)^5$
12. $(x^4)^n$
13. $(a^r \cdot a^s)^t$
14. $(x^{2m} \cdot x^{3n})^4$
15. $(2x^n)^n$

Perform the indicated multiplications and collect similar terms.

16. $(4x - 3)(3x + 6)$
17. $(x + 3)(2x - 5)$
18. $(5a + 2b)(5a - 2b)$
19. $(4x + 2y)(4x - 2y)$
20. $(3x - 2)(2x + 5)$
21. $(r^2 - s^2)(r - s)$
22. $(x - 2y)^2$
23. $(x^2 - xy + y^2)(x + y)$
24. $(2x + 3x^2 - 1)(3x - 2)$
25. $(x^2 - 2x - 2)(x + 2x^2 - 4)$
26. $(-xy + 2x^2 - 3y^2)(x^2 - 4y^2 + xy)$
27. $(x - 1)(x + 2)(x - 3)$
28. $(x - 2)(x + 3)(x + 2)$
29. $(x^4 + 2x^2y^2 + 4y^4)(x^2 - 2y^2)$
30. $(a^{2n} - 7a^n + 10)(a^n - 1)$
31. $(a^{2n} - 7a^n + 10)(a - 1)$
32. $(x - 2x^2 + 5 - x^3)(3x - 4 + x^2)$
33. $(x^{2n} + 2x^ny^n + y^{2n})(x^{2n} - 2x^ny^n + y^{2n})$
34. $(x^n - y^n)^3$
35. $(x - y)^4$

1-6 Division of algebraic expressions. Before we are able to divide one algebraic expression by another, we must establish an additional law of exponents. If $a \neq 0$ (a is not equal to zero) and n and m are positive integers,

$$\frac{a^n}{a^m} \equiv \underbrace{\frac{a \cdot a \cdot a \cdots a}{a \cdot a \cdot a \cdots a}}_{m \text{ factors } a}^{n \text{ factors } a}$$

$$\equiv a^{n-m} \quad \text{if } n \text{ is larger than } m \tag{1-13}$$

$$\equiv \frac{1}{a^{m-n}} \quad \text{if } m \text{ is larger than } n \tag{1-14}$$

$$\equiv 1 \quad \text{if } m = n. \tag{1-15}$$

ILLUSTRATION 1. $\dfrac{x^8}{x^3} \equiv x^5; \quad \dfrac{x^4}{x^{11}} \equiv \dfrac{1}{x^7}; \quad \dfrac{x^6}{x^6} \equiv 1.$

We are now prepared to divide any multinomial by a monomial. This is done by dividing each term of the multinomial by the monomial, and finding the algebraic sum of the resulting quotients.

EXAMPLE 1. Divide $12x^3y^4 + 18x^4y^2 - 36xy^3$ by $3x^2y^2$.
Solution.
$$\frac{12x^3y^4 + 18x^4y^2 - 36xy^3}{3x^2y^2} \equiv \frac{12x^3y^4}{3x^2y^2} + \frac{18x^4y^2}{3x^2y^2} - \frac{36xy^3}{3x^2y^2}$$
$$\equiv 4xy^2 + 6x^2 - \frac{12y}{x}.$$

We have defined (Article 1-4) an algebraic expression as the sum of algebraic terms. Any algebraic term is an *integral rational* term in certain literal numbers if it consists of the product of positive integral powers of these numbers multiplied by a factor not containing them (or if the term does not contain them). For example, ax^2y, $\sqrt{3}x^4y^3$, and $by^{3/4}$ are integral rational terms in x, and ax^2y and $\sqrt{3}x^4y^3$ are integral rational terms in y. Any multinomial in which each term is integral rational is called an *integral rational expression* or *polynomial*.

ILLUSTRATION 2. The multinomial $3 - 5x^3 + \frac{3}{5}xy^2$ is a polynomial in x and y. However, $-5/x^3$ is not integral in x, $5\sqrt{x}$ is not rational in x, and $5\sqrt{x} - 5/x^3$ is neither integral nor rational in x.

The degree of an *integral rational term* in some letter is defined to be the exponent of that letter. Thus ax^2y and $3x^4y^3$ are terms of the second and fourth degree in x, but first and third degree in y. The *degree of an integral rational term in two or more letters* is defined to be the sum of the exponents of those letters. Thus the degree of ax^2y in x and y is $2 + 1 = 3$, while the degree of $3x^4y^3$ in x and y is 7. The *degree of a polynomial* in certain letters is that of its term (or terms) of highest degree in those letters.

ILLUSTRATION 3. (a) $2x + 1$, $x^2 - 7x + 10$, and $3x^5 + 5x^2 - 6x$ are polynomials of the first, second and third degree in x, respectively. (b) $3x^2y^5 + 4xy^3 - 7x^3y^2$ is a polynomial of the third degree in x, fifth degree in y, and seventh degree in x and y.

Although the process for any division of algebraic expressions might be considered, we are primarily interested in dividing one polynomial by another. Our method will follow that used in arithmetic. If we wish to divide 28 by 9, the quotient is 3 and the remainder 1, and we write
$$\tfrac{28}{9} = 3 + \tfrac{1}{9}.$$

To divide one polynomial by another:
1. Arrange each polynomial in descending powers of some common letter.
2. Divide the first term of the dividend (the polynomial to be divided) by the first term of the divisor, giving the first term of the quotient.
3. Multiply the divisor by this first term of the quotient, and subtract this product from the dividend.
4. Using the remainder thus obtained as a new dividend, repeat this process, thus finding the second term of the quotient.
5. Continue this process until a remainder is obtained which is either zero or is of lower degree in the common letter than the divisor. If the remainder is zero, the division is exact.

The work should be arranged as shown in the example.

EXAMPLE 2. Divide $3x^3 - 4x^2y + 5xy^2 + 6y^3$ by $x^2 - 2xy + 3y^2$.

Solution.

(Dividend) $\quad 3x^3 - 4x^2y + 5xy^2 + 6y^3 \quad | \underline{x^2 - 2xy + 3y^2}$ (Divisor)
$\quad\quad\quad\quad\quad\;\, 3x^3 - 6x^2y + 9xy^2 \quad\quad\;\; | \overline{3x + 2y}$ (Quotient)

$\quad\quad\quad\quad\quad\quad\quad\quad 2x^2y - 4xy^2 + 6y^3$
$\quad\quad\quad\quad\quad\quad\quad\quad \underline{2x^2y - 4xy^2 + 6y^3}$
$\quad\quad\quad\quad\quad\quad\quad\quad\quad\quad\quad\quad\quad\quad 0 \quad$ (Remainder)

This result may be expressed

$$\frac{3x^3 - 4x^2y + 5xy^2 + 6y^3}{x^2 - 2xy + 3y^2} \equiv 3x + 2y.$$

EXAMPLE 3. Divide $5x^3 - 14x + 3$ by $x - 2$.

Solution. Since no x^2 term appears in the given dividend, we must supply it with a zero coefficient, $0x^2$

$\quad\quad\quad 5x^3 + 0x^2 - 14x + 3 \quad | \underline{x - 2}$
$\quad\quad\quad \underline{5x^3 - 10x^2} \quad\quad\quad\quad\quad\;\; | \overline{5x^2 + 10x + 6}$

$\quad\quad\quad\quad\quad\quad\; 10x^2 - 14x + 3$
$\quad\quad\quad\quad\quad\quad\; \underline{10x^2 - 20x}$

$\quad\quad\quad\quad\quad\quad\quad\quad\quad\quad 6x + 3$
$\quad\quad\quad\quad\quad\quad\quad\quad\quad\quad \underline{6x - 12}$
$\quad\quad\quad\quad\quad\quad\quad\quad\quad\quad\quad\quad\; 15$

This results in

$$\frac{5x^3 - 14x + 3}{x - 2} \equiv 5x^2 + 10x + 6 + \frac{15}{x - 2}.$$

A notation frequently used for a polynomial in x is the symbol $P(x)$, read "P of x" (this *does not* mean P times x); likewise, a polynomial in x and y might be denoted $Q(x,y)$. With this type of notation, if $P(x)$ denotes the dividend, $D(x)$ the divisor, $Q(x)$ the quotient, and $R(x)$ the remainder of a division of two polynomials in x, our result could be stated

$$\frac{P(x)}{D(x)} \equiv Q(x) + \frac{R(x)}{D(x)}, \tag{1-16}$$

or

$$P(x) \equiv Q(x) \cdot D(x) + R(x). \tag{1-17}$$

Similarly, for division of two polynomials in x and y,

$$P(x,y) \equiv Q(x,y) \cdot D(x,y) + R(x,y). \tag{1-18}$$

ILLUSTRATION 4. In Example 2 of this Article,

$P(x,y) = 3x^3 - 4x^2y + 5xy^2 + 6y^3,$ $D(x,y) = x^2 - 2xy + 3y^2,$
$Q(x,y) = 3x + 2y,$ $R(x,y) = 0.$

In Example 3,

$P(x) = 5x^3 - 14x + 3,$ $D(x) = x - 2,$
$Q(x) = 5x^2 + 10x + 6,$ $R(x) = 15.$

PROBLEMS

Divide:
1. $9xy^2 - 6x^3$ by $3x$
2. $4x^2y^3 - 24xy^4$ by $2xy^2$
3. $6x^3 - 9x^4y$ by $3xy$
4. $-15x^2y^3 + 20x^3y^2$ by $-5x^2y^2$
5. $3x^2y - 4xy^2 + 6x^3y^3$ by xy
6. $7x^3y^2 - 14x^5y^3 + 28x^8y^5 - 21x^7y^6$ by $7x^3y^2$

Divide, finding the quotient and remainder, and write out the result of each problem in the form of Eqs. (1-17) or (1-18):

7. $x^2 - 7x + 10$ by $x - 5$ 8. $2y^2 - 5y - 6$ by $2y - 1$
9. $3x^2 - 13x + 4$ by $x - 4$ 10. $2x^2 - 5x - 12$ by $2x + 3$
11. $2x^3 - 7x^2 + 11x - 4$ by $2x - 1$
12. $y^3 - 4y^2 - 2 + 5y$ by $y - 1$
13. $x^2y - 6x^3 - 12xy^2 - 6y^3$ by $2x - 3y$
14. $2x^3 - 11x^2y + 13xy^2 - 4y^3$ by $x - 4y$
15. $4x^3 + 5 + 4x^2 - 13x$ by $2x + 5$
16. $4x^3 + 5x - 6$ by $2x - 3$
17. $5x^3 - 2x^2 + 3x - 4$ by $x^2 - 2x + 1$
18. $4x^2 + x - 6x^3 + 3$ by $3x^2 + 5$
19. $x^6 - y^6$ by $x - y$
20. $x^7 - y^7$ by $x - y$

The process of division for polynomials in x (or any one letter), when the divisor is in the form $x - a$, may be greatly simplified. This process, known as *synthetic division*, will be illustrated by using the problem of Example 3. If the x's are omitted, writing only the coefficients, we would have

$$\begin{array}{rrrr|r}
5 + 0 & - 14 & 3 & & 1 - 2 \\
5 - 10 & & & & \overline{5 + 10 + 6} \\
\cline{1-2}
10 - 14 & & & & \\
10 - 20 & & & & \\
\cline{1-2}
& 6 & 3 & & \\
& 6 & - 12 & & \\
\cline{2-3}
& & 15 & &
\end{array}$$

We next omit those coefficients that are definite repetitions; the first term in the 2nd, 4th, 6th, $\cdots$ lines and the second term in the 3rd, 5th, 7th, $\cdots$ lines. Compressing the remaining terms, writing the first coefficient, 5, in the third line, and noting that the 1 of the divisor can be omitted, we have

$$\begin{array}{rrrr|r}
5 & +0 & -14 & 3 & -2 \\
  & -10 & -20 & -12 & \\
\hline
5 & 10 & 6 & 15 &
\end{array}$$

The coefficients of the quotient are also omitted, for they appear as the first three coefficients on the third line, while the remainder, 15, appears as the last number.

The final step of simplification is to replace the subtractions by additions, that is, to change the signs in the divisor (-2 to 2) and in the second line. Thus

$$\begin{array}{rrrr|r} 5 & +0 & -14 & 3 & \,2 \\ & 10 & 20 & 12 & \\ \hline 5 & 10 & 6 & 15 & \end{array}$$

This last arrangement represents the dividing of $5x^3 - 14x + 3$ by $x - 2$ by synthetic division to obtain the quotient $5x^2 + 10x + 6$ and the remainder 15.

Let us summarize this process of synthetic division. To divide a polynomial $P(x)$ by a binomial $x - a$, arrange on a line (in order of descending power) the coefficients of $P(x)$, inserting zero for the coefficient of any missing power of x, and write a on the right. Bring down the first coefficient of $P(x)$ in the first position on the third line. Multiply this first coefficient by a, writing the product in the second line under the second coefficient of $P(x)$. The sum of this product and second coefficient is placed in the third line. Multiply this sum by a, add this product to the next coefficient of $P(x)$, again writing this new sum on the third line, and so on, until a product has been added to the final coefficient of $P(x)$.

This last sum in the third line represents the remainder. The preceding numbers represent the coefficients, arranged in descending order, of the powers of x in the quotient, a polynomial of degree one less than the degree of $P(x)$.

EXAMPLE 4. By synthetic division, find the quotient and the remainder of $2x^4 + 3x^3 - 4x^2 + 5x + 6$ divided by $x + 3$.

Solution. In this example $x - a = x - (-3)$, so that $a = -3$.

$$\begin{array}{rrrrr|r} 2 & +3 & -4 & +5 & +6 & \,-3 \\ & -6 & 9 & -15 & +30 & \\ \hline 2 & -3 & 5 & -10 & 36 & \end{array}$$

Thus the quotient is $2x^3 - 3x^2 + 5x - 10$, and the remainder is 36.

The process of synthetic division will be most useful in later work.

Problems

By synthetic division, find the quotient and the remainder in each of the following divisions.

1. $3x^2 - 2x - 4$ by $x - 3$
2. $2x^3 + 3x^2 - 7$ by $x + 1$
3. $x^3 - 2x^2 + 9$ by $x + 2$
4. $x^3 + 4x - 7$ by $x - 3$
5. $x^4 - 2x^3 - 3x^2 - 4x - 8$ by (a) $x - 2$; (b) $x + 1$
6. $2x^4 - x^3 - 18x^2 - 7$ by (a) $x + 3$; (b) $x - 3$
7. $3x^4 - 7x - 20$ by (a) $x - 2$; (b) $x + 2$
8. $2x^4 - 3x^3 - 20x^2 - 6$ by (a) $x - 4$; (b) $x + 3$

Use synthetic division to find the quotient and remainder in the following divisions and express the answer in the form of Eq. (1–17).

9. $x^3 - 2x^2 + 3x - 4$ by $x - 3$
10. $2x^3 + x^2 - x + 4$ by $x + 1$
11. $x^4 - 5x^3 + x^2 - 6$ by $x - 1$
12. $x^3 + 3x^2 - 2x - 5$ by $x + 2$

1–7 Special products. There are certain special products which occur so frequently in algebra that they have been classified. These are given below. The letters involved in the formulas may stand for any algebraic expression. The reader should not only verify each by actual multiplication, but also memorize them, recognizing both the product from the factors and the factors from the product.

$$a(x + y) \equiv ax + ay. \qquad (1\text{–}19)$$

$$(x + y)(x - y) \equiv x^2 - y^2. \qquad (1\text{–}20)$$

$$(x \pm y)^2 \equiv x^2 \pm 2xy + y^2.^* \qquad (1\text{–}21)$$

$$(x + a)(x + b) \equiv x^2 + (a + b)x + ab. \qquad (1\text{–}22)$$

$$(ax + b)(cx + d) \equiv acx^2 + (ad + bc)x + bd. \qquad (1\text{–}23)$$

$$(x \pm y)^3 \equiv x^3 \pm 3x^2y + 3xy^2 \pm y^3. \qquad (1\text{–}24)$$

$$(x \pm y)(x^2 \mp xy + y^2) \equiv x^3 \pm y^3. \qquad (1\text{–}25)$$

In the following illustrations the reader should verify which of the above formulas are used.

---

*The sign $\pm$ is read "plus or minus." If the upper (lower) sign is used in the left member, it is also used in the right.

ILLUSTRATION 1. $(2x^2 - 3y)(2x^2 + 3y) \equiv (2x^2)^2 - (3y)^2$
$\equiv 4x^4 - 9y^2.$

ILLUSTRATION 2. $(x + 2)(x + 5) \equiv x^2 + (2 + 5)x + 10$
$\equiv x^2 + 7x + 10.$

ILLUSTRATION 3. $(3x + 4y)(2x - 3y) \equiv 6x^2 + (-9 + 8)xy - 12y^2$
$\equiv 6x^2 - xy - 12y^2.$

ILLUSTRATION 4. $(x + y - 1)^3$
$\equiv [(x + y) - 1]^3$
$\equiv (x + y)^3 - 3(x + y)^2 + 3(x + y) - 1$
$\equiv x^3 + 3x^2y + 3xy^2 + y^3 - 3x^2 - 6xy - 3y^2 + 3x + 3y - 1.$

Here $(x + y)$ is considered first as one term.

ILLUSTRATION 5. $(3x + 2y)(9x^2 - 6xy + 4y^2)$
$\equiv (3x + 2y)[(3x)^2 - (3x)(2y) + (2y)^2]$
$\equiv (3x)^3 + (2y)^3$
$\equiv 27x^3 + 8y^3.$

PROBLEMS

Find the following products.

1. $2a(3x - 4y)$
2. $-3x(2x + 7y)$
3. $-7xy(3x^2 + 4y)$
4. $4x^2yz(z^2 + xy + yz)$
5. $(2x - 3y)(2x + 3y)$
6. $(7x + 5y^2)(7x - 5y^2)$
7. $(x + 2y)(x - 2y)(x^2 + 4y^2)$
8. $(x - 3)^2$
9. $(2x + 7y)^2$
10. $(3x^2y - 5z^2)^2$
11. $(x - 2)(x - 5)$
12. $(2x + 3)(x - 5)$
13. $(xy^2 - z^2w)^2$
14. $(\tfrac{1}{2}x + \tfrac{2}{3}y)^2$
15. $(4x - 3y)(7x + 3y)$
16. $[(x + 1) - z][(x + 1) + z]$
17. $(2x + 3y + 3)(2x + 3y - 3)$
18. $(2x + 3y + 4z)^2$
19. $(x - 2y - z)^2$
20. $(2a + b)^3$
21. $(x + 2)(x^2 - 2x + 4)$
22. $(x - 3)(x^2 + 3x + 9)$
23. $(x + 3y + 2z - 4w)(x + 3y - 2z + 4w)$
24. $(4x - 2y - 3z + 3w)(4x + 2y + 3z + 3w)$

25. $(a - b + c - d)^2$ 26. $(2a + 3b - c - 4d)^2$
27. $[2(x + 2y) - 3][2(x + 2y) + 4]$
28. $[2(x - 3y) + 5][3(x - 3y) - 2]$
29. $(2x + 3y)^3$ 30. $(5x - 3y)^3$

1–8 Prime numbers, factors, and factoring. Returning our discussion to the integers, all positive integers except one may be classified into two types, composite numbers and primes. A positive integer is called *composite* if it can be expressed as the product of two or more smaller positive integers, which are its *factors*. In some cases some of these factors may be equal.

For example, 4, 6, 9, and 12 are composite for $4 = 2 \times 2$, $6 = 3 \times 2$, $9 = 3 \times 3$, and $12 = 3 \times 2 \times 2$. In fact, every even integer greater than 2 is composite.

A positive integer is called a *prime* if it is different from one and is not a composite. In other words, it can be expressed only as a product of two positive integers in the trivial way in which one factor is itself and the other the integer one. Examples of prime numbers are 2, 3, 5, and 7.

The *decomposition* of any composite number, that is, the expressing of such a number as a product of prime numbers is most important. Such a decomposition is always possible, since each factor which is composite can be expressed as the product of smaller factors, and ultimately the factors will all be prime. Thus, $60 = 12 \times 5 = 4 \times 3 \times 5 = 2 \times 2 \times 3 \times 5$. Moreover, such a decomposition is unique, although the proof of this fact is too advanced for this book.* Specifically, a composite number can be expressed as a product of prime factors in one and only one way, except for the order of the factors. Thus, also $60 = 15 \times 4 = 5 \times 3 \times 4 = 5 \times 3 \times 2 \times 2$. Two integers are called *relatively prime* or *prime to each other* if they contain no common prime factors.

The process of factoring an algebraic expression is similar to that of finding the factors of a composite number. This process, which is *usually* restricted at this elementary stage to factoring polynomials with rational coefficients and to factors completely free from irrational

---
* For the proof of this statement as well as a general discussion of prime numbers, see R. Courant and H. Robbins, *What Is Mathematics?*, New York: Oxford University Press, 1941, and Harriet Griffin, *Elementary Theory of Numbers*, New York: McGraw-Hill Book Company, 1954.

numbers, is performed frequently by reversing the processes of the preceding article. Such a factorization is considered complete when each algebraic factor is considered a *prime factor*, that is, an algebraic expression which cannot be factored without violating the above restrictions.

The more common types of factoring are illustrated.

EXAMPLE 1. Factor $2ax^2 - 4ay^2 + 8a^2x$.

Solution. This problem represents a polynomial with a common monomial factor.

$$2ax^2 - 4ay^2 + 8a^2x \equiv 2a(x^2 - 2y^2 + 4ax).$$

EXAMPLE 2. Factor $x(a + 2b) - 3y(a + 2b)$.

Solution. Each of the two expressions has the common term $(a + 2b)$. Therefore,

$$x(a + 2b) - 3y(a + 2b) \equiv (a + 2b)(x - 3y).$$

EXAMPLE 3. Factor $\dfrac{4x^2}{y^2} - (9a - b)^2$.

Solution. This expression is the difference between two perfect squares.

$$\frac{4x^2}{y^2} - (9a - b)^2 \equiv \left(\frac{2x}{y}\right)^2 - (9a - b)^2$$

$$\equiv \left[\frac{2x}{y} + (9a - b)\right]\left[\frac{2x}{y} - (9a - b)\right]$$

$$\equiv \left(\frac{2x}{y} + 9a - b\right)\left(\frac{2x}{y} - 9a + b\right).$$

EXAMPLE 4. Factor $9x^2 - 30xy + 25y^2$.

Solution. This algebraic expression is a perfect square.

$$9x^2 - 30xy + 25y^2 \equiv (3x - 5y)^2.$$

EXAMPLE 5. Factor $27x^3 + \dfrac{8}{y^3}$.

Solution. This algebraic expression is the sum of two cubes.
$$27x^3 + \frac{8}{y^3} \equiv \left(3x + \frac{2}{y}\right)\left(9x^2 - \frac{6x}{y} + \frac{4}{y^2}\right).$$

EXAMPLE 6. Factor $12x^2 + 7xy - 10y^2$.

Solution. This trinomial in the form of Eq. (1–23) is factored by trial and error. The result will be in the form of $(ax + by)(cx + dy)$, where $ac = 12$, $bd = -10$, and $ad + bc = 7$. Also a and c are both positive and b and d are different in sign. The correct combination, we find, is
$$12x^2 + 7xy - 10y^2 \equiv (4x + 5y)(3x - 2y).$$

EXAMPLE 7. Factor $6x^4 + 7x^2y^2 - 3y^4$.

Solution. This is the same type as Example 6.
$$6x^4 + 7x^2y^2 - 3y^4 \equiv (3x^2 - y^2)(2x^2 + 3y^2).$$

Although the first term on the right is the difference of two squares, it cannot be factored further, for such factorization would introduce irrational quantities.

PROBLEMS

Factor the following completely.

1. $4x - 20$
2. $10x + 15yz$
3. $3y^2 - 9y$
4. $4x^3y^2 + 6x^2y^3$
5. $xy^2z^3 - 3x^2yz^2 + 5xy^3z^2$
6. $a^2b^3c^4 - a^3b^4c^5 + 2a^2b^4c^4$
7. $3y(2x + 5) - 4x(2x + 5)$
8. $3y(4 - y) + 6x^2(4 - y)$
9. $2z^2(x + 3y) - 6xz(x + 3y)$
10. $3x(3 - 2y) - 2xy(3 - 2y)$
11. $9 - a^2$
12. $16x^2 - 9y^2$
13. $225a^8 - 64b^2$
14. $\dfrac{c^6}{d^8} - 121$
15. $x^3y^4 - 25xd^6$
16. $.01x^4 - 196y^8$
17. $(x + 2y)^2 - z^2$
18. $(3x - 2y)^2 - 25z^2$
19. $(a + b)^2 - (c + d)^2$
20. $9(2x - y)^2 - 4(2a + b)^2$
21. $81(4x - 3y)^2 - 25(3z + w)^2$
22. $x^2 + 6x + 9 - (y^2 + 4y + 4)$
23. $x^2 - 8x + 16$
24. $4a^2 - 12ab + 9b^2$

25. $66xy + 9x^2y^2 + 121$
26. $2x^3 - 28x^2 + 98x$
27. $5z^2 - 30wz + 45w^2$
28. $x^{2n} + 2x^n y^n + y^{2n}$
29. $(3 - x)^2 + 8(3 - x) + 16$
30. $25 - 30(2x - 3y) + 9(2x - 3y)^2$
31. $a^3 - 8$
32. $1 + \dfrac{8}{x^9}$
33. $8x^{6n} + 27y^{3m}$
34. $x^3 - \dfrac{y^3}{64}$
35. $27(x - y)^3 - 8(x + y)^3$
36. $5(a - 2b)^3 - 625(a - 2b)^3$
37. $x^2 - 7x + 12$
38. $y^2 - 2y - 8$
39. $a^2b^2 - ab - 20$
40. $2x^2 + 8x + 6$
41. $35x^2 - 24x + 4$
42. $3y^2 - y - 10$
43. $6a^2 + 7a - 20$
44. $2x^2 - 23xy - 39y^2$
45. $(x + y)^2 - 7(x + y) + 10$
46. $(y + z)^2 + (y + z) - 42$
47. $2(2x + y)^2 - (2x + y) - 10$
48. $6(x + y)^2 + 5(x + y)(y + z) - 6(y + z)^2$
49. $12(a + b)^2 - 14(a + b)(c + d) - 10(c + d)^2$
50. $4(x - 2)^2 + 5(x - 2)(y + 4) - 21(y + 4)^2$

There are many algebraic expressions which, by proper grouping, can be put into one of the forms listed in the previous examples, and then factored. This process will be illustrated.

EXAMPLE 8. Factor $ax - ay - bx + by$.

Solution. By grouping the first two terms together, and the last two together, and factoring out the common term, we transform the expression into the form of Example 2.

$$ax - ay - bx + by \equiv a(x - y) - b(x - y)$$
$$\equiv (x - y)(a - b).$$

EXAMPLE 9. Factor $4x^3 - 12x^2 - x + 3$.

Solution. Again we can group the first two terms and last two terms.

$$4x^3 - 12x^2 - x + 3 \equiv 4x^2(x - 3) - (x - 3)$$
$$\equiv (x - 3)(4x^2 - 1)$$
$$\equiv (x - 3)(2x + 1)(2x - 1).$$

In both these examples we could have grouped the first and third, and the second and fourth terms, and obtained the same result.

EXAMPLE 10. Factor $4x^2 - 12xy + 9y^2 + 4x - 6y - 3$.

Solution. Grouping the first three terms, the solution becomes clear.

$$\begin{aligned}4x^2 - 12xy + 9y^2 + 4x - 6y - 3 &\equiv (2x - 3y)^2 + 2(2x - 3y) - 3 \\ &\equiv [(2x - 3y) + 3][(2x - 3y) - 1] \\ &\equiv (2x - 3y + 3)(2x - 3y - 1).\end{aligned}$$

EXAMPLE 11. Factor $x^4 + 2x^2y^2 + 9y^4$.

Solution. If the coefficient of the second term were 6, the expression would be a perfect square. Therefore, adding (and subtracting) $4x^2y^2$, our solution becomes evident.

$$\begin{aligned}x^4 + 2x^2y^2 + 9y^4 &\equiv x^4 + 6x^2y^2 + 9y^4 - 4x^2y^2 \\ &\equiv (x^2 + 3y^2)^2 - (2xy)^2 \\ &\equiv (x^2 + 3y^2 + 2xy)(x^2 + 3y^2 - 2xy).\end{aligned}$$

PROBLEMS

Factor completely:

1. $ax - ay - by + bx$
2. $ax - 2ay - 6by + 3bx$
3. $x^3 - 2x^2 + 4x - 8$
4. $y^3 - 2y^2 + 5y - 10$
5. $2a - 6 - ab^2 + 3b^2$
6. $x^3 + 3x^2 - 9x - 27$
7. $x^2 - 2x + 1 - y^2$
8. $xy^3 + 2y^2 - xy - 2$
9. $4x^2 - y^2 + 4y - 4$
10. $x^6 - 7x^3 - 8$
11. $x^2 + 2xy + y^2 - z^2 + 2zw - w^2$
12. $4a^2 - x^2 + b^2 - y^2 - 4ab - 2xy$
13. $x^2 + 4xy + 4y^2 - x - 2y - 6$
14. $x^3 - 5x^2 - x + 5$
15. $x^4 - 7x^2y^2 + 9y^4$
16. $y^4 + y^2 + 25$
17. $a^4 + 2a^2b^2 + 9b^4$
18. $x^4 + 4y^4$
19. $a^8 - b^8$
20. $x^6 + 1$
21. $x^2 + 2xy - z^2 - 2yz$
22. $(x^2 + 2x - 3)^2 - 4$
23. $(x - y - 2z)^2 - (2x + y - z)^2$
24. $2(x + 2)^2(x - 3) + 3(x + 2)(x - 3)^2$

CHAPTER 2

FRACTIONS, EXPONENTS, AND RADICALS

In this chapter we shall continue our discussion of notions and processes fundamental to any study in mathematics. The ideas considered here will appear throughout this book, and must be thoroughly understood.

2–1 Simplification of fractions. We recall the *basic principle* with regard to fractions in algebra as well as arithmetic, which states that the value of a fraction is not changed if its numerator and denominator are both multiplied or both divided by the same quantity (not zero).

With this principle in mind, the *simplification* or reducing of a fraction to *lowest terms* is always possible. Factor both the numerator and denominator into their prime factors, and, using the basic principle, divide the numerator and denominator by all their common factors.

EXAMPLE 1. Reduce $\dfrac{8x^4y^7}{12x^6y^3}$ to lowest terms.

Solution. $\dfrac{8x^4y^7}{12x^6y^3} \equiv \dfrac{2^3 x^4 y^7}{2^2 \cdot 3 x^6 y^3}$.

By dividing both numerator and denominator by $2^2 x^4 y^3$,

$$\dfrac{8x^4y^7}{12x^6y^3} \equiv \dfrac{2y^4}{3x^2}.$$

EXAMPLE 2. Reduce $\dfrac{x^2 - 7x + 10}{2x^2 - x - 6}$ to lowest terms.

Solution. Factoring both numerator and denominator,

$$\dfrac{x^2 - 7x + 10}{2x^2 - x - 6} \equiv \dfrac{(x-5)(x-2)}{(2x+3)(x-2)}.$$

Dividing both numerator and denominator by $x - 2$,

$$\frac{x^2 - 7x + 10}{2x^2 - x - 6} \equiv \frac{x - 5}{2x + 3}.$$

The elimination of a common factor by dividing the numerator and denominator of a fraction by the same quantity is called *cancellation*. Such a process should be done *with great care*.

EXAMPLE 3. Reduce $\dfrac{12x^2 + 30x - 72}{52x - 8x^2 - 60}$ to lowest terms.

Solution. $\dfrac{12x^2 + 30x - 72}{52x - 8x^2 - 60} \equiv \dfrac{6(2x - 3)(x + 4)}{4(3 - 2x)(x - 5)} \equiv \dfrac{3(x + 4)}{2(5 - x)}.$

In dividing by $2x - 3$, this quantity divides into $3 - 2x$, minus one times, thus introducing a minus sign in the denominator. Recall for fractions $-\dfrac{a}{b} \equiv \dfrac{-a}{b} \equiv \dfrac{a}{-b}$.

PROBLEMS

Reduce to lowest terms:

1. $\dfrac{28}{63}$

2. $\dfrac{27x^3}{225x^5}$

3. $\dfrac{a^4 x^3 y}{a^2 x y^3}$

4. $\dfrac{a^2 + ab}{3a + 2a^3}$

5. $\dfrac{a^2 x - a^2 y}{ax^2 - ay^2}$

6. $\dfrac{24a^2}{6a^2 - 9a}$

7. $\dfrac{x^2 - 1}{x^2 - x}$

8. $\dfrac{x^2 - 4x + 4}{x^2 - 4}$

9. $\dfrac{x^2 - 16}{x^2 - 8x + 16}$

10. $\dfrac{a^2 - 3a - 4}{a^2 - a - 12}$

11. $\dfrac{y^2 - y - 6}{y^2 + 2y - 15}$

12. $\dfrac{2x^2 + 5x - 12}{4x^2 - 4x - 3}$

13. $\dfrac{6a^2 - 7a - 3}{4a^2 - 8a + 3}$

14. $\dfrac{ax + ay - bx - by}{am - bm - an + bn}$

15. $\dfrac{14x - 24 - 2x^2}{x^2 + x - 20}$

16. $\dfrac{(4x^2 - 9y^2)(18x - 12)}{(2x - 3y)(12x - 8)}$

17. $\dfrac{x^2 - 36}{x^3 - 216}$

18. $\dfrac{2x^2 - 14x + 20}{7x - 2x^2 - 6}$

19. $\dfrac{2(x^2 - y^2)xy + x^4 - y^4}{x^4 - y^4}$

20. $\dfrac{y^6 + 64}{y^4 - 4y^2 + 16}$

2–2 Addition of fractions. The algebraic sum of two or more fractions having the same denominator is the fraction with this common denominator. Its numerator is the algebraic sum of the numerators of the fractions considered.

ILLUSTRATION 1. $\dfrac{2x^2}{x-4} - \dfrac{3x}{x-4} + \dfrac{5}{x-4} \equiv \dfrac{2x^2 - 3x + 5}{x-4}$.

To find the algebraic sum of two or more fractions with different denominators, we must replace the fractions with equivalent fractions having the same denominators. This process of finding a common denominator, and preferably the *least common denominator* (L.C.D.) is important. The L.C.D. of two or more fractions is that algebraic expression consisting of the product of all the prime factors, appearing in any of the denominators, each with an exponent equal to the largest exponent of any such prime factor.

EXAMPLE 1. Find the L.C.D. of the fractions

$$\dfrac{3x}{x^2 - 4x + 4}, \quad \dfrac{5x^2}{3(x^2 - 4)}, \quad \dfrac{2}{2x^2 - x - 6}.$$

Solution. Factoring each denominator, we have

$x^2 - 4x + 4 \equiv (x-2)^2, \qquad 3(x^2 - 4) \equiv 3(x+2)(x-2),$
$\qquad 2x^2 - x - 6 \equiv (2x+3)(x-2).$

The L.C.D. is $3(x+2)(x-2)^2(2x+3)$.

We are now able to find the algebraic sum of two or more fractions with different denominators. After the L.C.D. has been determined, each equivalent fraction may be formed with this same denominator by dividing the L.C.D. of the given fraction by the

ADDITION OF FRACTIONS

denominator of the fraction in question, and then multiplying both numerator and denominator of this fraction by the result. These equivalent fractions may now be added as in Illustration 1.

EXAMPLE 2. Change the following fractions to equivalent ones with their L.C.D. as denominator, and find their sum:

$$\frac{4}{x+2}, \quad \frac{x+3}{x^2-4}, \quad \frac{2x+1}{x-2}.$$

Solution. The L.C.D. is $(x+2)(x-2)$. Therefore,

$$\frac{4}{x+2} \equiv \frac{4(x-2)}{(x+2)(x-2)}, \quad \frac{x+3}{x^2-4} \equiv \frac{x+3}{(x+2)(x-2)},$$

$$\frac{2x+1}{x-2} \equiv \frac{(2x+1)(x+2)}{(x+2)(x-2)}$$

and

$$\frac{4}{x+2} + \frac{x+3}{x^2-4} + \frac{2x+1}{x-2} \equiv \frac{4(x-2)}{(x+2)(x-2)} + \frac{x+3}{(x+2)(x-2)}$$

$$+ \frac{(2x+1)(x+2)}{(x+2)(x-2)}$$

$$\equiv \frac{(4x-8) + (x+3) + (2x^2+5x+2)}{x^2-4}$$

$$\equiv \frac{2x^2+10x-3}{x^2-4}.$$

PROBLEMS

Reduce to a single fraction and simplify:

1. $\frac{2}{3} + \frac{5}{6} - \frac{3}{10}$
2. $5 - \frac{4}{9} - \frac{7}{15}$
3. $\frac{3x}{4y} - \frac{4y}{3x}$
4. $\frac{a^2}{b} - \frac{b^2}{a}$
5. $\frac{2x+3}{6} - \frac{4x-7}{9}$
6. $\frac{3x-1}{5} + \frac{4-5x}{6}$
7. $x + y + \frac{x^2}{x-y}$
8. $\frac{x+1}{x+2} - \frac{x+3}{x}$

9. $\dfrac{3x - 2y}{5x - 3} + \dfrac{2x - y}{3 - 5x}$

10. $\dfrac{2}{12x^2 - 3} + \dfrac{3}{2x - 4x^2}$

11. $\dfrac{5}{x} - \dfrac{4}{y} + \dfrac{3}{z}$

12. $\dfrac{4}{x^2 - 4x - 5} + \dfrac{2}{x^2 - 1}$

13. $\dfrac{2x - 1}{4 - x} + \dfrac{x + 2}{3x - 12}$

14. $\dfrac{x + 5}{x^2 + 7x + 10} - \dfrac{x - 1}{x^2 + 5x + 6}$

15. $\dfrac{x - 1}{2x^2 - 13x + 15} + \dfrac{x + 3}{2x^2 - 15x + 18}$

16. $\dfrac{2x + 3}{3x^2 + x - 2} - \dfrac{3x - 4}{2x^2 - 3x - 5}$

17. $\dfrac{3}{a - 3} + \dfrac{a^2 + 2}{a^3 - 27}$

18. $\dfrac{2xy}{x^3 + y^3} - \dfrac{x}{x^2 - xy + y^2}$

19. $\dfrac{2}{x^2 + 3x + 2} - \dfrac{3}{x^2 + 5x + 6} - \dfrac{4}{x^2 + 4x + 3}$

20. $x + 6 + \dfrac{5x + 1}{12x^2 + 5x - 2} - \dfrac{x}{3x + 2}$

21. $2y - 3 + \dfrac{y - 2}{4y^2 - 12y + 9} + \dfrac{y + 2}{2y^2 - y - 3}$

22. $\dfrac{1}{(x - y)(y - z)} + \dfrac{1}{(y - z)(z - x)} + \dfrac{1}{(z - x)(x - y)}$

23. $\dfrac{x}{(x - y)(y - z)} + \dfrac{y}{(y - z)(z - x)} + \dfrac{z}{(z - x)(x - y)}$

24. $\dfrac{2x - 1}{2x^2 - x - 6} + \dfrac{x + 3}{6x^2 + x - 12} - \dfrac{2x - 3}{3x^2 - 10x + 8}$

2–3 Multiplication and division of fractions. In algebra, as in arithmetic, *the product of two or more fractions* is the fraction whose numerator is the product of their numerators, and whose denominator is the product of their denominators. In obtaining any such product, the process of dividing out factors common to the numerator and denominator may be used, and any result should also be reduced to its simplest form.

ILLUSTRATION 1. $\dfrac{x - 4}{2x + 8} \cdot \dfrac{4x + 8}{x^2 - 16} \equiv \dfrac{(x - 4) \cdot 2^2 \cdot (x + 2)}{2(x + 4)(x + 4)(x - 4)}$

$\equiv \dfrac{2(x + 2)}{(x + 4)^2}.$

The *reciprocal* of any number is 1 divided by the number. Specifically, the reciprocal of $a\,(a \neq 0)$ is $1/a$, and the reciprocal of any fraction b/c is $1/(b/c)$. If we multiply both numerator and denominator of this fraction by c/b, we obtain

$$\frac{1}{\dfrac{b}{c}} \cdot \frac{\dfrac{c}{b}}{\dfrac{c}{b}} \equiv \frac{\dfrac{c}{b}}{1} \equiv \frac{c}{b}.$$

Thus, the reciprocal of any fraction is that fraction inverted. We may now state the rule for division of fractions. The *quotient of two fractions* is the fraction formed by multiplying the dividend by the reciprocal of the divisor. Explain how this rule makes use of the basic principle stated in Article 2–1.

ILLUSTRATION 2.

$$\frac{3x - 15}{x + 3} \div \frac{12x + 18}{4x + 12} \equiv \frac{3x - 15}{x + 3} \cdot \frac{4x + 12}{12x + 18}$$

$$\equiv \frac{3(x - 5)}{x + 3} \cdot \frac{4(x + 3)}{6(2x + 3)}$$

$$\equiv \frac{2(x - 5)}{2x + 3}.$$

PROBLEMS

Find the reciprocal of each of the following:

1. 4
2. 7/10
3. $2\frac{3}{7}$
4. .68
5. $a + b$
6. $\dfrac{x + 2}{3x - 4}$

Perform the following operations and reduce the result to its simplest form:

7. $\dfrac{3x^3}{4y^2} \cdot \dfrac{5y}{x^2}$
8. $\dfrac{7a}{12b^3} \cdot \dfrac{20b^5}{35a^3}$
9. $\dfrac{40x^3y^2}{24xy^4} \div \dfrac{27xy}{8x^2y^3}$
10. $\dfrac{xy^3}{yz} \div x^2z$
11. $\dfrac{x^2 - y^2}{x^3 - y^3} \div \dfrac{(x + y)}{x}$
12. $\dfrac{x^2 - 2x + y^2}{x^3 - y^3} \cdot \dfrac{x^2 + xy + y^2}{x - y}$

13. $\dfrac{x^2 - 6x + 9}{x^2 - 7x + 12} \cdot \dfrac{x^3 - 4x^2 + 9x - 36}{x^4 - 81}$

14. $\dfrac{y^2 - 2y - 15}{y^2 - 9} \div \dfrac{12 - 4y}{y^2 - 6y + 9}$

15. $\dfrac{x^4 - y^4}{(x - y)^2} \cdot \dfrac{y^2}{x^2 + y^2} \cdot \dfrac{x - y}{xy + y^2}$

16. $(a^2 - b^2) \div \left[\dfrac{a^2 + ab}{b^2 + ab} \div \dfrac{a^2 - ab}{b^2 - ab}\right]$

17. $\dfrac{9x^2 + 6x - 8}{6x^2 + 5x - 4} \cdot \dfrac{2x^2 - 7x - 4}{2x^2 - 5x - 12} \cdot \dfrac{4x^2 + 4x - 3}{6x^2 - x - 2}$

18. $\left[\dfrac{y^3 + 4y^2 - 5y}{y^2 - 2y + 1} \div \dfrac{y^2 + y - 2}{y^4 + 8y}\right] \cdot \dfrac{y - 4}{y^2 - 2y + 4}$

19. $\left[\dfrac{2x}{x - 1} + \dfrac{x^2}{x^2 - 1}\right] \div \dfrac{x^3}{1 - x}$

20. $\left[\dfrac{3x}{x - 3} - \dfrac{3x + 2}{x^2 - 6x + 9}\right] \cdot \left[\dfrac{x + 2}{x + 3} - \dfrac{x}{x^2 + 6x + 9}\right]$

Most of the fractions considered have been simple fractions. Any fraction which contains one or more other fractions in either numerator or denominator, or in both, is called a *complex fraction*. Such a complex fraction may be simplified by reducing the numerator and denominator to single fractions, and then performing the division.

EXAMPLE 1. Simplify $\dfrac{a - \dfrac{1}{a}}{a - \dfrac{1}{a^2}}$.

Solution. $\dfrac{a - \dfrac{1}{a}}{a - \dfrac{1}{a^2}} \equiv \dfrac{a^2 - 1}{a} \div \dfrac{a^3 - 1}{a^2}$

$\equiv \dfrac{(a + 1)(a - 1)}{a} \cdot \dfrac{a^2}{(a - 1)(a^2 + a + 1)}$

$\equiv \dfrac{a(a + 1)}{a^2 + a + 1}.$

Problems

Reduce each of the following to a single fraction in simplest form.

1. $\dfrac{3 + \dfrac{4}{5}}{\dfrac{2}{3} - 1}$

2. $\dfrac{\dfrac{a}{b} + \dfrac{a}{c}}{ab + ac}$

3. $\dfrac{x - \dfrac{x}{y}}{z - \dfrac{z}{y}}$

4. $\dfrac{\dfrac{4x}{5} + \dfrac{2y}{3}}{\dfrac{3x}{5} - \dfrac{3y}{4}}$

5. $\dfrac{\dfrac{2}{x} + \dfrac{5}{y}}{\dfrac{2}{x} - \dfrac{5}{y}}$

6. $\dfrac{\dfrac{1}{x} + \dfrac{1}{y}}{\dfrac{1}{x^2} - \dfrac{1}{y^2}}$

7. $\dfrac{x^2 - \dfrac{1}{x}}{x + 1 + \dfrac{1}{x}}$

8. $\dfrac{\dfrac{1}{2} - \dfrac{4}{x^3}}{\dfrac{1}{x^2} + \dfrac{1}{4} + \dfrac{1}{2x}}$

9. $\dfrac{9x^2 - 4y^2}{\dfrac{x-y}{y-2x} - 1}$

10. $\dfrac{\dfrac{x}{y} + \dfrac{y^2}{x^2}}{\dfrac{y}{x^2} - \dfrac{1}{x} + \dfrac{1}{y}}$

2-4 Integral and zero exponents. We discussed the meaning of $a^n (a \neq 0)$, where n was any positive integer, in Article 1–5. From the laws of exponents [Eqs. (1–10)–(1–15)], we may generalize our definitions, in order to define a^n where n is any integer, positive, negative, or zero. Recalling Eq. (1–15), $a^n/a^n \equiv 1$. If Eq. (1–13) were permissible, we would have

$$a^{n-n} \equiv a^0 \equiv 1. \qquad (2\text{–}1)$$

To be consistent with these laws of exponents, we *define* $a^0 \equiv 1$. Moreover, we *define* a^{-n}, where n is a positive integer, as

$$a^{-n} \equiv \dfrac{1}{a^n}. \qquad (2\text{–}2)$$

We therefore have, for any $a \neq 0$, and any integers (positive, negative, or zero) n and m,

$$a^n a^m \equiv a^{n+m}, \tag{2-3}$$

$$(ab)^n \equiv a^n b^n, \tag{2-4}$$

$$(a^n)^m \equiv a^{nm}, \tag{2-5}$$

and

$$\frac{a^n}{a^m} \equiv a^{n-m}. \tag{2-6}$$

ILLUSTRATION 1.

$$5^0 = 1, \quad 2x^0 \equiv 2 \cdot 1 = 2, \quad (2x^2y)^0 \equiv 1,$$
$$x^{-2} \equiv 1/x^2, \quad (x^3)^4 \equiv x^{12}, \quad x^3/x^5 \equiv x^{-2} \equiv 1/x^2.$$

PROBLEMS

Remove the negative and zero exponents and simplify the following expressions.

1. $8x^0$
2. $(8x)^0$
3. $3x^{-2}y^4$
4. $5y^{-2}x^3z^0$
5. $\dfrac{2x^3y^{-2}}{3x^{-2}y^3}$
6. $\dfrac{4x^{-2}y^4}{6x^{-5}y^{-2}}$
7. $\dfrac{a^{-1}+b^{-1}}{(cd)^{-1}}$
8. $\dfrac{x^{-2}+y^{-2}}{x^{-1}+y^{-1}}$
9. $\dfrac{a^{-1}+b^{-1}}{(a+b)^{-1}}$
10. $(a^{-1}+b^{-1})^{-1}$
11. $(x^n y^2)^m$
12. $(-1)^n \cdot (-1)^m (-1)^1$

2–5 Rational exponents. As was the case in the last article when we extended our definitions to zero and negative exponents, we shall further extend the definitions to rational exponents. Any number a whose nth power (n, any positive integer) is equal to b and which satisfies the equation $a^n = b$, is called an nth root of b. We recall, for example, that either 2 or -2 are square roots of 4, since $2^2 = 4$ and $(-2)^2 = 4$. Similarly, since $(-2)^3 = -8$, -2 is a cube root of -8. In general, every number except zero has exactly n distinct nth roots, although most or all of them may be imaginary numbers.*

---

* See Chapter 15.

2–5] RATIONAL EXPONENTS

In many cases it is convenient to have a principal nth root defined. The *principal nth root* of a positive number is the positive root. The principal nth root of a negative number is the negative root if n is odd. If n is even and the number is negative, no principal nth root is defined, for no such real value exists. (Consider for example, the equation $x^2 = -4$). The symbol $\sqrt[n]{b}$ means the principal nth root, where n is called the *index* and b the *radicand* of the *radical*.

ILLUSTRATION 1. $\sqrt{9} = 3$, $\sqrt[3]{-8} = -2$, $\sqrt[4]{81} = 3$ are the principal square root, cube root, and fourth root of 9, -8, and 81 respectively.

We are now able to extend the laws of exponents to rational values p/q, where p and q are any two relatively prime positive integers. If $n = 1/q$, where q is any positive integer, and Eq. (1–10) is to be satisfied,

$$\underbrace{b^{1/q} \cdot b^{1/q} \cdot b^{1/q} \cdots b^{1/q}}_{q \text{ factors } b^{1/q}} \equiv b^{q/q} \equiv b. \qquad (2\text{–}7)$$

Hence $b^{1/q}$ is defined as a qth root of b and in fact, the symbol $b^{1/q}$ means the principal qth root. Likewise, if Eq. (1–11) is to be satisfied,

$$(b^{1/q})^p \equiv b^{p/q}. \qquad (2\text{–}8)$$

This identity states that $b^{1/q}$ is a pth root of $b^{p/q}$ and that $b^{p/q}$ is the pth power of the principal qth root of b. Also we can now show that

$$\underbrace{b^{p/q} \cdot b^{p/q} \cdot b^{p/q} \cdots b^{p/q}}_{q \text{ factors } b^{p/q}} \equiv b^p, \qquad (2\text{–}9)$$

and thus $b^{p/q}$ is also the principal qth root of b^p. Therefore we have, from (2–8), (2–9), and the previous notation,

$$b^{p/q} \equiv (b^p)^{1/q} \equiv \sqrt[q]{b^p}* \qquad (2\text{–}10)$$

and also

$$b^{p/q} \equiv (b^{1/q})^p = (\sqrt[q]{b})^p. \qquad (2\text{–}11)$$

It can now be shown that the five laws of exponents, Eqs. (2–2) through (2–6), hold for all rational exponents, as well as integers.

---

* We must exclude the case where b is negative and q is even, as we did in our definition of principal value.

ILLUSTRATION 2. $8^{2/3} = \sqrt[3]{8^2} = \sqrt[3]{64} = 4$, or
$= (\sqrt[3]{8})^2 = 2^2 = 4$.

ILLUSTRATION 3. $81^{-3/4} = \dfrac{1}{(\sqrt[4]{81})^3} = \dfrac{1}{3^3} = \dfrac{1}{27}$.

ILLUSTRATION 4. $x^{1/4} \cdot x^{2/3} \equiv x^{1/4 + 2/3} \equiv x^{11/12} \equiv \sqrt[12]{x^{11}}$.

ILLUSTRATION 5. $x^{1/4} \div x^{2/3} \equiv x^{1/4 - 2/3} \equiv x^{-5/12} \equiv \dfrac{1}{\sqrt[12]{x^5}}$.

ILLUSTRATION 6. $(x^{-4})^{-3/4} \equiv x^3$.

Problems

Find the numerical value of the following.

1. $25^{1/2}$ 2. $81^{3/4}$ 3. $(\frac{16}{49})^{1/2}$
4. $(\frac{8}{125})^{-1/3}$ 5. $(\frac{64}{27})^{2/3}$ 6. $32^{-4/5}$
7. $(2^{10})^{-3/5}$ 8. $(2^{-6})^{2/3}$ 9. $3^{1/2} \cdot 3^{1/2}$

Remove the negative exponents, simplify, and express the result in radical form.

10. $x^{1/4} \cdot x^{1/5}$ 11. $x^{1/4} \div x^{1/5}$
12. $(x^{1/4})^{1/5}$ 13. $x^{1/4} \cdot x^{-1/5}$
14. $(x^{1/4})^{-1/5}$ 15. $(x^{-1/4})^{-1/5}$

Remove the negative and zero exponents and simplify:

16. $(9x^{-4}y^2)^{1/2}$ 17. $(2x^{1/6}y^{5/6})^{-6}$
18. $(2x^{-3}y^4)^3$ 19. $\left(\dfrac{125x^4y^3}{27x^{-2}y^6}\right)^{1/3}$
20. $\left(\dfrac{5^0 x^4 y^3 z}{16x^{-6}yz^5}\right)^{-1/2}$ 21. $(a^{1/2} + b^{1/2})^2$
22. $(a^{1/2} + b^{1/2})(a^{1/2} - b^{1/2})$ 23. $(x^{1/3} + y^{1/3})(x^{2/3} - x^{1/3}y^{1/3} + y^{2/3})$
24. $(x + y)^{-2}(x^{-2} - y^{-2})$ 25. $(x + y^{-1})^2$

2–6 Radicals. In many cases it is more advantageous to express a quantity in terms of radicals rather than rational exponents. The laws of radicals follow directly from the definitions. If m and n are positive integers,

$$(\sqrt[n]{a})^n \equiv a. \tag{2-12}$$

$$\sqrt[n]{ab} \equiv (ab)^{1/n} \equiv a^{1/n} \cdot b^{1/n} \equiv \sqrt[n]{a} \cdot \sqrt[n]{b}. \tag{2-13}$$

$$\sqrt[n]{\frac{a}{b}} \equiv \left(\frac{a}{b}\right)^{1/n} \equiv \frac{a^{1/n}}{b^{1/n}} \equiv \frac{\sqrt[n]{a}}{\sqrt[n]{b}}. \tag{2-14}$$

$$\sqrt[m]{\sqrt[n]{a}} \equiv (a^{1/n})^{1/m} \equiv a^{1/nm} \equiv \sqrt[nm]{a}. \tag{2-15}$$

ILLUSTRATION 1. $\sqrt{50} = \sqrt{25 \cdot 2} = \sqrt{25} \cdot \sqrt{2} = 5\sqrt{2}$.

ILLUSTRATION 2. $\sqrt[3]{\frac{4}{27}} = \frac{\sqrt[3]{4}}{\sqrt[3]{27}} = \frac{\sqrt[3]{4}}{3}$.

ILLUSTRATION 3. $\sqrt[6]{27} = \sqrt{\sqrt[3]{27}} = \sqrt{3}$.

In simplifying any radicals by use of these laws, we must have:
1. No factors which are perfect nth powers under a radical whose index is n.
2. No fractions under the radical sign.
3. The index of the radical as small as possible.

Any radical that satisfies these conditions is said to be in *simplest form*. In addition to the three illustrations above, which are in simplest form, we might illustrate further.

EXAMPLE 1. Simplify $\sqrt[3]{81x^5y^7}$.

Solution. In this expression it will be necessary to remove from the radicand all factors which are perfect cubes.

$$\sqrt[3]{81x^5y^7} \equiv \sqrt[3]{27x^3y^6 \cdot 3x^2y} = \sqrt[3]{27x^3y^6} \cdot \sqrt[3]{3x^2y} = 3xy^2\sqrt[3]{3x^2y}.$$

EXAMPLE 2. Simplify $\sqrt{\frac{3}{2}}$.

Solution. In any radical with the radicand appearing as a fraction, it is important to eliminate the fraction because of computational reasons. This process enables the radical to be approximated by only one root extraction and a simple division, rather than by two root extractions and a more complicated division. Our process introduces a perfect square in the denominator, so that it may be removed from under the radical.

$$\sqrt{\frac{3}{2}} = \sqrt{\frac{3 \cdot 2}{2 \cdot 2}} = \frac{\sqrt{6}}{2}.$$

EXAMPLE 3. Simplify $\sqrt[4]{\frac{64x^2y^4}{z^2}}.$

Solution. Problems of this type are usually clearer if rational exponents are introduced. We also must eliminate the denominator under the radical.

$$\sqrt[4]{\frac{64x^2y^4}{z^2}} \equiv \left(\frac{2^6 x^2 y^4}{z^2}\right)^{1/4} \equiv \frac{2^{3\frac{1}{2}} x^{1/2} y}{z^{1/2}} \cdot \frac{z^{1/2}}{z^{1/2}}$$

$$\equiv \frac{2y}{z} \sqrt{2xz}.$$

PROBLEMS

Simplify each of the following:

1. $\sqrt{8}$
2. $\sqrt{98}$
3. $\sqrt{\frac{75}{12}}$
4. $\sqrt[3]{40}$
5. $\sqrt[3]{-625}$
6. $\sqrt[4]{32}$
7. $\sqrt{27x^3y^5}$
8. $\sqrt{192a^3b^7}$
9. $\sqrt[3]{81z^4x^6y^5}$
10. $\sqrt{\frac{3}{5}}$
11. $\sqrt{\frac{125}{63}}$
12. $\sqrt{\frac{1}{4} + \frac{1}{9}}$
13. $\sqrt{a^2b^2 + b^2c^2}$
14. $\sqrt{a^{-2} + b^{-2}}$
15. $\sqrt{\frac{3x}{y^3}}$
16. $\sqrt[3]{\frac{x^4y}{z^2}}$
17. $\sqrt[4]{\frac{x^7y^6}{243}}$
18. $\sqrt{x - 2 + \frac{1}{x}}$
19. $\sqrt[4]{25}$
20. $\sqrt[6]{49x^4}$
21. $\sqrt[3]{\frac{5}{3x^2}}$
22. $\sqrt[6]{\frac{9}{16}}$
23. $\sqrt[4]{\frac{169x^6z^2}{y^4}}$
24. $\sqrt[4]{1 - \frac{4}{x} + \frac{4}{x^2}}$

2-7 Addition and subtraction of radicals. In any problems involving addition or subtraction of radicals, all similar radicals (that is, those which result in the same index and radicand) are combined into single terms. Consider the example.

EXAMPLE. Simplify by combining similar terms,
$$4\sqrt{12} + 5\sqrt{8} - \sqrt{50} - 7\sqrt{48}.$$

Solution.
$$4\sqrt{12} + 5\sqrt{8} - \sqrt{50} - 7\sqrt{48}$$
$$= 4\sqrt{4 \cdot 3} + 5\sqrt{4 \cdot 2} - \sqrt{25 \cdot 2} - 7\sqrt{16 \cdot 3}$$
$$= 8\sqrt{3} - 28\sqrt{3} + 10\sqrt{2} - 5\sqrt{2}$$
$$= 5\sqrt{2} - 20\sqrt{3}.$$

PROBLEMS

Simplify by combining similar terms:

1. $4\sqrt{3} - 5\sqrt{12} + 2\sqrt{75}$
2. $\sqrt[3]{2} + \sqrt[3]{16} - \sqrt[3]{54}$
3. $5\sqrt{2} - \sqrt[4]{64} + 2\sqrt{32}$
4. $\sqrt{x^3} + \sqrt{25x^3} + \sqrt{9x}$
5. $\sqrt{4(x+y)} - 2\sqrt{9(x+y)} + 3\sqrt{x+y}$
6. $3\sqrt{18} - 3\sqrt{32} + 3\sqrt{12} - 3\sqrt{3}$
7. $\sqrt{a^3bc^5} + \sqrt{ab^7c^3} + \sqrt{a^9b^5c}$
8. $\sqrt{\dfrac{a-b}{a+b}} - \sqrt{\dfrac{a+b}{a-b}} + \sqrt{\dfrac{a^2}{a^2-b^2}}$
9. $8\sqrt{\dfrac{1}{3}} + \dfrac{3}{2}\sqrt{108} - \sqrt[4]{9}$
10. $\sqrt{\left(\dfrac{x}{y} - \dfrac{y}{x}\right)\dfrac{1}{xy}} - \sqrt{\dfrac{x+y}{x-y}}$
11. $\dfrac{7 \pm \sqrt{49 - 4 \cdot 10}}{2}$
12. $\dfrac{4 \pm \sqrt{16 - 4 \cdot 2}}{2}$

2-8 Multiplication and division of radicals. The multiplication of two or more radicals is accomplished by using Eq. (2-13). If the radicals have the same index the result follows immediately. If, however, the radicals are of different indices, they must be converted to radicals with the same index before the multiplication takes place. This is always possible, and is usually carried out with the aid of equivalent expressions with rational exponents.

38   FRACTIONS, EXPONENTS, AND RADICALS   [CHAP. 2

EXAMPLE 1. Find the product of $\sqrt{15ax^3}$ and $\sqrt{45a^2xy^3}$ and simplify.

Solution. Since both radicals have the index 2, we use Eq. (2–13) immediately.

$$\sqrt{15ax^3} \cdot \sqrt{45a^2xy^3} \equiv \sqrt{15^2 \cdot 3a^3x^4y^3}$$
$$\equiv 15ax^2y\sqrt{3ay}.$$

EXAMPLE 2. Multiply $\sqrt{6x^3}$ by $\sqrt[3]{4x^4y^2}$ and simplify.

Solution. By converting the radicals to rational exponents, we have

$$\sqrt{6x^3} \cdot \sqrt[3]{4x^4y^2} \equiv (3 \cdot 2x^3)^{1/2} \cdot (2^2x^4y^2)^{1/3}$$
$$\equiv (3 \cdot 2x^3)^{3/6} \cdot (2^2x^4y^2)^{2/6}$$
$$\equiv (3^3 \cdot 2^3 \cdot x^9 \cdot 2^4x^8y^4)^{1/6}$$
$$\equiv \sqrt[6]{3^3 \cdot 2 \cdot 2^6x^{12} \cdot x^5y^4}$$
$$\equiv 2x^2\sqrt[6]{54x^5y^4}.$$

Division of two radicals is handled in a similar manner, by using Eq. (2–14). Again, if the radicals are of different indices, they must be converted to the same index.

EXAMPLE 3. Divide $6\sqrt[3]{5}$ by $2\sqrt{3}$ and simplify.

Solution. Since $\sqrt[3]{5} = \sqrt[6]{25}$ and $\sqrt{3} = \sqrt[6]{27}$, we have

$$\frac{6\sqrt[3]{5}}{2\sqrt{3}} = 3\sqrt[6]{\frac{25}{27}}.$$

Recalling that a radical in simplest form has a denominator free of radicals, we must multiply the numerator and denominator by $\sqrt[6]{27}$. Thus

$$\frac{6\sqrt[3]{5}}{2\sqrt{3}} = 3\sqrt[6]{\frac{25 \cdot 27}{27 \cdot 27}} = 3\sqrt[6]{\frac{25 \cdot 27}{3^6}} = \frac{3}{3}\sqrt[6]{25 \cdot 27} = \sqrt[6]{675}.$$

This process of eliminating radicals from the denominator, as was illustrated in Example 3, as well as Example 2 of Article 2–6, is called *rationalizing the denominator*. The problem of eliminating all radicals in a denominator may be more complicated.

EXAMPLE 4. Rationalize the denominator of $\dfrac{2 + \sqrt{3}}{\sqrt{5} - \sqrt{3}}$.

Solution. Since we must eliminate both radicals in the denominator, we must find an expression (*rationalizing factor*) which, when multiplied by $\sqrt{5} - \sqrt{3}$, will give a result free of any radicals. Since $(\sqrt{5} - \sqrt{3})(\sqrt{5} + \sqrt{3}) = 5 - 3$, we multiply both numerator and denominator by $\sqrt{5} + \sqrt{3}$.

$$\frac{2 + \sqrt{3}}{\sqrt{5} - \sqrt{3}} \cdot \frac{\sqrt{5} + \sqrt{3}}{\sqrt{5} + \sqrt{3}} = \frac{2\sqrt{5} + \sqrt{15} + 2\sqrt{3} + 3}{5 - 3}$$

$$= \frac{2\sqrt{5} + \sqrt{15} + 2\sqrt{3} + 3}{2}.$$

PROBLEMS

Perform the following multiplications, expressing the result in simplest form.

1. $\sqrt{5} \cdot \sqrt{13}$
2. $\sqrt{14} \cdot \sqrt{21}$
3. $\sqrt[3]{4} \cdot \sqrt[3]{26}$
4. $\sqrt[3]{3x^2} \cdot \sqrt{2x}$
5. $\sqrt{x^2 - y^2} \cdot \sqrt{x - y}$
6. $\sqrt{3x^2y^3} \cdot \sqrt{12x^5y}$
7. $\sqrt{x^3 + y^3} \cdot \sqrt{x + y}$
8. $\sqrt[3]{9x} \cdot \sqrt[6]{27x^4}$
9. $\sqrt{2} \cdot \sqrt[3]{3} \cdot \sqrt[4]{4}$
10. $\sqrt{a} \cdot \sqrt[3]{a} \cdot \sqrt[4]{a}$
11. $\sqrt{2}\,(\sqrt{6} + \sqrt{14})$
12. $(2 + \sqrt{3})(2 - \sqrt{3})$
13. $(\sqrt{3} + \sqrt{5})(\sqrt{3} - \sqrt{5})$
14. $(2\sqrt{3} - 3\sqrt{2})^2$
15. $(\sqrt{5} + 2\sqrt{3})(\sqrt{5} - 3\sqrt{3})$
16. $\sqrt{3 + 2\sqrt{2}} \cdot \sqrt{3 - 2\sqrt{2}}$
17. $\left(\dfrac{\sqrt{5} - 1}{2}\right)^2$
18. $\left(\dfrac{\sqrt{6} - \sqrt{2}}{4}\right)^2$

Perform the following divisions, expressing the result in simplest form.

19. $4\sqrt{28} \div 3\sqrt{7}$
20. $\sqrt[3]{\frac{4}{5}} \div \sqrt[3]{\frac{108}{25}}$
21. $\sqrt[6]{12} \div \sqrt{3}\sqrt[3]{2}$
22. $(2\sqrt{6} + 3\sqrt{14}) \div \sqrt{2}$
23. $\sqrt[4]{24a^3b} \div \sqrt[4]{8ab^3}$
24. $\sqrt{xy^2} \div \sqrt[3]{x^2y}$

Rationalize the denominator of each of the following:

25. $\dfrac{2\sqrt{3}}{4\sqrt{5}}$
26. $\dfrac{2\sqrt[3]{3}}{4\sqrt[3]{5}}$

27. $\dfrac{4}{\sqrt[3]{16}}$
28. $\dfrac{3}{2 + \sqrt{3}}$

29. $\dfrac{5}{\sqrt{7} - \sqrt{3}}$
30. $\dfrac{\sqrt{2} + \sqrt{3}}{\sqrt{2} - \sqrt{3}}$

31. $\dfrac{x}{x + \sqrt{y}}$
32. $\dfrac{y}{\sqrt{x} - \sqrt{y}}$

33. $\dfrac{2\sqrt{7} + \sqrt{3}}{3\sqrt{7} - 5\sqrt{3}}$

34. $\dfrac{1}{\sqrt[3]{a} + \sqrt[3]{b}}$ [Hint: $x^3 + y^3 \equiv (x + y)(x^2 - xy + y^2)$]

35. $\dfrac{2}{2 - \sqrt[3]{3}}$
36. $\dfrac{1}{\sqrt[3]{9} + \sqrt[3]{6} + \sqrt[3]{4}}$

37. $\dfrac{1}{\sqrt{2} + \sqrt{3} + \sqrt{5}}$ [Hint: Multiply both numerator and denominator by $(\sqrt{2} + \sqrt{3}) - \sqrt{5}$.]

CHAPTER 3

COORDINATE SYSTEMS, FUNCTIONS, AND GRAPHICAL REPRESENTATION

There are certain notions and concepts which are basic to any systematic study of algebra and trigonometry. In this chapter we shall introduce and briefly discuss several of the most important of these.

3-1 A one-dimensional coordinate system. The method of associating numbers with points on a line is of considerable help in mathematics and has resulted in great progress in the application of mathematics to science. Any scale which measures quantities, such as a yardstick or thermometer, makes use of an association of this kind. To each numerical value assumed by the physical quantity there corresponds a position on the scale, and to each position on the scale there corresponds a number. Such a correspondence establishes a coordinate system. The simplest and most useful coordinate system in one dimension employs this one-to-one correspondence between the real numbers and the points on a straight line. Let us consider such a system.

On a fixed straight line of reference of unlimited length, we choose any point O, called the *origin*, and lay off equal divisions* of arbitrary length in both directions from O (see Fig. 3-1). We now associate zero with the origin, the positive integers with the successive points on one side, and the negative integers with the successive points on the other. The usual convention on such a horizontal line is to consider the integers to the right as positive and those to the left as negative.

The point associated with any rational number can be determined by the simple geometric construction used to divide any line segment into b equal parts. Thus the number $\frac{3}{4}$ is represented by a point three-fourths of the way from 0 to the point identified with 1. Also, $-1\frac{3}{8}$ is represented by a point at a distance $1\frac{3}{8}$ units to the left of 0.

---
* There are systems where the subdivisions are not equal. Consider the slide rule scale.

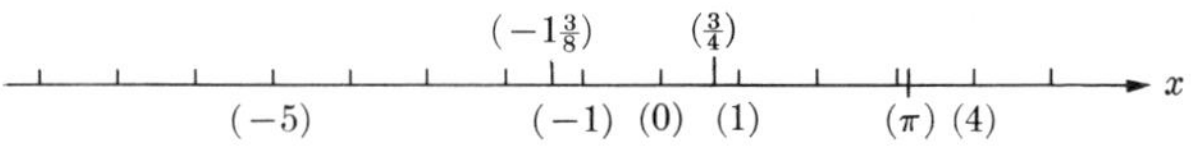

```
         (-1 3/8)   (3/4)
—┴—┴—┴—┴—┴—┴—┼┼—┴—┴—┴—┼┼—┴—┴→ x
 (-5)        (-1) (0) (1)   (π) (4)
```

FIGURE 3-1

The points associated with some of the irrational numbers may also be found by geometric construction. For example, the point associated with $\sqrt{2}$ may be located, because $\sqrt{2}$ is the hypotenuse of an isosceles right triangle, with each leg one unit in length. Although this type of geometric construction is not possible for all real numbers, we shall assume that the correspondence can be extended to all real numbers. This is done by associating every line segment with a real number which represents its length. Thus to each real number we have associated one point on the line and, conversely, to each point on the line there is associated one real number.

The *coordinate* of a point is defined to be the number associated with that point. It is written (x), and will be referred to as "the point x."

There are certain properties of this system which we should observe. It gives us a graphic interpretation of the relative magnitude of numbers. Thus $5 > 3$ (read 5 is greater than 3)* corresponds to the fact that (5) lies to the right of (3), while $-5 < -3$ (read -5 is less than -3) corresponds to the fact that (-5) is to the left of (-3). The notation $x < y < z$ indicates that y is greater than x but less than z.

An expression for the distance between any two points in this system is important and useful. We need only to subtract the coordinate of the left-hand point from the coordinate of the right-hand point. Thus, if $x_1 < x_2$, the distance between (x_1) and (x_2) is $x_2 - x_1$, but if $x_2 < x_1$, the distance must be $x_1 - x_2$, since we wish to have the distance always positive. We can avoid the inconvenience of having to distinguish between the two points by employing the notion of absolute value.

The absolute value of x, denoted by $|x|$, indicates its size or magnitude without regard to its sign. For example, $|3| = 3$, and

---

* If x and y are any two real numbers, by definition,
$x > y$ is equivalent to $x - y$ is positive,
$x < y$ is equivalent to $x - y$ is negative.

$|-3| = 3$. Specifically, the *absolute value* of a number x is defined as

$$|x| = \begin{cases} x \text{ if } x > 0, \\ -x \text{ if } x < 0, \\ 0 \text{ if } x = 0. \end{cases} \quad (3\text{-}1)$$

In this connection we should recall the definition of the square root sign, $\sqrt{}$. For any positive number a, $\sqrt{a}$ denotes the positive square root of a; that is, for any real number x, x^2 is positive (or zero if $x = 0$) and

$$\sqrt{x^2} = \begin{cases} x \text{ if } x > 0, \\ -x \text{ if } x < 0, \\ 0 \text{ if } x = 0. \end{cases} \quad (3\text{-}2)$$

Thus $\sqrt{7^2} = 7$, and $\sqrt{(-7)^2} = -(-7) = 7$. Since (3-1) and (3-2) define the same values, either expression can be used to indicate absolute value.

We are now able to give the general expression for the distance between any two points P_1 and P_2 with coordinates (x_1) and (x_2):

$$d = \overline{P_1P_2} = |x_1 - x_2| = \sqrt{(x_1 - x_2)^2}. \quad (3\text{-}3)$$

For example, the distance between the two points (5) and (-3) is given either by the expression

$$d = \sqrt{[5 - (-3)]^2} = 8,$$

or the expression

$$d = \sqrt{(-3 - 5)^2} = 8.$$

Problems

1. Arrange the following numbers in ascending order of magnitude, and plot them on a linear coordinate system such as that of Fig. 3-1: 2.3, 0.333, 2^3, $\sqrt{4}$, $1/3$, -5, -1, 0, -6.5.

2. Choose two negative numbers, a and b, such that $a > b$. Show that $a - b$ is a positive number.

3. State in words the geometrical interpretation of the following:
   (a) $a < b$
   (b) $a < 2$
   (c) $a > b$
   (d) $a > b > c$
   (e) $a - b = 1$
   (f) $3.14 < \pi < 3.15$

(g) $1.41 < \sqrt{2} < 1.42$
(h) $|5 - 2| > |1 - 3|$
(i) $|a - b| > 0$
(j) $|x - 2| < 3$
(k) $|x - 1| > 4$
(l) $-1 < x < 1$

4. If the coordinates of two points, P_1 and P_2, on a line are (2) and (8) respectively, show that the coordinate of the mid-point of the segment P_1P_2 is (5).

5. Find the coordinate of the mid-point of the line joining (4) and (-4), (3) and (-5), (-1.7) and (3.7), $(\sqrt{2})$ and $(\sqrt{3})$, (x_1) and (x_2).

6. Solve the following equations for x:

(a) $x = |10|$
(b) $x = |2 - 5|$
(c) $x = \sqrt{3^2}$
(d) $x = \sqrt{(-4)^2}$
(e) $x = |-3/2|$
(f) $x = \sqrt{(-1)^2}$
(g) $x = |1/3 - 5/3|$
(h) $x = |1/3| + |-5/3|$

7. Solve the following equations for all possible values of x:

(a) $|x| = 2$
(b) $|x| = \sqrt{5}$
(c) $\sqrt{x^2} = 3$
(d) $\sqrt{x^2} = 1/4$
(e) $|x - 2| = 5$
(f) $|x - 4| = 0$
(g) $|3 - x| = 6$
(h) $\sqrt{(x - 1)^2} = 5$
(i) $\sqrt{(2 - x)^2} = 4$
(j) $|x - 2| = -3$
(k) $\sqrt{(x - 4)^2} = -1$
(l) $\sqrt{(x - 5)^2} = 3$

8. Recall from plane geometry that a circle is the locus (totality) of points at a given distance from a given point, where the distance is called the radius and the given point is called the center. In this geometry of one dimension, how many points are at a given distance from a fixed point? Of how many points would a "circle" consist?

9. If (1) is a given point, and 2 a given distance, explain how $|x - 1| = 2$ would be the condition that any point (x) must be 2 units distant from (1). This is the condition that the point (x) lie on the "circle" with center (1) and radius 2, and is called the equation of the "circle."

10. In terms of "circles in one dimension," give the geometric significance of each of the equations in Problem 7.

11. Give the equation of a "circle in one dimension" with its center at the point (a) and of radius r.

3–2 A two-dimensional coordinate system. In the last article we observed a useful coordinate system in one dimension which not only enabled us to view the relative magnitudes of numbers in a graphic way, but also allowed us to represent differences in magnitude by the use of distance. But for many purposes this is not enough. One of the more important concepts in mathematics is the study of the relation or dependence of two sets of numbers. Since the corresponding values of two such related sets might be considered as pairs of numbers, any system which produces an association between a point and a pair of numbers would be most advantageous in studying such a relationship. This is easily accomplished in two dimensions.

The most frequently used system that sets up an association between each point in a plane and a pair of real numbers is the rectangular cartesian system of coordinates. In 1637 René Descartes, a French mathematician and philosopher, discovered this method of associating numbers with points, and by this unifying of algebra and geometry, great progress in mathematics and the application of mathematics in science was made possible.

Let us construct two perpendicular straight lines and, for convenience, let one of them be horizontal. We shall call these *coordinate axes*. Using the point of intersection as the *origin O*, set up on each line a one-dimensional system. Ordinarily the same unit of length is used on both lines, although in some cases it is convenient to do otherwise. We now denote by the symbol $(x,0)$ the point on the horizontal line corresponding to the number x in its one-dimensional system. Similarly, we denote the point on the vertical line corresponding to the number y in its one-dimensional system by the symbol $(0,y)$. The horizontal line is called the *x-axis* or *axis of abscissas*, while the vertical line is referred to as the *y-axis* or *axis of ordinates*. As is customary, the point on the y-axis $(0,y)$ is above the x-axis when y is positive.

In this reference system of axes, shown in Fig. 3–2, consider any pair of values of x and y, say x_1 and y_1. To find the point corresponding to this pair of values, we draw lines parallel to the axes through the point $(x_1,0)$ on the x-axis and the point $(0,y_1)$ on the y-axis. These lines intersect at a point P, a distance x_1 from the y-axis (to the right or left, depending upon whether x_1 is positive or

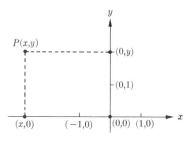

FIGURE 3–2

negative) and a distance y_1 from the x-axis (above or below, depending upon whether y_1 is positive or negative). This point P, determined by the pair of values x_1 and y_1, is denoted by the symbol (x_1,y_1), where x_1 and y_1 are called the *coordinates* of P. As might be expected, the value x is called the *abscissa* of P and the y-value is called its *ordinate*. Clearly, there is only one point determined by any pair of values (x,y). Conversely, for each point there is only one pair of values (x,y), for the point has unique directed distances from the axes. Thus the one-to-one correspondence is established between all the points in the plane and the set of all number pairs (x,y).

The two coordinate axes divide the plane into four parts, called the *first, second, third,* and *fourth quadrants*. It is helpful to verify that the coordinates of points located in the different quadrants have the signs shown in the table:

Quadrants	abscissa	ordinate
I	+	+
II	−	+
III	−	−
IV	+	−

Let us illustrate in Fig. 3–3 the plotting of several points.

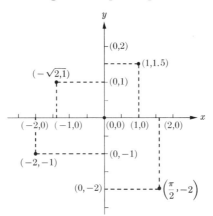

FIGURE 3–3

Problems

1. Plot the following points:
   (a) with abscissa 4 and ordinate 3,
   (b) $(4,-3)$,
   (c) with $x = -4$ and $y = 3$.
2. Plot the following points:
   (a) $(2,6)$, $(-1,4)$, $(3,-2)$, $(-1,-3)$;
   (b) $(4,0)$, $(-4,0)$, $(0,4)$, $(0,-4)$, $(0,0)$.
3. What are the coordinates of a point (a) three units to the right of the y-axis, and two above the x-axis? (b) four units to the left of the y-axis, and six above the x-axis? (c) five units to the right of the y-axis, on the x-axis?
4. (a) What is the abscissa of any point on the y-axis? (b) What is the ordinate of any point on the x-axis?
5. Without plotting, indicate the quadrant in which each of the following points lies: $(-1,2)$, $(2,-4)$, $(-3,-7)$, $(4,6)$, $(-5,2)$, $(28,-2)$.
6. (a) Give the coordinates of four points which are the vertices of a rectangle. (b) Give the coordinates of three points which are the vertices of a right triangle. (c) Give the coordinates of four points on a circle with its center at $(2,3)$ and of radius 4.
7. In each of the following, three vertices of a parallelogram are given. Give the three possible sets of coordinates for the fourth vertex:
   (a) $(0,0)$, $(2,4)$, and $(6,0)$;
   (b) $(-2,1)$, $(1,2)$, and $(0,-3)$.
8. Three vertices of a parallelogram are (a,b), $(0,0)$, and $(c,0)$. What are the possible coordinates of the fourth vertex?
9. Indicate in a rectangular coordinate system the location of the set of all the points (x,y) which satisfy the following conditions:
   (a) $x = 2$ (b) $y = -3$
   (c) $x > 2$ (d) $y > 4$
   (e) $x < -1$ (f) $x = y$
   (g) $x > 2$ (h) $x > y$
   $y = 3$
   (i) $x < y$ (j) $x > 2$
   $y < 4$
   (k) $x = 2$ (l) $x = 2$
   $y < -1$ $y = 3$

3–3 The distance formula. We are now prepared to obtain a formula which has many applications in the material dealt with in this book. To obtain an expression for the distance d between any

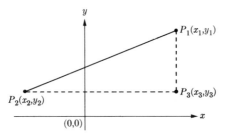

FIGURE 3-4

two points $P_1(x_1,y_1)$ and $P_2(x_2,y_2)$, where the same unit lengths are used on both axes, we make use of the famous theorem of Pythagoras. Considering $P_1(x_1,y_1)$ and $P_2(x_2,y_2)$ as any two points in the plane, construct a right triangle, as in Fig. 3-4, with P_1P_2 the hypotenuse, and the two legs parallel to the axes. Call their point of intersection, where the right angle is formed, $P_3(x_3,y_3)$. Since $x_3 = x_1$ and $y_3 = y_2$, the distance between P_2 and P_3 is

$$P_2P_3 = \sqrt{(x_1 - x_2)^2}$$

and the distance between P_1 and P_3 is

$$P_1P_3 = \sqrt{(y_1 - y_2)^2}.$$

Recalling the theorem of Pythagoras, which states

$$\overline{P_1P_2}^2 = \overline{P_2P_3}^2 + \overline{P_1P_3}^2,$$

we obtain

$$\overline{P_1P_2}^2 = (x_1 - x_2)^2 + (y_1 - y_2)^2.$$

This result may be written:

$$d = P_1P_2 = \sqrt{(x_1 - x_2)^2 + (y_1 - y_2)^2}. \tag{3-4}$$

EXAMPLE 1. The distance between the points $P_1(-4,2)$ and $P_2(3,-1)$ is

$$P_1P_2 = \sqrt{[3 - (-4)]^2 + (-1 - 2)^2}$$
$$= \sqrt{58}.$$

EXAMPLE 2. The distance between the origin (0,0) and any point (x,y) is
$$d = \sqrt{(x-0)^2 + (y-0)^2}$$
$$= \sqrt{x^2 + y^2}.$$

EXAMPLE 3. The triangle with the points $P_1(-5,-1)$, $P_2(2,3)$, and $P_3(3,-2)$ as vertices forms an isosceles triangle.
$$P_1P_2 = \sqrt{(-5-2)^2 + (-1-3)^2}$$
$$= \sqrt{49+16} = \sqrt{65}.$$
$$P_1P_3 = \sqrt{(-5-3)^2 + (-1+2)^2}$$
$$= \sqrt{64+1} = \sqrt{65}.$$

PROBLEMS

In each of the following exercises draw the figure on coordinate paper.

1. Find the distance between the given points:
(a) (3,2) and (6,7)
(b) (−4,3) and (5,−2)
(c) (5/2,−3/4) and (7/4,−3/2)
(d) (0,0) and (5,−12)
(e) (−3,7) and (5,7)
(f) (−1,3) and (x,y)

2. By proving that two sides of the triangle are equal, show that the triangle whose vertices are (2,1), (5,5), and (−2,4) is an isosceles triangle.

3. Show that the points (8,1), (−6,−7), and (2,7) are the vertices of an isosceles triangle.

4. Show that the points (6,1), (5,6), (−4,3), and (−3,−2) are the vertices of a parallelogram.

5. Prove that the points (2,3), (−4,−3), and (6,−1) are the vertices of a right triangle. Notice that we must use the converse of the theorem of Pythagoras to prove this.

6. Show that the points (12,9), (20,−6), (5,−14), and (−3,1) are the vertices of a square. What is the length of a diagonal?

7. Test algebraically to see whether or not the following triples of points are collinear (lie on the same line): (6,2), (1,1), (−4,0); (−6,5), (3,−10), (−2,−2).

8. Find the point on the y-axis which is equidistant from the points (−4,4) and (4,10).

9. If two vertices of an equilateral triangle are $(-4,-3)$ and $(4,1)$, find the remaining vertex.

10. Find those points whose ordinates are -5 and whose distance from the origin is 13.

11. Draw the square whose diagonals lie along the coordinate axes and the length of whose side is a. What are the coordinates of the four vertices?

12. If a circle had its center at the point $(2,3)$ and passed through $(8,-5)$, what would be its radius? Would it pass through $(-6,9)$?

13. Consider the circle with its center at the origin and with a radius of 1. Through which of the following points does it pass: $(1,0)$, $(0,-1)$, $(1,1)$, $(1/\sqrt{2}, 1/\sqrt{2})$, $(1/2, 1/2)$, $(-1/2, \sqrt{3}/2)$?

*14. By giving the expression for the distance between the origin and the point (x,y), and equating this distance to 1, we have stated the algebraic condition on x and y which must be satisfied by the coordinates of any point (x,y) lying on the circle whose center is $(0,0)$ and whose radius is 1. Show that this condition, when simplified, becomes $x^2 + y^2 = 1$. This is called the equation of the unit circle in the plane.

15. If a point lies on a curve, its coordinates must satisfy the equation representing that curve. Check the results of Problem 13 by determining whether the coordinates of the points satisfy the equation of the unit circle obtained in Problem 14, namely, $x^2 + y^2 = 1$.

3-4 The circle and arc length. In any rectangular coordinate system a geometric figure, locus, or curve, such as a circle, is considered as a set of points. The coordinates of each point in this set satisfy the equation of this curve and, conversely, if the coordinates of any point satisfy this equation, the point must lie on the curve. Thus an *equation of any curve* is a statement of the condition which the coordinates of each of the points of the curve, and only these points, must satisfy.

We recall the definition of a *circle* as the locus (totality) of points in the plane which are at a constant distance from a fixed point. Using the distance formula, it is now possible to obtain the general equation of a circle.

With $C(h,k)$ as the center and r as the radius (Fig. 3-5), the

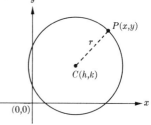

FIGURE 3-5

condition that any point $P(x,y)$ lying on the circle must satisfy is
$$CP = r,$$
which is the same as having the coordinates of P satisfy
$$\sqrt{(x-h)^2 + (y-k)^2} = r$$
from Eq. (3–4). Conversely, if $CP = r$, then P is on the circle. We therefore have
$$(x-h)^2 + (y-k)^2 = r^2, \tag{3-5}$$
which is the general equation of a circle with center (h,k) and with radius r.

EXAMPLE 1. An equation of the circle with its center at $(2,-3)$ and radius 4 is
$$(x-2)^2 + (y+3)^2 = 16.$$

EXAMPLE 2. An equation of the circle with its center at the origin and a radius of 1 is
$$(x-0)^2 + (y-0)^2 = 1,$$
or
$$x^2 + y^2 = 1.$$
This is called an equation of the unit circle, an important special case in our study. Recall Problem 14 of the last section.

We are interested not only in the equation of a circle, but also in the notion of length of portions of its circumference. The length of a circular arc is most useful in discussing and measuring angles.

By considering a circular arc $\overset{\frown}{AB}$, as in Fig. 3–6, it is possible to assign to this arc $\overset{\frown}{AB}$ a length s, although it is impossible to measure the length directly. By denoting several points on $\overset{\frown}{AB}$ by C, D, and so on, the length of the polygonal line joining these points is the sum of the lengths of the appropriate chords, obtained by using formula (3–4).

If we now consider other points on the arc $\overset{\frown}{AB}$, such as E, the new polygonal line $ACDEB$ has a greater length than $ACDB$,

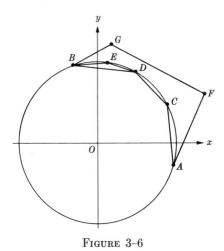

Figure 3-6

since

$$BD < DE + EB.$$

Continuing this process of inserting points, we shall obtain polygonal lines whose lengths are greater than the preceding ones. However, the length of our polygonal line cannot become infinite, for it can be shown that the length of any line obtained in this way is always less than the length of any polygonal line which joins A and B outside the circle, such as $AFGB$. Therefore, we have a number of polygonal lines, each with a length greater than the last, and yet smaller than a fixed number. If the number of chords is increased, and their lengths become arbitrarily close to zero, it can be shown that the length of this polygonal line will approach a definite value. This definite value, s, is called the *arc length* of $\stackrel{\frown}{AB}$.

EXAMPLE. The circumference or arc length of the entire circle with radius r is equal to $2\pi r$, or approximately 6.28 times the radius. Compare this with the two regular geometric figures shown in Fig. 3-7. Since the inscribed hexagon has a perimeter of $6r$ and the circumscribed square has a perimeter of $8r$, the inequality $6r < 2\pi r < 8r$ would be expected.

Problems

1. Write an equation of the following circles:
   (a) center at (3,1) and radius 5,
   (b) center at (4,−2) and radius 3,
   (c) center at (−1,3) tangent to the x-axis,
   (d) center at (2,−4) and passing through (5,−8).
2. Write an equation of the circle with its center at the origin and radius r.
3. Describe the set of points (x,y) which satisfy:
   (a) $(x - 1)^2 + (y - 1)^2 \leq 1$,
   (b) $(x - 2)^2 + (y + 3)^2 \geq 4$,
   (c) $(x - 3)^2 + (y + 1)^2 = 0$.
4. What is the length of the perimeter of the square inscribed in a circle of radius r?
5. Determine the length of the perimeter of the regular hexagon circumscribed about a circle of radius r.
*6. Expressing your answer in terms of π, what is the length of the circumference of the unit circle (a circle with a radius of 1)? What is the length of an arc one-fourth the distance around the circle? What is the length of an arc one-sixth the distance around the circle?

3–5 Angles. In geometry an angle has usually been defined as the configuration consisting of two half-lines (rays) radiating from a point. However, in trigonometry we generalize this definition by stating that an angle thus defined by two half-lines has a measure which corresponds to the amount of rotation required to move a ray from the position of one of these lines to the other. Consider Fig. 3–8,

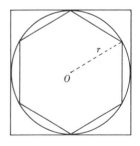

Figure 3–7

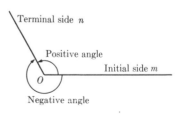

Figure 3–8

with the two lines m and n intersecting at O and lying in a plane perpendicular to our line of vision. If we consider m as the *initial side* and n as the *terminal side* of the angle with O as its vertex, there are two possible directions of rotation of the initial side m. The angle is said to be *positive* if the rotation is counterclockwise, but *negative* if clockwise. A curved arrow will indicate the direction of rotation.

Let us now consider a ray m which issues from the origin of a rectangular coordinate system and coincides with the positive x-axis (Fig. 3–9). As this ray rotates, any point P on m will trace out part or all of the circumference of a circle of radius OP. In fact, the circumference may be traced several times. After the rotation, OP will be in some position OP', where the circular arc $\overparen{PP'}$, denoted by s, may be used to measure the angle POP'. An angle such as POP' is said to be in *standard position*, and to be in the quadrant in which its terminal side OP' is located.

The most logical units for measuring the magnitude of an angle would seem to be the number of revolutions due to the rotation from the initial to the terminal side of the angle. Since the number of revolutions of any angle is determined by the ratio of the intercepted circular arc length s to the circumference of the circle, we define the magnitude of an angle in revolutions as

$$\text{Angle* in revolutions} = \frac{s}{2\pi r}. \qquad (3\text{--}6)$$

For example, if P traces out an arc one-half the circumference, the corresponding angle is one of one-half a revolution. Likewise, if the arc is twice the circumference, the angle measure is two revolutions.

Consider the two concentric circles at O, in Fig. 3–10, with $\overparen{PP'}$ an arc of length s on the circle of radius r, and $\overparen{QQ'}$ an arc of length s' on the circle of radius r'. By using the theorem that similar triangles have proportional sides, and recalling the definition of arc length from Article 3–4, it can be proved that

$$\frac{s'}{r'} = \frac{s}{r},$$

---

* Since we so frequently are considering angles in terms of their magnitudes, we use the word "angle" in place of "measure of angle."

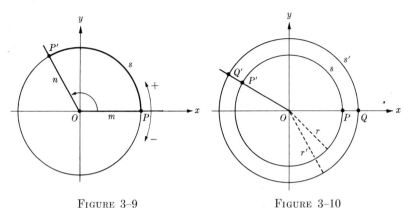

FIGURE 3-9   FIGURE 3-10

and, therefore,
$$\frac{s'}{2\pi r'} = \frac{s}{2\pi r}.$$

The magnitude of any angle is thus independent of the length of its initial or terminal side.

Although the use of revolutions is the most natural method for measuring angles, there are other more convenient systems.

The system most commonly used in elementary work such as surveying and navigation is the *Sexagesimal System*, in which the degree* is the fundamental unit. In this system one revolution $= 360°$, $1° = 60'$ (minutes), and $1' = 60''$ (seconds). Thus:

$$\text{Angle in degrees} = (\text{revolutions})\,(360°). \qquad (3\text{-}7)$$

For example, one-half a revolution is $180°$, or an angle of two revolutions is $720°$.

The other important system used in calculus and other more advanced mathematics is the *Radian System*. Recalling that the circumference of the unit circle is 2π, we define one revolution to be 2π radians and have

$$\text{Angle in radians} = (\text{revolutions})\,(2\pi). \qquad (3\text{-}8)$$

---

* The origin of the use of the degree for measurement is discussed by O. Neugebauer in *Studies of the History of Science*. Philadelphia: University of Pennsylvania Press, 1941. Bicentennial Conference. Chapter on Ancient Astronomy, p. 16.

For example, one-half a revolution is π radians, or an angle of two revolutions is 4π radians. When no unit of measure is designated, radian measure is understood.

We have two immediate results from these definitions. Solving (3–7) and (3–8) for (revolutions) and equating, we find the relationship between an angle expressed in degrees and radians to be

$$\frac{\text{Angle in degrees}}{360°} = \frac{\text{Angle in radians}}{2\pi}. \qquad (3\text{–}9)$$

EXAMPLE 1. An angle of 45° is equal to

$$\frac{45°}{360°} 2\pi = \frac{\pi}{4} \text{ radians.}$$

EXAMPLE 2. An angle of $5\pi/6$ radians is equal to

$$\left(\frac{5\pi/6}{2\pi}\right) 360° = 150°.$$

EXAMPLE 3. To express an angle of one radian in degrees, the same method is used:

$$1 \text{ radian} = \frac{1}{2\pi} 360° = \frac{360°}{2\pi} = 57°18' \text{ (approximately).} \qquad (3\text{–}10)$$

Similarly, for an angle of one degree,

$$1° = 2\pi \left(\frac{1°}{360°}\right) = .01745 \text{ radian (approximately).} \qquad (3\text{–}11)$$

EXAMPLE 4. Transform an angle of 194°23′ to radian measure.

Solution. In Table I, the angles are given in both degrees and radians. We may use this table to change from one system to the other. Since

$$194°23' = 180° + 14°23',$$

we work with 14°23′.

$$14°20' = .2502 \text{ radian,}$$

and

$$14°30' = .2531 \text{ radian.}$$

Since 23' is .3 of the difference between 30' and 20', the radian measure of the required angle is greater than .2502 by .3 of the difference between .2531 and .2502, or

$$14°23' = .2511 \text{ radian.}$$

Therefore

$$194°23' = (\pi + .2511) \text{ radians}$$
$$= 3.3927 \text{ radians (approximately).}$$

The second result shows the advantage of the radian system in measuring angles. Writing (3–8) with the substitution from (3–6), we have

$$\text{Angle in radians} = \frac{s}{2\pi r}(2\pi) = \frac{s}{r}.$$

Therefore the length of the circular arc s cut by a central angle θ (measured in radians) in a circle of radius r is given by

$$s = r\theta. \tag{3-12}$$

EXAMPLE. A circle has a radius of 40 inches. (a) How long is the arc subtended by a central angle of 36°?

Solution. Since the number of radians corresponding to 36° is $\pi/5$,

$$s = 40\left(\frac{\pi}{5}\right) = 8\pi \text{ inches.}$$

Using an approximation for π, this answer can be written to any desired degree of accuracy.

(b) How large is the central angle that subtends an arc of 15 inches?

Solution. Again using $s = r\theta$, $15 = 40\theta$ or $\theta = \frac{3}{8}$ radian. If the result is desired in degrees, we merely change $\frac{3}{8}$ radian into degrees, obtaining 21°30'.

PROBLEMS

1. In a rectangular coordinate system, locate the following angles in standard position, showing the initial and terminal sides. Use a curved arrow to indicate the direction in which the angle is measured.

(a) 1/4 rev.   (b) −3/4 rev.   (c) 3 rev.   (d) 3/8 rev.
(e) −1/6 rev. (f) −5/4 rev.   (g) 5/6 rev. (h) −5/3 rev.

2. Repeat Problem 1 for the following angles expressed in degrees:
(a) 45° (b) 135° (c) −225° (d) −300°
(e) 240° (f) 450° (g) 720° (h) −120°

3. Repeat Problem 1 for the following angles expressed in radians:
(a) $\pi/6$ (b) $2\pi/3$ (c) $\pi/4$ (d) $4\pi/9$
(e) $-3\pi/2$ (f) $-5\pi/6$ (g) $5\pi/12$ (h) -5π

4. Express the angles given in Problems 1 and 2 in radian measure, leaving the answer in terms of π.

5. Express the angles given in Problems 2 and 3 in revolutions.

6. Express the angles given in Problems 1 and 3 in degrees.

7. Transform the following angles to radians, using Table I if needed, and giving the answer to four decimal places.
(a) 27° (b) 156°20′ (c) 47°
(d) 189°32′ (e) 253°10′ (f) −378°49′

8. Transform the following angles to degrees and minutes, using Table I if needed, and giving the answer to the nearest minute.
(a) $\pi/8$ (b) $-2\pi/13$ (c) .2443 (d) −1.3730
(e) 1.8600 (f) 9/4 (g) −1.2900 (h) 5.7200

9. Express each of the following as a single angle in degrees and minutes from 0′ to 59′.
(a) 15°27′ + 32°14′ (b) 18°41′ + 15°12′
(c) 13°32′ + 37°28′ (d) 142°5′ + 8°55′
(e) 29°43′ + 51°38′ (f) 61°19′ + 23°58′
(g) 180° − 15°13′ (h) 90° − 47°38′
(i) 360° − 147°23′ (j) 270° − 63°48′
(k) $\frac{1}{2}(18°47′ + 56°29′)$ (l) $\frac{1}{2}(56°28′ - 47°36′)$

10. If an arc 20 ft long subtends an angle of 2 radians at the center of a circle, find its radius.

11. If a wheel of radius 2 ft rolls 3 ft, how many radians has it turned? How many degrees has it turned?

12. In a circle of radius 14 inches, how long an arc does a central angle of 82° intersect?

13. How many degrees are there between the minute and hour hands of a clock at 4:00 o'clock? At 1:00? At 9:15? At 5:47?

14. Assuming that the earth's radius is 3960 miles, find the distance on the surface of the earth from Columbus, Ohio to the equator. The latitude of Columbus is 40°.

15. If a point on the circumference of a wheel whose diameter is 20 inches travels 3000 ft a minute, through how many radians does the wheel turn in one second?

16. For small angles, the intercepted arc and chord are approximately the same length. Assuming that the earth moves around the sun in a circle of 93,000,000 miles radius, find the sun's diameter if it subtends an angle of 32' at the earth.

*17. In the unit circle ($r = 1$), formula (3–12) becomes $s = \theta$. Explain the meaning of this extremely important formula. What would be the length of the arc of the unit circle intercepted by a central angle of 5 radians? 1 radian? θ radians?

3–6 Functions. In mathematics, the concept of function is most important and useful. In considering the unit circle, discussed in Example 2, Article 3–4, we found its equation to be $x^2 + y^2 = 1$. By subtracting x^2 from both members of this equality, and taking the positive square root, we have

$$y = \sqrt{1 - x^2}, \qquad (3\text{–}13)$$

the equation of the "top" half of the circle (Fig. 3–11). Either from the figure or equation (3–13) we see that for a particular point to lie on the curve, or have its coordinates satisfy the equation, there must be a certain value of y which corresponds to any specific value of x. For example, if $x = 0$, y must be 1; if $x = 1$, y must be 0; or if $x = 1/2$, y must be $\sqrt{3}/2$. This equation (3–13) defines a pairing of numbers in the sense that for each value of x, there is a specified value of y. The x and y are called *variables*. The x is called the *independent variable* since it represents any number substituted in the equation, and y is called the *dependent variable*, since it results from the original choice of the substituted value of x. We express

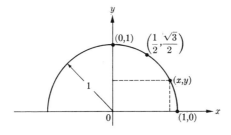

FIGURE 3–11

this relationship between the variables x and y by calling y a *function* of x.

Let us define this basic concept in general. *If two variables x and y are so related that whenever a value is assigned to x, one or more values of y* are determined by some rule or rules, then y is said to be a function of x.*

In this book the rule for determining the value of y will usually be expressed by an equation, a table of values, or the use of a graph. The customary method for designating y as a function of x is to write

$$y = f(x),$$

and to read this "y is a function of x" or "y equals f of x." The symbol $f(a)$ will denote the value of the function when a is substituted for x. Thus for $y = f(x) = \sqrt{1 - x^2}, f(0) = 1, f(1) = 0$, and $f(\frac{1}{2}) = \sqrt{3}/2$.

The set of x values for which the function is defined is called the *domain* of x, while the corresponding values of y form a set of y-values, called the *range* of y. In the case of the function $y = \sqrt{1 - x^2}$, the domain of x is all real values of x which satisfy $-1 \leq x \leq 1$, while the range of y is $0 \leq y \leq 1$.

When it is necessary to refer to more than one function in some discussion, other expressions such as $g(x)$, $F(x)$, $P(x)$ may also be used.

To summarize, in considering the concept of function there are three important parts: the set of x values, the set of y values, and the rule, relationship, or correspondence which associates these two sets.

ILLUSTRATION 1. The equation $y = f(x) = 2x - 6$ expresses y as a first degree function of x. The values of x may be any real numbers, as may the values of y. Specifically $f(0) = -6, f(1) = -4, f(3) = 0, f(4) = 2$, and so on.

ILLUSTRATION 2. The area A of a circle of radius r is given by the expression $A = \pi r^2$. Since π is a quantity which remains fixed in value during the discussion (called a *constant*),

---

* If for any value of x, only one value of y is determined, we call the function *single-valued*.

$$A = f(r) = \pi r^2, \quad \text{for } r \geq 0.$$

For example $f(0) = 0$, $f(1) = \pi$, $f(2) = 4\pi$, and so on. For this function r has for its domain all positive values.

ILLUSTRATION 3. Let $y = f(x)$ be defined

$$f(x) = \begin{cases} x & \text{for } x \geq 0 \\ -x & \text{for } x < 0. \end{cases}$$

This function defined by different expressions over different domains of x, we recall [Eq. (3-1)], may be written $y = |x|$, and was discussed.

ILLUSTRATION 4. Let $f(\theta)$ be defined for any real number θ as the ordinate of the point reached by starting at $(1,0)$ and measuring an arc along the unit circle $x^2 + y^2 = 1$, of length $|\theta|$ in the counterclockwise direction if $\theta > 0$, but in the clockwise direction if $\theta < 0$. This function is defined by a rule, and is not expressible by an algebraic equation. Verify that

$$f(0) = 0, \qquad f(\pi/2) = 1,$$
$$f(\pi) = 0, \qquad f(3\pi/2) = -1.$$

This is a most important function and will be discussed in detail in Chapter 4.

PROBLEMS

1. If $f(x) = 2x - 5$, find $f(0), f(1), f(3), f(-1)$.
2. If $f(x) = x^2 - 7x + 10$, find $f(2), f(5), f(3), f(0)$.
3. If $f(x) = \dfrac{1}{x-3}$, find $f(4), f(2), f(-1)$. What can be said about $f(3)$?
4. If $f(x) = 2^x$, find $f(0), f(1), f(5), f(-1), f(-5)$.
5. If $f(x) = x^{1/2}$, find $f(0), f(2), f(4)$. What domain does x have in order to have real values of $f(x)$?
6. Define $f(x)$ so that $10^{f(x)} = x$. Find $f(1), f(10), f(100), f(\frac{1}{10}), f(\frac{1}{100})$.
7. If $y = f(x) = |x - 2|$, find $f(0), f(2), f(4), f(-2)$. Give the domain for x and the range for y.
8. If x is the length of one side of a square, express the perimeter P as a function of x. Express the area A as a function of x.

9. If x is the length of one side of an equilateral triangle, express the perimeter P as a function of x. Express the area A as a function of x. [*Hint:* In a 30°–60° triangle, the hypotenuse is double the shorter leg.] Expressing the altitude in terms of x, $A = \frac{1}{2}$ base $\times$ altitude.

10. With s measured in feet and t in seconds, the function $s = f(t) = -16t^2 + 32t$ expresses the height of a ball above the ground after t seconds, if it were thrown upward with a velocity of 32 ft/sec. Find $f(0)$, $f(\frac{1}{2})$, $f(1)$, $f(\frac{3}{2})$, $f(2)$, and explain the result.

11. Verify the values of the function given in Illustration 4.

*12. For the same θ as that described in Illustration 4, define $g(\theta)$ as the abscissa of the point on the unit circle, rather than the ordinate. Find $g(0)$, $g(\pi/2)$, $g(\pi)$, $g(3\pi/2)$, $g(2\pi)$.

*13. If $f(x) = x^2$, show that $f(-x) \equiv f(x)$. Any function satisfying the condition $f(-x) \equiv f(x)$ is called an *even* function. Give another example of such a function.

*14. If $f(x) = x^3$, show that $f(-x) \equiv -f(x)$. Any function satisfying this condition is called an *odd* function. Give another example of an odd function.

3–7 Graphical representation of functions. Through the use of the rectangular coordinate system discussed in Article 3–2, we are able to exhibit the relationship between x and y (or any two variables) in the case of any particular function. Although all the points (x,y) whose coordinates satisfy a given relation $y = f(x)$ cannot be plotted, usually a sufficient number may be, so that a good approximation to a picture of the function may be obtained. The aggregate of *all* such points forms the *graph* of the function or the *curve* which represents the function.

EXAMPLE 1. Draw the graph of the function $y = f(x) = 3x - 4$.

Solution. By assigning arbitrary values for x, and computing the corresponding y values, we can obtain any number of points (x,y) whose coordinates satisfy the equation $y = 3x - 4$. The points, arranged in the table, are then plotted, and joined by a smooth curve. This function is represented by a straight line (Fig. 3–12).

x	0	1	2	3
y	-4	-1	2	5

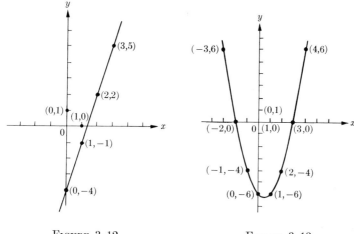

FIGURE 3-12   FIGURE 3-13

EXAMPLE 2. Draw the graph of the function $y = f(x) = x^2 - x - 6$.

Solution. Again, assign values to x, compute the corresponding y values, and arrange the results in a table.

x	0	1	2	3	4	-1	-2	-3
y	-6	-6	-4	0	6	-4	0	6

Drawing a smooth curve through these points (x,y) starting with the point whose abscissa is -3, then -2, -1, and so on, we obtain the result shown in Fig. 3-13. Often additional points such as $(\frac{1}{2}, -6\frac{1}{4})$ must be found, whose coordinates satisfy the relation in order to correctly complete the graph. This curve, called a *parabola*, is the required graph.

We have noticed in both Examples 1 and 2 that the graph of the function crosses the x-axis. At any such point, the y coordinate is zero. The x value at such points is called a zero of the function. In Example 1, the zero of the function is $x = \frac{4}{3}$; in Example 2, there are two zeros of the function, -2 and 3. In general, a *zero of a function* is the abscissa of a point where the graph of the function crosses or touches the x-axis.

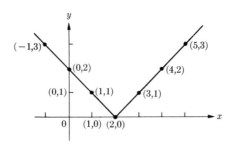

FIGURE 3–14

EXAMPLE 3. Draw the graph of the function $y = f(x) = |x - 2|$.

Solution. We construct the table,

x	0	1	2	3	4	5	-1
y	2	1	0	1	2	3	3

plot the points, and draw the graph. (See Fig. 3–14.) Notice $x = 2$ is the zero of this function, although the curve does not cross the x-axis.

PROBLEMS

Draw the graph of each of the following functions on ruled paper, showing the scales on both axes and giving the zeros of the function in each case:

1. $y = 2x + 5$
2. $y = 6 - 3x$
3. $y = x^2$
4. $y = -4x^2$
5. $y = x^2 - 7x + 10$
6. $y = -x^2 - x + 30$
7. $y = |x - 1|$
8. $y = |2x - 3|$
9. $y = \sqrt{x - 1}$ [Hint: y is never negative.]
10. $y = x^3$
11. $y = x^3 - x$
12. $y = \sqrt{x(2 - x)}$. What is the domain for x and the range for y?

3–8 Graphical representation of empirical data. Certain functional relations are expressed most clearly by a table of statistics or scientific data. Often such material cannot even be expressed by a

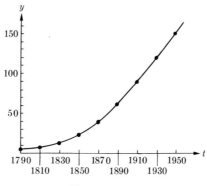

FIGURE 3–15

mathematical formula, but an adaptation of the method of graphing may be used. Consider the example, showing how our population has grown. The table shows the total population at intervals of twenty years since 1790, when the first census was taken.

Year (t)	1790	1810	1830	1850	1870	1890	1910	1930	1950
Population (y) (in millions)	3.9	7.2	12.9	23.2	39.8	62.9	92.0	122.8	150.7

This functional relationship is in no sense mathematical, but we can use an adaptation of our method of graphing to obtain a clear representation of the data. Choosing the t-axis as horizontal, the points are plotted and a smooth curve is drawn through these points to indicate the general trend. This graph is shown in Fig. 3–15.

PROBLEMS

Construct a graph representing the data given in each of the following problems. Indicate the scales clearly in each case.

1. This table lists the percentage of the population of the United States which is foreign-born at ten-year intervals since 1890.

Year (t)	1890	1900	1910	1920	1930	1940	1950
Percentage (y)	14.7	13.6	14.7	13.2	11.6	8.7	6.7

2. This table lists the total horsepower hours of energy (in billions) used from mineral fuel and water power in the United States since 1860.

Year (t)	1860	1880	1900	1920	1940	1950
Energy (y)	25	40	78	190	280	410

3. This table lists the buying power of per capita yearly income for certain years since 1839 in the United States. The figures are adjusted to represent modern purchasing power.

(t)	1839	1859	1879	1899	1919	1929	1933	1941
(y)	198	296	309	482	620	681	495	792

4. This table lists the public school attendance (in millions) of students in the United States since 1880.

(t)	1880	1890	1900	1910	1920	1930	1940	1950
(y)	6.1	8.2	10.6	12.8	16.2	21.3	22.0	21.7

5. This table lists the number of teachers (in units of 10,000) in the United States since 1880.

(t)	1880	1890	1900	1910	1920	1930	1940	1950
(n)	28.7	36.4	42.3	52.3	67.9	85.4	87.5	85.9

CHAPTER 4

THE CIRCULAR FUNCTIONS

4-1 Trigonometry. In about 1600 A.D. Bartholomaus Pitiscus, a professor of mathematics at Heidelberg, wrote the first textbook to bear the title of *Trigonometry*. What he had in mind was exactly what the name implies: triangle measurement. Actually, however, trigonometry had its origin in early historical times. This was a part of the attempt to study and describe the celestial sphere in which the sun, moon, and stars were supposed to move. The two most prominent men interested in these developments were the Greek astronomers, Hipparchus of Nicaea (Second Century B.C.) and Claudius Ptolemy (Second Century A.D.). As a consequence, one often gains the impression that the principle, if not the sole application, of trigonometry is the solving of triangles, and thus that the application of trigonometry lies in the fields of astronomy, navigation, and surveying. This may have been true 2000 or even 400 years ago, but it is certainly not the case today.

With the development of trigonometry, the general study of the circular functions has progressed. In fact, we now define trigonometry as that branch of mathematics which is concerned with the properties and applications of the circular or trigonometric functions.

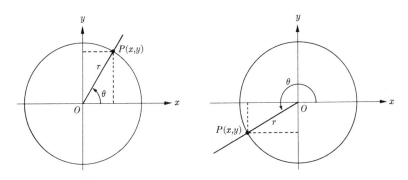

FIGURE 4-1

4–2 Definitions of the circular functions. Consider an angle θ in standard position which has been generated by the portion of the rotating ray of length r (Fig. 4–1). Let (x,y) be the coordinates of P, the point on the terminal side of the angle, a distance r from the origin. We define the *sine, cosine,* and *tangent functions of* θ in terms of these coordinates and of r, the radius of the generated circle. Using the usual abbreviations, we have

$$\sin \theta = \frac{y}{r}, \tag{4-1}$$

$$\cos \theta = \frac{x}{r}, \tag{4-2}$$

and

$$\tan \theta = \frac{y}{x} \quad (x \neq 0). \tag{4-3}$$

These functions are uniquely determined for any specific value of θ [except in (4–3) where, for certain values of θ, $\tan \theta$ is not defined because $x = 0$] and in no way depend upon the distance of P from the origin. To clarify this statement, consider the two concentric circles with r and r' as radii and $P(x,y)$ and $P'(x',y')$ lying on the terminal side of angle θ, as in Fig. 4–2. Since the right

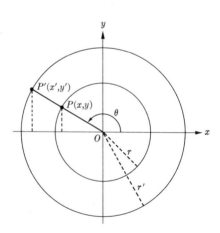

Figure 4–2

triangles formed with the lengths of the coordinates x, y, x', and y' as legs are similar, the corresponding ratios which define these functions are equal.

Notice the relationship between these functions. Since $\tan \theta = y/x \equiv (y/r)/(x/r)$, we have

$$\tan \theta \equiv \frac{\sin \theta}{\cos \theta}, \qquad (4\text{-}4)$$

which is defined for all values of θ except where $\cos \theta = 0$. In this case $\tan \theta$ is undefined (see Article 1–1).

From Article 3–2, we recall the signs of the coordinates of points in the different quadrants. In the same way, we determine in which quadrants the circular functions are positive or negative. Since $r = \sqrt{x^2 + y^2} > 0$ in every quadrant, we have the following table, which the student should verify.

Quadrant	$\sin \theta = y/r$	$\cos \theta = x/r$	$\tan \theta = y/x$
I	+	+	+
II	+	−	−
III	−	−	+
IV	−	+	−

EXAMPLE 1. Let θ be an angle in standard position with its terminal side passing through $(-3,4)$. Find $\sin \theta$, $\cos \theta$, and $\tan \theta$.

Solution. Since $x = -3$ and $y = 4$, $r = \sqrt{(-3)^2 + 4^2} = \sqrt{9+16} = \sqrt{25} = 5$. Thus, $\sin \theta = 4/5$, $\cos \theta = -3/5$, and $\tan \theta = -4/3$.

EXAMPLE 2. Find the value of $\cos \theta$ and $\tan \theta$ if $\sin \theta = -5/13$ and $\tan \theta > 0$.

Solution. Since $\sin \theta < 0$ and $\tan \theta > 0$, θ terminates in the third quadrant. Moreover, with $\sin \theta = -5/13$, we may assume that the terminal side of θ passes through $(-12,-5)$. (Why?) Therefore $\cos \theta = -12/13$, and $\tan \theta = 5/12$.

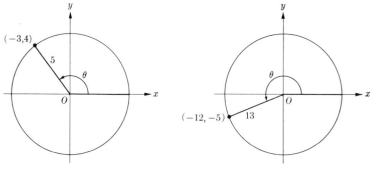

FIGURE 4–3   FIGURE 4–4

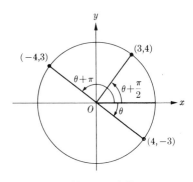

FIGURE 4–5

EXAMPLE 3. Let θ be an angle in standard position with its terminal side passing through $(4,-3)$. Find the sine, cosine, and tangent of $\theta + \pi/2$, and $\theta + \pi$.

Solution. By similar right triangles the terminal side of $\theta + \pi/2$, in standard position, passes through $(3,4)$, while that of $\theta + \pi$ passes through $(-4,3)$. Since $r = 5$ in all cases, we have

$$\sin(\theta + \pi/2) = 4/5, \quad \sin(\theta + \pi) = 3/5,$$
$$\cos(\theta + \pi/2) = 3/5, \quad \cos(\theta + \pi) = -4/5,$$
$$\tan(\theta + \pi/2) = 4/3, \quad \tan(\theta + \pi) = -3/4.$$

If θ were increased or decreased by an integral multiple of 2π (or 360°) the terminal side of the new angle would coincide with the original terminal side, so that P would have the same coordinates. Therefore

$$\sin [\theta + k(2\pi)] \equiv \sin \theta, \qquad (4\text{--}5)$$

and

$$\cos [\theta + k(2\pi)] \equiv \cos \theta, \qquad (4\text{--}6)$$

where k is any integer. Hence, if we know the values of $\sin \theta$ and $\cos \theta$ in the range $0 \leqslant \theta \leqslant 2\pi$, we know their values for all θ. The sine and cosine functions are called *periodic functions* with a period of 2π.*

The tangent function differs from the sine and cosine functions. Since $\tan \theta = y/x = -y/-x$, and the point $P'(-x,-y)$ lies on the terminal side of the angle $\theta + \pi$,

$$\tan (\theta + k\pi) \equiv \tan \theta, \qquad (4\text{--}7)$$

where k is any integer. The tangent function is periodic with a period of π.

There are three other circular functions of less importance than those already defined. The *cosecant, secant,* and *cotangent functions* are defined in terms of the coordinates of $P(x,y)$ as follows:

$$\csc \theta = \frac{r}{y} \quad (y \neq 0), \qquad (4\text{--}8)$$

$$\sec \theta = \frac{r}{x} \quad (x \neq 0), \qquad (4\text{--}9)$$

and

$$\cot \theta = \frac{x}{y} \quad (y \neq 0). \qquad (4\text{--}10)$$

It will be noticed immediately by using Eqs. (4–1,2,3) that

---

* In general, any function of θ is said to be *periodic* with period p, provided the function of $\theta + p$ is equal to the function of θ, and p is the smallest constant for which this is true. Note how $k = 1$ in Eq. (4–5) gives this smallest constant, 2π.

and
$$\sin \theta \csc \theta \equiv 1, \tag{4-11}$$
$$\cos \theta \sec \theta \equiv 1, \tag{4-12}$$
$$\tan \theta \cot \theta \equiv 1. \tag{4-13}$$

Because of this type of relationship, these three functions are called the *reciprocal functions*. If $\sin \theta$, $\cos \theta$, and $\tan \theta$ are known, any of these other three functions can be found immediately. For this reason we shall not emphasize them.

Problems

Which of the following expressions are positive? Which are negative?

1. $\sin 164°$
2. $\cos 158°$
3. $\tan 195°$
4. $\cos 327°$
5. $\tan 227°$
6. $\sin 264°$
7. $\sin (-38°)$
8. $\cot (-125°)$
9. $\tan (-213°)$
10. $\cos 2\pi/3$
11. $\tan 5\pi/4$
12. $\sin (-11\pi/6)$
13. $\cot 9\pi/7$
14. $\sec (-7\pi/8)$
15. $\csc (-\pi/8)$

Assuming that θ is in standard position, determine the quadrants in which θ may lie under the following conditions:

16. $\sin \theta > 0$
17. $\cos \theta > 0$
18. $\tan \theta > 0$
19. $\sin \theta < 0$
20. $\cos \theta < 0$
21. $\sec \theta > 0$
22. $\sin \theta > 0$ and $\cos \theta > 0$
23. $\cos \theta > 0$ and $\sin \theta < 0$
24. $\tan \theta > 0$ and $\sin \theta < 0$
25. $\sin \theta > 0$ and $\cos \theta < 0$
26. $\sin \theta < 0$ and $\cos \theta < 0$

In each of the following the terminal side of the angle, in standard position, passes through the indicated point. Sketch and find the circular functions of each angle.

27. $(3,4)$
28. $(-5,12)$
29. $(-4,3)$
30. $(24,-7)$
31. $(-8,-15)$
32. $(10,-8)$
33. $(0,-2)$
34. $(-1,7)$
35. $(6,0)$

Find the values of $\sin \theta$, $\cos \theta$, and $\tan \theta$ under the following conditions:

36. $\sin \theta = 5/13$, θ in the first quadrant.
37. $\cos \theta = -4/5$, θ in the third quadrant.

4-3] THE UNIT CIRCLE 73

38. $\tan \theta = -1/3$, θ in the second quadrant.
39. $\sin \theta = 2/3$, θ not in the first quadrant.
40. $\cot \theta = 4/3$, θ not in the first quadrant.
41. $\cos \theta = 5/6$, θ not in the first quadrant.
42. $\tan \theta = 1/2$, and $\sin \theta$ is positive.
43. $\sin \theta = -3/5$, and $\tan \theta$ is positive.
44. $\cos \theta = -7/9$, and $\tan \theta$ is negative.
45. $\tan \theta = 2/3$.
46. $\sec \theta = -5/3$.
47. $\csc \theta = 12/7$.

In Problems 48 through 52, the terminal side of the angle θ, in standard position, passes through (8,15).

48. Sketch and find the circular functions of $\theta + \pi/2$.
49. Sketch and find the circular functions of $\theta + 180°$.
50. Sketch and find the circular functions of $\theta + 270°$.
51. Sketch and find the circular functions of $\theta - \pi$.
52. Sketch and find the circular functions of $\theta - 90°$.

*53. For any angle θ in standard position with its terminal side intersecting the circle of radius r at P, show that P has coordinates $(r \cos \theta, r \sin \theta)$. This is a most important concept, and should be emphasized.

4-3 The unit circle. We have mentioned the *unit circle* with its center at the origin and a radius of one. Since the circular functions are independent of the radius of the circle, many general properties can be obtained through the use of this circle.

Consider the angle θ in standard position (Fig. 4-6), with its terminal side intersecting the unit circle at $P(x,y)$. Since $r = 1$, we have $\sin \theta = y$ and $\cos \theta = x$. In other words, P is the point with coordinates $(\cos \theta, \sin \theta)$. For certain angles the values of their circular functions are immediately apparent. The terminal side of the angle $0° = 0$ radians coincides with the initial side, so that the

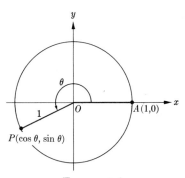

FIGURE 4-6

coordinates of P are $(1,0)$. Hence,

$$\sin 0 = 0, \qquad \tan 0 = \frac{0}{1} = 0,$$

$$\cos 0 = 1, \qquad \sec 0 = \frac{1}{1} = 1,$$

while cot 0 and csc 0 (i.e., 1/0) are undefined.

The angle $90° = \pi/2$ has its terminal side intersecting the unit circle at the point $(0,1)$. Using this fact, the values of circular functions of $90°$ are found. The functional values of the other *quadrantal angles* (any angle which is a multiple of $90°$) are also obtained by noting the coordinates of the point where the terminal side of the angle intersects the unit circle.

It is now clear how the values of these functions change as θ increases from 0 to 2π. Verify the following table:

Quadrant	θ varies from	Value of $\sin \theta$ varies from	Value of $\cos \theta$ varies from
I	0 to $90°$ (0 to $\pi/2$)	0 to 1	1 to 0
II	$90°$ to $180°$ ($\pi/2$ to π)	1 to 0	0 to -1
III	$180°$ to $270°$ (π to $3\pi/2$)	0 to -1	-1 to 0
IV	$270°$ to $360°$ ($3\pi/2$ to 2π)	-1 to 0	0 to 1

The fact that the values of the sine and cosine function are never larger than one is apparent from the table. That this must be so is clear, since they are the coordinates of some point on the unit circle. The tangent, however, can have any positive or negative real number for its value.

As the point $P(\cos \theta, \sin \theta)$ is thought of as moving around the unit circle, we have observed the changes in both $\sin \theta$ and $\cos \theta$. If we wish to consider the behavior of $\sin \theta$ alone, a study of the graph of $y = \sin \theta$ is most convenient, with values of θ plotted on the horizontal axis, and values of y on the vertical axis. Such a graph may be obtained directly from the unit circle.

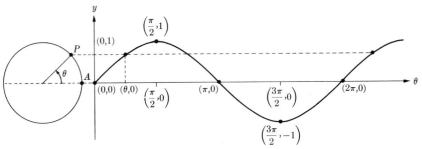

FIGURE 4-7

Consider a rectangular coordinate system with intervals $\pi/2$ in length laid off along the θ-axis, as in Fig. 4-7. A unit circle is drawn with its center on the θ-axis. To obtain the graph of $y = \sin \theta$, the ordinate of P is carried over to the point whose abscissa is θ. As many points as may be desired on the graph are obtained by repeated bisection of the four quadrants. By drawing a smooth curve through the points located in this way, we have a graph of $y = \sin \theta$.* The curve continues indefinitely to the right and left, but need only be drawn from 0 to 2π. (Why?)

With the use of radian measure and the unit circle, the value of θ corresponds exactly to the length of its subtended arc. Thus in Fig. 4-7, the length of the arc $\overset{\frown}{AP}$ is equal to the distance from 0 to θ on the θ-axis. Notice how clearly the following properties are shown by the graph of $y = \sin \theta$.

1. This function is periodic with period 2π. [Recall Eq. (4-5).]
2. In the first quadrant this function is an *increasing function*.† In what other quadrant is this the case?
3. The values of $\sin \theta$ are positive for θ, in the first and second quadrants, and negative in the third and fourth quadrants.
4. The values of $\sin \theta$ lie between -1 and $+1$.
5. The zeros of $\sin \theta$ are at multiples of π,

$$\sin k\pi = 0, \quad k = 0, \pm 1, \pm 2, \cdots.$$

---

* Such a graph may also be plotted using points, whose coordinates are obtained from Table I in this book. A further discussion of graphing appears in Chapter 10.

† If $\theta_1 > \theta_2$, then $\sin \theta_1 > \sin \theta_2$.

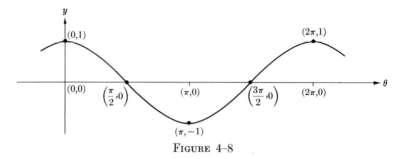

FIGURE 4–8

6. This function has the property that $f(\theta) \equiv -f(-\theta)$, specifically $\sin \theta \equiv -\sin(-\theta)$ [see Eq. (5–7)]. Thus the sine function is odd. Recall Problem 14, Article 3–6.

In Fig. 4–8, the graph of $y = \cos \theta$ is shown, which may be constructed by a procedure similar to the one just described. Using this figure, it is helpful to make a list of some of the properties of $y = \cos \theta$.

We have already noticed certain relations between the various functions such as (4–4,11,12,13). Others can also be derived. Since $\cot \theta \equiv 1/\tan \theta$ [see (4–13)], and $\tan \theta \equiv \sin \theta / \cos \theta$ [see (4–4)],

$$\cot \theta \equiv \frac{\cos \theta}{\sin \theta}. \quad (4\text{–}14)$$

More important, however, are the following results. Since the coordinates of any point on the unit circle are $(\cos \theta, \sin \theta)$, and these coordinates must satisfy the equation of the unit circle, $x^2 + y^2 = 1$, we have

$$\sin^2 \theta + \cos^2 \theta \equiv 1 \quad (4\text{–}15)$$

for any value of θ, where $\sin^2 \theta$ is the usual notation for $(\sin \theta)^2$, and so on. By dividing each term of (4–15) by $\cos^2 \theta$, we get

$$\frac{\sin^2 \theta}{\cos^2 \theta} + 1 \equiv \frac{1}{\cos^2 \theta},$$

or

$$\tan^2 \theta + 1 \equiv \sec^2 \theta. \quad (4\text{–}16)$$

By dividing each term of (4–15) by $\sin^2 \theta$, we get

$$1 + \frac{\cos^2 \theta}{\sin^2 \theta} \equiv \frac{1}{\sin^2 \theta},$$

or
$$1 + \cot^2 \theta \equiv \csc^2 \theta. \tag{4-17}$$

EXAMPLE. Find all six functions of θ if $\cos \theta = 3/5$.

Solution. This example is similar to Example 2, Article 4–2. We can now solve it by a second method. Using (4–12), we have $\sec \theta = 5/3$. From (4–15) $\sin \theta \equiv \pm\sqrt{1 - \cos^2\theta} = \pm\sqrt{1 - 9/25} = \pm\sqrt{16/25} = \pm 4/5$. The sign depends on the quadrant of θ. Since $\cos \theta > 0$, we know θ is in the first or fourth quadrant. If it were in the first, the plus sign would appear before the $4/5$; if it were in the fourth, the minus sign would be chosen. From (4–4), $\tan \theta \equiv \sin \theta / \cos \theta = \pm(4/5)/(3/5) = \pm 4/3$. Using (4–11), $\cot \theta = \pm 3/4$, and (4–8) gives us $\csc \theta = \pm 5/4$.

PROBLEMS

1. What are the values of the six circular functions of θ when $\theta = \pi$? Recall that the coordinates of P in this case are $(-1,0)$.

2. What are the values of the six circular functions of $\theta = 270°$?

3. What are the values of the six circular functions of 2π? How do these values compare with those for $\theta = 0$?

4. What are the values of the six circular functions of $-\pi/2$?

5. What are the values of the six circular functions of the following angles?

(a) 3π (b) $7\pi/2$ (c) $9\pi/2$
(d) 16π (e) 15π (f) 100π
(g) $450°$ (h) $720°$ (i) $900°$

6. What are the values of the six circular functions of the following angles?

(a) -5π (b) $-\pi/2$ (c) $-3\pi/2$
(d) -6π (e) $-11\pi/2$ (f) -40π
(g) $-180°$ (h) $-630°$ (i) $-450°$

7. Verify the table in this article for the variation of $\sin \theta$ and $\cos \theta$.

8. Make a similar table for the other four functions.

9–20. Do Problems 36 through 47 of Article 4–2 by the method outlined in the example of this article.

21. Show that $\sin (k\pi) = 0$ and $\cos (k\pi) = (-1)^k$ for any integer k (positive, negative, or zero).

*22. Construct a graph of $y = \cos \theta$, as in Fig. 4–8, indicating the procedure used. From the graph list properties for $y = \cos \theta$, similar to those listed for $y = \sin \theta$.

23. In the same rectangular coordinate system sketch both the curve $y = \sin \theta$ and $y = \cos \theta$, one curve superimposed on the other. Although the following facts will be proved in Chapter 5, could they be deduced from the figure?

(a) $\sin (\pi/2 + \theta) \equiv \sin (\pi/2 - \theta)$,
(b) $\sin (-\theta) \equiv -\sin \theta$,
(c) $\cos (\pi/2 - \theta) \equiv -\cos (\pi/2 + \theta)$,
(d) $\cos (-\theta) \equiv \cos \theta$,
(e) $\cos \theta \equiv \sin (\pi/2 - \theta)$,
(f) $\sin \theta \equiv \cos (\pi/2 - \theta)$.

*24. Obtain an expression in terms of θ for the length of any chord (in a unit circle) whose corresponding arc is subtended by the central angle θ. *Hint.* In Fig. 4–6, the coordinates of A are $(1,0)$ and of P are $(\cos \theta, \sin \theta)$. Using the distance formula (3–4),

$$AP = \sqrt{(1 - \cos \theta)^2 + \sin^2 \theta}$$
$$= \sqrt{1 - 2\cos \theta + \cos^2 \theta + \sin^2 \theta},$$

or

$$\text{Length of Chord} = \sqrt{2 - 2\cos \theta}. \qquad (4\text{–}18)$$

25. Find the length of the chord in a unit circle, if the subtending angle is (a) $\pi/2$, (b) π.

26. In Fig. 4–9, the terminal side of angle θ intersects the unit circle at P. Draw the tangent to the circle at A and extend the terminal side from the origin O through P. The point at which these lines intersect can be called R. Labeling the point $(0,1)$ on the circle B, draw the tangent to the circle through this point. This tangent line will intersect the line from O through P

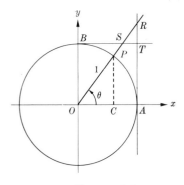

FIGURE 4–9

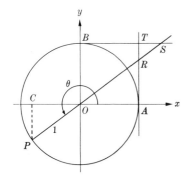

FIGURE 4–10

4–4]   VALUES OF SPECIAL ANGLE FUNCTIONS   79

and the tangent line to the circle at A. Call these points S and T respectively. (a) Verify for this figure that $\tan \theta$ is equal in magnitude to the length of the line segment AR. Draw a line from P perpendicular to the x-axis. Calling the foot of this perpendicular C, triangle OPC is similar to ORA. Use the ratio of similar sides of these triangles to obtain the result. (b) Verify that $\cot \theta$ is equal in magnitude to the length of the line segment BS.

27. Figure 4–10 has been drawn in the same manner as Fig. 4–9 except that P is in the third quadrant. In a similar manner, construct two other figures, one where θ is in the second quadrant and one where θ is in the fourth quadrant.

28. Employing the figures from Problem 27, Fig. 4–9 and Fig. 4–10, and the results of Problem 26, verify that the following definitions are equivalent to the original definitions: (a) $\tan \theta$ is the y-coordinate of the point R, (b) $\cot \theta$ is the x-coordinate of the point S.

*29. By using the unit circle and the definition for $\tan \theta$ given in Problem 28, construct a graph of $y = \tan \theta$ (see Fig. 4–11).

30. From the graph of $y = \tan \theta$ given in Fig. 4–11, list properties for this function similar to those listed for $y = \sin \theta$.

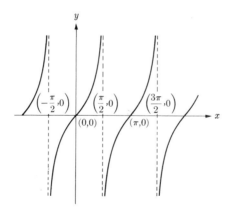

FIGURE 4–11

4–4 Values of special angle functions.

In addition to the quadrantal angles, the circular functions of certain other angles can be found exactly. Consider the angle $\pi/6 = 30°$, located in standard

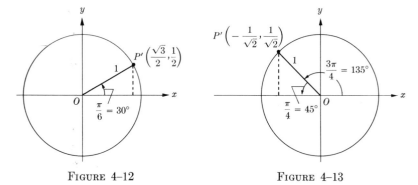

FIGURE 4-12   FIGURE 4-13

position with the unit circle in Fig. 4-12. In plane geometry, it is shown that in a right triangle with an angle of 30°, the hypotenuse is twice as long as the side opposite this angle. Since the point P' lies on the unit circle and has $1/2$ for its y-coordinate, substituting in $x^2 + y^2 = 1$, the equation of the circle, we have $x^2 + (1/2)^2 = 1$ or $x^2 = 3/4$. Thus, in the first quadrant, $x = \sqrt{3}/2$, and

$$\sin 30° = 1/2, \qquad \cot 30° = \sqrt{3},$$
$$\cos 30° = \sqrt{3}/2, \qquad \sec 30° = 2/\sqrt{3},$$
$$\tan 30° = 1/\sqrt{3}, \qquad \csc 30° = 2.$$

By properly placing the 30°-60° triangle in the unit circle, the coordinates of a point on the terminal side of any nonquadrantal angle of size $k \cdot 30°$, with k an integer, can be found. From these the values of the circular functions of any such angle result.

Again consider $\theta = 3\pi/4 = 135°$. Since $135° = 90° + 45°$ in Fig. 4-13, the right triangle, with OP' as hypotenuse and also as the terminal side of the angle 135°, is isosceles. Since P' lies on the unit circle and has x- and y-coordinates equal in size, with $y = -x$, on substituting, we have $x^2 + x^2 = 2x^2 = 1$, or $x = -1/\sqrt{2}$. Hence,

$$\sin 135° = 1/\sqrt{2}, \qquad \cot 135° = -1,$$
$$\cos 135° = -1/\sqrt{2}, \qquad \sec 135° = -\sqrt{2},$$
$$\tan 135° = -1, \qquad \csc 135° = \sqrt{2}.$$

The values of the circular functions of any other odd multiple of 45° can be obtained by similar methods.

Problems

Draw a figure showing each angle in the unit circle, and verify the following by finding the exact values.

1. (a) $\sin \pi/3 = \cos \pi/6$ (b) $\sin 30° = \cos 60°$
2. (a) $\sin \pi/6 = \sin 5\pi/6$ (b) $\cos 150° = -\cos 30°$
3. (a) $\tan 45° = \tan 225°$ (b) $\cot 135° = \cot 315°$
4. (a) $\sec 11\pi/6 = \sec \pi/6$ (b) $\csc 2\pi/3 = -\csc 4\pi/3$
5. (a) $\sin 120° = \sin(-240°)$ (b) $\cos 7\pi/6 = \cos(-5\pi/6)$
6. (a) $\sin 60° = 2 \sin 30° \cos 30°$
   (b) $\sin \pi/2 = 2 \sin \pi/4 \cos \pi/4$
7. (a) $\sin \pi/6 = \sqrt{\dfrac{1 - \cos \pi/3}{2}}$ (b) $\cos \pi/6 = \sqrt{\dfrac{1 + \cos \pi/3}{2}}$
8. $\tan 30° = \dfrac{1 - \cos 60°}{\sin 60°} = \dfrac{\sin 60°}{1 + \cos 60°}$
9. $\cos 60° = \cos^2 30° - \sin^2 30° = 2\cos^2 30° - 1$

Find the exact numerical values of the following:

10. (a) $\sin^2 \pi/6 + \cos^2 \pi/6$ (b) $\sin^2 0 + \cos^2 0$
11. (a) $\sec^2 \pi/3 - \tan^2 \pi/3$ (b) $\sec^2 5\pi/4 - \tan^2 5\pi/4$
12. (a) $\csc^2 315° - \cot^2 315°$ (b) $\csc^2 135° - \cot^2 135°$
13. $\sin 2\pi/3 + \cos 7\pi/6 + \tan 5\pi/3$
14. $\tan 5\pi/4 + \cot 7\pi/4 - \sec 5\pi/6$
15. $\csc 150° - \cos 240° + \tan 120°$
16. $\sin 120° \cos 150° + \cos 120° \sin 150°$
17. $\cos 3\pi/4 \cos \pi/4 - \sin 3\pi/4 \sin \pi/4$
18. $\sin 330° \cos 120° \tan 135°$
19. $(\cos 11\pi/6 + \sin \pi/3)(\tan \pi/6 + \cot 4\pi/3)$
20. $(\tan 5\pi/4 + \sin 3\pi/2) \cos 5\pi/6$

Find all the angles between 0° and 360° that satisfy each of the following equations, and express your answers in degrees and in radians:

21. $\sin \theta = 1/2$
22. $\cos \theta = -1/2$
23. $\tan \theta = 1/\sqrt{3}$
24. $\sin \theta = -\sqrt{3}/2$
25. $\tan \theta = -1$
26. $\cos \theta = -\sqrt{2}/2$
27. $\sin \theta = \sqrt{2}/2$
28. $\sec \theta = 2$
29. $\cot \theta = -\sqrt{3}$
30. $\csc \theta = 2/\sqrt{3}$

31. By drawing a figure for each of the following angles in the unit circle, make a table giving the angle in both degrees and radians, and the six circular functions of each: 0°, 30°, 45°, 60°, 90°, 120°, 135°, 150°, 180°, 210°, 225°, 240°, 270°, 300°, 315°, 330°, 360°.

32. Using formula (4–18) find the length of the chord in a unit circle if the subtending angle is (a) $\pi/6$, (b) $\pi/3$, (c) $2\pi/3$.

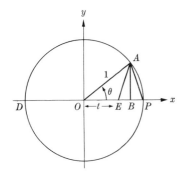

FIGURE 4–14

4–5 Exact values of the functions for $\theta = \pi/5$. There is one other special value of an angle which is of interest. Although a slightly more complicated construction is required for $36° = \pi/5$, we are able to find exact values for the circular functions of this angle, and thus have exact values for the functions of π, $\pi/2$, $\pi/3$, $\pi/4$, $\pi/5$, and $\pi/6$. Consider the unit circle with center O and radius OP (see Fig. 4–14). Locate the point E on the radius OP, so that

$$\frac{\text{Length } OP}{\text{Length } OE} = \frac{\text{Length } OE^*}{\text{Length } EP}.$$

By letting t denote the length of OE, we have the relationship

$$\frac{1}{t} = \frac{t}{1-t}. \qquad (4\text{–}19)\dagger$$

---

* This location is possible with the use of ruler and compass, and is essentially the same as that used for constructing a regular pentagon.

† The ratio defined by Eq. (4–19) has been considered since the time of the Greek mathematicians, and is called the *golden section*. Much has been written on this subject which would be of interest to the student as additional reading. For example:

1. H. V. Baravalle, The Geometry of the Pentagon and the Golden Section, *Mathematics Teacher*, Jan. 1948.

2. W. W. Rouse Ball, *Mathematical Recreations and Essays*, rev. by H. S. M. Coxeter, 11th Ed. London: Macmillan and Company, Limited, 1940.

3. R. Courant and H. Robbins, *What is Mathematics?* New York: Oxford University Press, 1941.

4. Jay Hambridge, *The Elements of Dynamic Symmetry*. New York: Brentano's, 1926.

Solving this quadratic equation* for the positive value of t, we find $t = (\sqrt{5} - 1)/2$.

Locate A so that the length of the chord AP is equal to t. Therefore the triangles OPA and AEP are similar, and thus triangle AEP is isosceles, so that the length of $AE = t$, and triangle OEA is also isosceles. From this it follows directly that $\angle OPA$ is twice $\angle AOP$ (see Problems 3, 4, and 5). Since $\angle OPA$ is measured by $\frac{1}{2}\overset{\frown}{AD}$,† $\angle AOP$ is measured by $\frac{1}{4}\overset{\frown}{AD}$. Also $\angle AOP$ is measured by $\overset{\frown}{AP}$. Thus $\overset{\frown}{AD} = 4\overset{\frown}{AP}$. But $\overset{\frown}{AD} + \overset{\frown}{AP} = \pi$. As a direct result, $\overset{\frown}{AP} = \pi/5$ and $\theta = \pi/5 = 36°$.

By considering AB perpendicular to OP at B, we easily find the exact lengths of OB and AB, that is, the cosine and sine of $\pi/5$. The length of $EP = 1 - t = 1 - \dfrac{\sqrt{5} - 1}{2} = \dfrac{3 - \sqrt{5}}{2}$; thus the length of EB is $\dfrac{3 - \sqrt{5}}{4}$. Therefore the length of $OB = \dfrac{\sqrt{5} - 1}{2} + \dfrac{3 - \sqrt{5}}{4} = \dfrac{\sqrt{5} + 1}{4}$. Also, in the right triangle OAB, (length of $AB)^2 = 1 - \left(\dfrac{\sqrt{5} + 1}{4}\right)^2 = \dfrac{5 - \sqrt{5}}{8}$. Thus

$$\sin \frac{\pi}{5} = \sqrt{\frac{5 - \sqrt{5}}{8}}, \qquad \cos \frac{\pi}{5} = \frac{\sqrt{5} + 1}{4}. \qquad (4\text{--}20)$$

Should the numerical values of the other functions be desired, they can now be found.

Problems

Referring to Fig. 4–14, prove the following in detail:

1. Triangle AEP is isosceles.
2. Triangle OEA is isosceles.

---

* The solution of quadratic equations is discussed in Article 6–3. See Problem 3 (second list).

† Use is made of the theorem in plane geometry which states: an angle inscribed in a circle is measured by one-half the intercepted arc.

3. Angles AOP, OAE, and EAP are equal.
4. $\angle OAP = \angle OPA$.
5. $\angle APO = $ twice $\angle AOP$.

6. Using formula (4–18), find the length of the chord in a unit circle which is subtended by the angle $\pi/5$. How does this value compare with the value of t?

CHAPTER 5

CIRCULAR FUNCTIONS INVOLVING MORE THAN ONE ANGLE

This chapter deals with two additional general types of circular function relations, which differ from those studied thus far in that they contain functions of more than one angle. Not only do we frequently encounter the problem of expressing a circular function of θ plus some multiple of $\pi/2$ as a function of θ only, but often we wish to consider circular functions of two angles in general. Such functions, we shall see, can be expressed in terms of functions of each angle separately. More specifically, we know that $\sin(\pi/4 + \pi/3)$ is not equal to $\sin \pi/4 + \sin \pi/3$, since the latter is equal to $1/\sqrt{2} + \sqrt{3}/2$ (a value greater than one), an impossible value for the sine function. However, $\sin(\pi/4 + \pi/3)$ can be expressed in terms of functions of $\pi/4$ and $\pi/3$, and its value readily obtained. We find it convenient to develop first a formula for $\cos(\alpha - \beta)$.

5-1 Proof of the formula for $\cos(\alpha - \beta)$. We wish to prove one of the most basic formulas of Trigonometry,

$$\cos(\alpha - \beta) \equiv \cos \alpha \cos \beta + \sin \alpha \sin \beta. \tag{5-1}$$

As in Fig. 5-1, let α and β be any two angles in standard position in

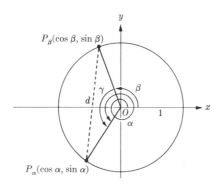

FIGURE 5-1

the unit circle. Denote by P_α and P_β the points where the terminal sides of the angles intersect the unit circle. Since the coordinates of P_α are (cos α, sin α) and of P_β are (cos β, sin β), the distance d between these two points [Eq. (3–4)] is given by

$$d = \sqrt{(\cos \alpha - \cos \beta)^2 + (\sin \alpha - \sin \beta)^2}$$
$$= \sqrt{\cos^2 \alpha - 2 \cos \alpha \cos \beta + \cos^2 \beta + \sin^2 \alpha - 2 \sin \alpha \sin \beta + \sin^2 \beta}$$
$$= \sqrt{2 - 2(\cos \alpha \cos \beta + \sin \alpha \sin \beta)}.$$

By recalling (4–18), the expression for the length of any chord whose subtending angle is θ in the unit circle, we have another expression for d. Since d represents the length of the chord whose subtending angle is γ, where $\gamma = (\alpha - \beta) \pm$ some integral multiple of 360°,

$$d = \sqrt{2 - 2 \cos \gamma}$$
$$= \sqrt{2 - 2 \cos (\alpha - \beta)}.$$

Upon equating these two expressions for d and simplifying, we see that
$$\cos (\alpha - \beta) \equiv \cos \alpha \cos \beta + \sin \alpha \sin \beta, \qquad (5\text{–}1)$$

which was to be proved. It should be pointed out, and emphasized, that this formula expresses the cosine of $\alpha - \beta$ in terms of the functions of α and β themselves, and holds for any value of α or β. It is in this respect that the formula becomes important.

5–2 Special reduction formulas. As is probably somewhat evident by this time, any circular function of any angle can be expressed as a function of an angle between zero and $\pi/4$. This is shown by using certain reduction formulas derived from (5–1) with special values for α or β.

If we replace α by 90° in (5–1), we obtain

$$\cos (90° - \beta) \equiv \cos 90° \cos \beta + \sin 90° \sin \beta.$$

Since cos 90° = 0 and sin 90° = 1, we have

$$\cos (90° - \beta) \equiv \sin \beta. \qquad (5\text{–}2)$$

It should be emphasized that these relations hold for all angles α or β, and the reader should observe this as the formulas are derived. If

5-2] SPECIAL REDUCTION FORMULAS

in (5-2) we replace β by $90° - \beta$, the relation

$$\cos [90° - (90° - \beta)] \equiv \sin (90° - \beta)$$

holds for all β. Simplifying, this becomes

$$\sin (90° - \beta) \equiv \cos \beta. \tag{5-3}$$

From (5-2) and (5-3), we immediately see that

$$\tan (90° - \beta) \equiv \cot \beta, \tag{5-4}$$

and

$$\cot (90° - \beta) \equiv \tan \beta. \tag{5-5}$$

Although true for all values, these four results are especially useful for acute angles when computation is involved. For example, $\sin 56° = \cos 34°$, $\tan 81° = \cot 9°$, or $\cos 72° = \sin 18°$.

We have noticed that circular functions have names which can be paired. In each pair, one function is the *cofunction* of the other. The sine is the cofunction of the cosine, and the cosine is the cofunction of the sine, and so on. With this concept of cofunction, the four relations above can be stated: The cofunction of any angle equals the function of the *complementary* angle.

The relationship between the functions of any angle and its negative are most useful, and also follow from (5-1). Letting $\alpha = 0°$,

$$\cos (0° - \beta) \equiv \cos 0° \cos \beta + \sin 0° \sin \beta.$$

Since $\cos 0° = 1$ and $\sin 0° = 0$,

$$\cos (-\beta) \equiv \cos \beta. \tag{5-6}$$

Also, if we replace β by $-\beta$ in (5-2),

$$\cos (90° + \beta) \equiv \sin (-\beta).$$

Since $90° + \beta = \beta - (-90°)$,

$$\sin (-\beta) \equiv \cos [\beta - (-90°)]$$
$$\equiv \cos \beta \cos (-90°) + \sin \beta \sin (-90°)$$
$$\equiv (\cos \beta)(0) + (\sin \beta)(-1),$$

or

$$\sin (-\beta) \equiv -\sin \beta. \tag{5-7}$$

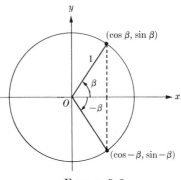

FIGURE 5-2

It follows directly that

$$\tan(-\beta) \equiv -\tan\beta. \tag{5-8}$$

For example, $\sin(-15°) = -\sin 15°$, and $\tan(-176°) = -\tan 176°$, but $\cos(-279°) = \cos 279°$. Notice the significance of (5-6) and (5-7) in Fig. 5-2.

5-3 General addition formulas. We shall now derive the general formulas for the sine, cosine, and tangent of the sum or difference of two angles. Since

$$\cos(\alpha + \beta) \equiv \cos[\alpha - (-\beta)]$$
$$\equiv \cos\alpha\cos(-\beta) + \sin\alpha\sin(-\beta),$$

we have

$$\cos(\alpha + \beta) \equiv \cos\alpha\cos\beta - \sin\alpha\sin\beta. \tag{5-9}$$

Also,

$$\sin(\alpha + \beta) \equiv \cos[90° - (\alpha + \beta)]$$
$$\equiv \cos[(90° - \alpha) - \beta]$$
$$\equiv \cos(90° - \alpha)\cos\beta + \sin(90° - \alpha)\sin\beta$$

and we have

$$\sin(\alpha + \beta) \equiv \sin\alpha\cos\beta + \cos\alpha\sin\beta. \tag{5-10}$$

Replacing β by $-\beta$ in (5-10), we immediately find

$$\sin(\alpha - \beta) \equiv \sin\alpha\cos\beta - \cos\alpha\sin\beta. \tag{5-11}$$

GENERAL ADDITION FORMULAS

The tangent formulas,

$$\tan(\alpha+\beta) \equiv \frac{\tan\alpha + \tan\beta}{1 - \tan\alpha\tan\beta} \qquad (5\text{-}12)$$

and

$$\tan(\alpha-\beta) \equiv \frac{\tan\alpha - \tan\beta}{1 + \tan\alpha\tan\beta}, \qquad (5\text{-}13)$$

result from the fact that $\tan\theta \equiv \sin\theta/\cos\theta$.

By using (5-9) and (5-10),

$$\begin{aligned}\tan(\alpha+\beta) &\equiv \frac{\sin(\alpha+\beta)}{\cos(\alpha+\beta)} \\ &\equiv \frac{\sin\alpha\cos\beta + \cos\alpha\sin\beta}{\cos\alpha\cos\beta - \sin\alpha\sin\beta} \\ &\equiv \frac{\dfrac{\sin\alpha\cos\beta}{\cos\alpha\cos\beta} + \dfrac{\cos\alpha\sin\beta}{\cos\alpha\cos\beta}}{\dfrac{\cos\alpha\cos\beta}{\cos\alpha\cos\beta} - \dfrac{\sin\alpha\sin\beta}{\cos\alpha\cos\beta}} \\ &\equiv \frac{\tan\alpha + \tan\beta}{1 - \tan\alpha\tan\beta}.\end{aligned}$$

Formula (5-13) may be derived in a similar way, or by using (5-8).

EXAMPLE 1. Compute $\sin 7\pi/12$ from the functions of $\pi/3$ and $\pi/4$.

Solution. $\sin 7\pi/12 = \sin(\pi/3 + \pi/4)$
$= \sin\pi/3 \cos\pi/4 + \cos\pi/3 \sin\pi/4$
$= \sqrt{3}/2 \cdot \sqrt{2}/2 + 1/2 \cdot \sqrt{2}/2$
$= \sqrt{6}/4 + \sqrt{2}/4 = (\sqrt{6} + \sqrt{2})/4.$

EXAMPLE 2. Compute $\cos 15°$.

Solution. $\cos 15° = \cos(60° - 45°)$
$= \cos 60° \cos 45° + \sin 60° \sin 45°$
$= 1/2 \cdot \sqrt{2}/2 + \sqrt{3}/2 \cdot \sqrt{2}/2$
$= \sqrt{2}/4 + \sqrt{6}/4 = (\sqrt{2} + \sqrt{6})/4.$

90   CIRCULAR FUNCTIONS OF MORE THAN ONE ANGLE   [CHAP. 5

EXAMPLE 3. Given $\sin \alpha = 12/13$, with α in the first quadrant, and $\cos \beta = 3/5$, with β in the first quadrant. Find in what quadrant $\alpha + \beta$ lies.

Solution. Since α and β are both in the first quadrant, $\alpha + \beta$ is either in the first or second quadrant. Since the cosine is positive in the first and negative in the second, it will suffice to find $\cos(\alpha + \beta)$. We are able to find $\cos \alpha$ and $\sin \beta$ as in Chapter 4: $\cos \alpha = 5/13$, and $\sin \beta = 4/5$. Thus

$$\cos(\alpha + \beta) \equiv \cos \alpha \cos \beta - \sin \alpha \sin \beta$$
$$= 5/13 \cdot 3/5 - 12/13 \cdot 4/5$$
$$= 15/65 - 48/65 = -33/65.$$

Since $\cos(\alpha + \beta)$ is negative, it follows that $\alpha + \beta$ is in the second quadrant.

EXAMPLE 4. Show that $\sin(270° - \theta) \equiv -\cos \theta$.

Solution. In (5–11) let $\alpha = 270°$ and $\beta = \theta$. Therefore,

$$\sin(270° - \theta) \equiv \sin 270° \cos \theta - \cos 270° \sin \theta$$
$$\equiv (-1) \cos \theta - (0) \sin \theta$$
$$\equiv -\cos \theta.$$

EXAMPLE 5. Express $3 \sin \theta + 4 \cos \theta$ in the form $k \sin(\theta + \theta_1)$, where θ_1 is in the first quadrant.

Solution. By multiplying and dividing the expression by $\sqrt{3^2 + 4^2} = 5$, we have

$$5(\tfrac{3}{5} \sin \theta + \tfrac{4}{5} \cos \theta) \equiv 5(\sin \theta \cdot \tfrac{3}{5} + \cos \theta \cdot \tfrac{4}{5}).$$

Letting $\cos \theta_1 = 3/5$, we find that $\sin \theta_1 = 4/5$; thus the above expression becomes

$$5(\sin \theta \cos \theta_1 + \cos \theta \sin \theta_1) \equiv 5 \sin(\theta + \theta_1),$$

where θ_1 has its sine equal to $4/5$ and its cosine equal to $3/5$.

PROBLEMS

1. Find the exact value of the sine, cosine, and tangent of $5\pi/12 = 75°$ by setting $\alpha = 45°$ and $\beta = 30°$ in the appropriate one of the formulas previously derived.

GENERAL ADDITION FORMULAS

2. Find the exact value of sin 15° and tan 15° by taking 15° = 60° − 45°.
3. Find the exact value of cos $7\pi/12$ and tan $7\pi/12$, as done in Example 1.
4. Find the exact values of the sine, cosine, and tangent of $11\pi/12 = 165°$ by setting $\alpha = 3\pi/4$ and $\beta = \pi/6$.

It is interesting to note that with the results of Problems 1–4, and previously known results, the exact values of the circular functions of any integral multiple of $\pi/12$ have been or can be readily found.

5. If sin $\alpha = 4/5$, and sin $\beta = 12/13$, and both α and β are in the first quadrant, find

   (a) sin $(\alpha + \beta)$ (b) cos $(\alpha + \beta)$ (c) tan $(\alpha + \beta)$
   (d) sin $(\alpha - \beta)$ (e) cos $(\alpha - \beta)$ (f) tan $(\alpha - \beta)$

6. If cos $\alpha = -24/25$, tan $\beta = 9/40$, α is in the second quadrant, and β is in the third quadrant, find

   (a) sin $(\alpha + \beta)$ (b) cos $(\alpha + \beta)$ (c) tan $(\alpha + \beta)$
   (d) sin $(\alpha - \beta)$ (e) cos $(\alpha - \beta)$ (f) tan $(\alpha - \beta)$

7. (a) In what quadrant is $\alpha + \beta$ in Problem 5?   (b) In what quadrant is $\alpha - \beta$ in Problem 6?

Find sin $(\alpha + \beta)$ and cos $(\alpha + \beta)$, given that:

8. tan $\alpha = 3/4$, sec $\beta = 13/5$, and neither α nor β is in the first quadrant.
9. tan $\alpha = -15/8$, sin $\beta = -7/25$, and neither α nor β is in the fourth quadrant.

Express each of the following in terms of functions of θ only:

10. cos $(\pi/4 + \theta)$
11. tan $(\theta + \pi/6)$
12. sec $(\theta - 45°)$
13. cot $(\pi/4 + \theta)$
14. sin $(\theta + 60°)$
15. csc $(\theta - 30°)$
16. cos $(\pi/6 - \theta)$
17. sin $(\theta - 45°)$

Show that the following statements are true.

18. tan $(\theta + \pi/4) - \tan (\theta - 3\pi/4) \equiv 0$
19. sin $(\theta - \pi/6) + \cos (\theta - \pi/3) \equiv \sqrt{3} \sin \theta$
20. cot $(\alpha + \beta) \equiv \dfrac{\cot \alpha \cot \beta - 1}{\cot \alpha + \cot \beta}$
21. tan $(\theta + \pi/4) \equiv (1 + \tan \theta)/(1 - \tan \theta)$
22. sin $(\alpha + \beta)/\cos \alpha \cos \beta \equiv \tan \alpha + \tan \beta$
*23. (a) sin $(\alpha + \beta) + \sin (\alpha - \beta) \equiv 2 \sin \alpha \cos \beta$
    (b) sin $(\alpha + \beta) - \sin (\alpha - \beta) \equiv 2 \cos \alpha \sin \beta$

24. (a) $\cos(\alpha + \beta) + \cos(\alpha - \beta) \equiv 2\cos\alpha\cos\beta$
    (b) $\cos(\alpha + \beta) - \cos(\alpha - \beta) \equiv -2\sin\alpha\sin\beta$

*25. By letting $\alpha + \beta = x$, and $\alpha - \beta = y$, and dividing the respective members of Problems 23(a) and (b), prove
$$\frac{\sin x - \sin y}{\sin x + \sin y} \equiv \frac{\tan\frac{1}{2}(x - y)}{\tan\frac{1}{2}(x + y)}.$$

26. (a) $\sin(\alpha + \beta)\sin(\alpha - \beta) \equiv \sin^2\alpha - \sin^2\beta$
    (b) $\cos(\alpha + \beta)\cos(\alpha - \beta) \equiv \cos^2\alpha - \sin^2\beta$

27. $\cos(\alpha + \beta)\cos\beta + \sin(\alpha + \beta)\sin\beta \equiv \cos\alpha$

28. $\sin(\alpha - \beta)\cos\beta + \cos(\alpha - \beta)\sin\beta \equiv \sin\alpha$

*29. (a) $\cos\theta \equiv \cos(\theta/2 + \theta/2) \equiv \cos^2(\theta/2) - \sin^2(\theta/2) \equiv 1 - 2\sin^2(\theta/2)$
    (b) $\cos\theta \equiv 2\cos^2(\theta/2) - 1$

*30. $\sin\theta \equiv 2\sin\frac{\theta}{2}\cos\frac{\theta}{2}$

*31. $\sin\frac{\theta}{2} \equiv \pm\sqrt{\frac{1 - \cos\theta}{2}}$. *Hint:* Use Problem 29(a).

*32. $\cos\frac{\theta}{2} \equiv \pm\sqrt{\frac{1 + \cos\theta}{2}}$. *Hint:* Use Problem 29(b).

Express the following in the form $k\sin(\theta + \theta_1)$, where θ_1 is between $-\pi/2$ and $\pi/2$.

33. $5\sin\theta + 12\cos\theta$
34. $15\sin\theta + 8\cos\theta$
35. $4\sin\theta - 3\cos\theta$
36. $24\sin\theta + 7\cos\theta$
37. $\sin\theta + \cos\theta$
38. $2\sin\theta - 5\cos\theta$

5–4 General reduction formulas. It is often necessary to express the circular functions of a given angle in terms of functions of an acute angle. We are able to do this by using reduction formulas obtained from (5–1). If we wish to reduce any angle by multiples of 90°, in order to work with the acute angle we recall (Problem 21, Article 4–3) that $\sin(2k \cdot 90°) = 0$ and $\cos(2k \cdot 90°) = (-1)^k$ for any integer k (positive, negative, or zero). Since
$$\sin(2k \cdot 90° + \beta) \equiv \sin(2k \cdot 90°)\cos\beta + \cos(2k \cdot 90°)\sin\beta,$$
we have
$$\sin(2k \cdot 90° + \beta) \equiv (-1)^k \sin\beta, \qquad (5\text{–}14)$$
and similarly,
$$\cos(2k \cdot 90° + \beta) \equiv (-1)^k \cos\beta. \qquad (5\text{–}15)$$

These two relationships are for β increased or decreased by even multiples of 90°. The function is not changed although the sign may be. For odd multiples of 90°, the function is changed to its cofunction, and again the sign may also change, for

$$\sin [(2k + 1) 90° + \beta]$$
$$\equiv \sin [90° + (2k \cdot 90° + \beta)]$$
$$\equiv \sin 90° \cos (2k \cdot 90° + \beta) + \cos 90° \sin (2k \cdot 90° + \beta)$$
$$\equiv \cos (2k \cdot 90° + \beta),$$

and therefore, by (5–15),

$$\sin [(2k + 1) 90° + \beta] \equiv (-1)^k \cos \beta. \quad (5\text{–}16)$$

Moreover,

$$\cos [(2k + 1) 90° + \beta]$$
$$\equiv \cos [90° + (2k \cdot 90° + \beta)]$$
$$\equiv \cos 90° \cos (2k \cdot 90° + \beta) - \sin 90° \sin (2k \cdot 90° + \beta)$$
$$\equiv -(-1)^k \sin \beta,$$

and thus,
$$\cos [(2k + 1) 90° + \beta] \equiv (-1)^{k+1} \sin \beta. \quad (5\text{–}17)$$

The special case of (5–14) and (5–15), where k has the value 2, is important, since

$$\sin (\beta + 2\pi) \equiv \sin \beta, \quad (5\text{–}18)$$

and

$$\cos (\beta + 2\pi) \equiv \cos \beta. \quad (5\text{–}19)$$

It was for this reason that these functions were called *periodic* functions, with periods of 2π. [Recall Eqs. (4–5, 6).]

EXAMPLE 1. Express sin 624° as a function of a positive acute angle less than 45°.

Solution. Since $624° = 6 \cdot 90° + 84°$, by using (5–14) we have $\sin 624° = -\sin 84°$. From (5–2), $-\sin 84° = -\cos 6°$, and we have $\sin 624° = -\cos 6°$.

EXAMPLE 2. Express cos 1243° as a function of a positive acute angle less than 45°.

Solution. Again $1243° = 13 \cdot 90° + 73°$. Thus, using (5–17), $\cos 1243° = -\sin 73°$, and by (5–2), we have $\cos 1243° = -\cos 17°$.

EXAMPLE 3. Repeat Examples 1 and 2 for $\cos(-497°)$.

Solution. Since $-497° = -5 \cdot 90° - 47°$, we have $\cos(-497°) = -\sin 47° = -\cos 43°$.

Having done several exercises of this type, one should be able to write the answer without the use of Formulas (5–14) through (5–17). Draw a figure, choose the sign of the function value to be simplified, depending upon the quadrant, and with this sign write the corresponding acute angle. In considering Example 1, since 624° is in the third quadrant, sin 624° is negative, and the corresponding acute angle is 84°. Thus $\sin 624° = -\sin 84°$.

PROBLEMS

Express each of the following as a function of a positive acute angle less than 45°:

1. $\sin 196°$
2. $\cos 147°$
3. $\sin 319°$
4. $\cos 254°$
5. $\tan 294°$
6. $\cos 728°$
7. $\sin(-625°)$
8. $\cos(-435°)$
9. $\tan 1004°$
10. $\sin 248°25'$
11. $\cos 106°18'$
12. $\tan 163°17'$
13. $\cos 204°46'$
14. $\tan 136°34'$
15. $\sin 156°39'$

By using (5–1) or (5–9) through (5–13) show that each of the following statements is true. Check those involving sines or cosines by (5–14) through (5–17).

16. $\sin(180° + \theta) \equiv -\sin \theta$
17. $\cos(180° + \theta) \equiv -\cos \theta$
18. $\sin(180° - \theta) \equiv \sin \theta$.
19. $\cos(180° - \theta) \equiv -\cos \theta$
20. $\tan(270° - \theta) \equiv \cot \theta$
21. $\cos(\theta - 180°) \equiv -\cos \theta$
22. $\cos(270° + \theta) \equiv \sin \theta$
23. $\sin(\theta + 270°) \equiv -\cos \theta$
24. $\tan(180° + \theta) \equiv \tan \theta$
25. $\cos(360° - \theta) \equiv \cos \theta$
26. $\sin(360° - \theta) \equiv -\sin \theta$

27. In Problem 24, we showed that $\tan(\theta + \pi) \equiv \tan \theta$. This was proved as a special case of (4–7) with $k = 1$. What is the period of $\tan \theta$?

28. What are the periods of the cotangent, secant, and cosecant functions?

5-5 Function values of any angle. In Chapter 4 and Article 5-3 of this chapter we computed the values of the circular functions for certain angles. Considering $60° = \pi/3$ radians, we found $\cos 60° = .5000$, an exact decimal value, while $\sin 60° = \sqrt{3}/2$ and $\tan 60° = \sqrt{3}$ could be approximated as decimal values to whatever accuracy we wished. Such approximations from exact values, expressed in radicals, can be obtained only in special cases.

To find the values of the functions for any angle, we might draw the angle using a protractor and find approximate values with a scale to measure distances. Because of inaccuracies in drawing and measuring, such a method has definite limitations. The general method for finding the values of the functions to any degree of accuracy, based on calculus, is beyond the scope of this book.

We are able to find exact values of the functions for angles of every 3°, however, by a completely elementary process, and although this method is not an efficient way of obtaining the values for a table, it has sufficient mathematical interest to warrant this discussion. By combining the exact values of the functions for $\theta = 30°$, 36°, 45°, 60°, and 90°, which we found in Chapter 4, and using some of the relations of Article 5-3, we obtain our results.

Let us find exact values for sin 3° and cos 3°. By applying the values found for cos 36° (Article 4-5) to Problems 31 and 32, Article 5-3, we have

$$\sin 18° = \sqrt{\frac{1 - \cos 36°}{2}} = \sqrt{\frac{3 - \sqrt{5}}{8}},$$

and

$$\cos 18° = \sqrt{\frac{1 + \cos 36°}{2}} = \sqrt{\frac{5 + \sqrt{5}}{8}}.$$

Also, from Example 2 and Problem 2, Article 5-3, we have

$$\sin 15° = \frac{\sqrt{6} - \sqrt{2}}{4},$$

and

$$\cos 15° = \frac{\sqrt{6} + \sqrt{2}}{4}.$$

Since $18° - 15° = 3°$, using (5–11),

$$\sin 3° = \sin(18° - 15°)$$
$$= \sin 18° \cos 15° - \cos 18° \sin 15°$$
$$= \sqrt{\frac{3 - \sqrt{5}}{8}} \frac{\sqrt{6} + \sqrt{2}}{4} - \sqrt{\frac{5 + \sqrt{5}}{8}} \frac{\sqrt{6} - \sqrt{2}}{4}$$
$$= \sqrt{\frac{(3 - \sqrt{5})(8 + 4\sqrt{3})}{128}} - \sqrt{\frac{(5 + \sqrt{5})(8 - 4\sqrt{3})}{128}},$$

so that

$$\sin 3°$$
$$= \frac{\sqrt{12 - 2\sqrt{15} - 4\sqrt{5} + 6\sqrt{3}} - \sqrt{20 - 2\sqrt{15} + 4\sqrt{5} - 10\sqrt{3}}}{8}.$$
(5–20)

In a similar way, using (5–1),

$$\cos 3°$$
$$= \frac{\sqrt{20 + 2\sqrt{15} + 4\sqrt{5} + 10\sqrt{3}} + \sqrt{12 + 2\sqrt{15} - 4\sqrt{5} - 6\sqrt{3}}}{8}.$$
(5–21)

The values of the other functions can be found from these exact values. Of course, approximated decimal values can also be obtained to any degree of accuracy desired.

The functions of other angles could be handled similarly. For example, $6° = 36° - 30°$, $9° = 45° - 36°$, $12° = 30° - 18°$, and so on. The purpose of this discussion has been to demonstrate how a workable table could be constructed by a completely elementary process.

It is also interesting to note, although the proof will not be given, that the same angle, the angles of n degrees, with n an integral multiple of 3, are exactly those which are constructible with ruler and compass.

For the general angle, we have tables, and can use them either to find the approximate value of the functions of any given angle, or to find the angle when the value of some function of that angle is given. Table I, in the back of the book, lists the values of the sine, cosine, tangent, and cotangent, approximated to four decimal places, of angles at 10' intervals in the first quadrant. From the values of the functions for these angles, given in both degree and radian measure, it is possible to find the functions of angles of any size.

Since the circular function of any angle is the same as the cofunction of the complementary angle, Table I is efficiently constructed so that angles from 0° to 45° are at the left, while angles from 45° to 90° are at the right. Moreover, the circular functions listed at the top go with the angles at the left, while those at the bottom go with the angles at the right.

For example, opposite 24°10' and in the column under sine, we find .4094, which is sin 24°10'. The value for sin 68°20' is .9293, which we find in the column above sine, since 68°20' is given at the right.

By assuming that the graph of the circular functions approximates a straight line, we may use the method of *linear interpolation* to find the circular functions of other angles. If x and $x + 10'$ are consecutive angles in the table, and r is an integer between 0 and 10, we may approximate sin $(x + r')$ by interpolating by proportion:

$$\sin(x + r') = \sin x + \frac{r}{10}[\sin(x + 10') - \sin x]. \quad (5\text{-}22)$$

The other functions are found by similar formulas.

EXAMPLE 1. Find sin 24°16'.

Solution. Since the angle 24°16' lies between 24°10' and 24°20',

$$\begin{aligned}\sin 24°16' &= \sin 24°10' + \tfrac{6}{10}[\sin 24°20' - \sin 24°10'] \\ &= .4094 + \tfrac{6}{10}(.4120 - .4094) \\ &= .4094 + .0016 \\ &= .4110.\end{aligned}$$

EXAMPLE 2. Find cos 57°42'.

Solution. With 57°42' between 57°40' and 57°50', we have

$$\begin{aligned}\cos 57°42' &= \cos 57°40' + \tfrac{2}{10}[\cos 57°50' - \cos 57°40'] \\ &= .5348 + \tfrac{2}{10}(.5324 - .5348) \\ &= .5348 - .0005 \\ &= .5343.\end{aligned}$$

Table I may also be used to find the angle between 0° and 90° if the value of a circular function is given. For example, if $\tan \theta = .9435$, by looking in the tangent column we find the entry .9435 opposite 43°20′, and thus $\theta = 43°20'$.

The table may also be interpolated to find an angle, approximated to the nearest minute, when the value of a function of this angle lies between two entries in the table. If $\sin \theta$ is given, we find two consecutive entries in the sine column between which the given value lies. Thus, letting $\theta = x + r'$, where r is some integer between 0 and 10, and x and $x + 10'$ are the consecutive entries in the table, we find r by using the approximation

$$\frac{r}{10} = \frac{\sin (x + r') - \sin x}{\sin (x + 10') - \sin x}. \qquad (5\text{–}23)$$

Again, other functions follow the same pattern.

EXAMPLE 3. Find the angle θ between 0° and 90° if $\sin \theta = .6231$.

Solution. We locate .6231 in the sine column between $\sin 38°30' = .6225$ and $\sin 38°40' = .6248$. Thus

$$\frac{r}{10} = \frac{.6231 - .6225}{.6248 - .6225},$$

or

$$r = 10 \left(\frac{.0006}{.0023} \right)$$
$$= 3,$$

so that $\theta = 38°33'$, approximated to the nearest minute.

EXAMPLE 4. Find the angle θ, if $\cos \theta = .5741$.

Solution. We find $\cos 54°50' = .5760$, and $\cos 55° = .5736$. Therefore

$$\frac{r}{10} = \frac{.5741 - .5760}{.5736 - .5760},$$

so that

$$r = 10 \left(\frac{.0019}{.0024} \right) = 8,$$

and our result is $\theta = 54°58'$.

FUNCTION VALUES OF ANY ANGLE

If the angle is larger than 90° the reduction formulas of the last article must be used, before the table evaluation. Moreover, it should be clear that other angles greater than 90° satisfy the conditions of Examples 3 and 4. The finding of such angles is discussed in Chapters 6 and 9.

Problems

1. Establish (5–21) of this section.

2. Express 21°, 24°, and 27° by using combinations similar to those above, which might be used to find their sines or cosines.

3. Do the same for 33°, 39°, and 42°.

4. If you have access to a computing machine, find the decimal approximation for sin 3° and cos 3°, using (5–20) and (5–21). Check your results with the value given in Table I.

5. Find the value of each of the following by using Table I:

(a) sin 14°20′
(b) tan 52°40′
(c) cos 28°50′
(d) sin 63°30′
(e) tan 21°10′
(f) cos 72°20′
(g) sin 115°30′
(h) cos 161°10′

6. Find the approximate value of each of the following by using Table I and interpolation:

(a) sin 72°43′
(b) cos 28°46′
(c) tan 51°29′
(d) cos 63°23′
(e) tan 39°18′
(f) sin 128°36′
(g) cos 153°17′
(h) sin 8°9′

7. Find the angle θ between 0° and 90° using Table I if:

(a) $\sin \theta = .4253$
(b) $\tan \theta = 1.1237$
(c) $\cos \theta = .8857$
(d) $\sin \theta = .8450$
(e) $\tan \theta = .2156$
(f) $\cos \theta = .2447$
(g) $\sin \theta = .3475$
(h) $\cos \theta = .7844$

8. Find the approximate value between 0° and 90° of θ to the nearest minute, by using Table I and interpolation, if:

(a) $\tan \theta = .8172$
(b) $\sin \theta = .5331$
(c) $\cos \theta = .2717$
(d) $\cos \theta = .9392$
(e) $\sin \theta = .7531$
(f) $\tan \theta = .8083$
(g) $\cos \theta = .5386$
(h) $\sin \theta = .9648$

CHAPTER 6

LINEAR AND QUADRATIC FUNCTIONS

6–1 The linear function. The circular functions considered in the last two chapters were not defined by algebraic expressions but rather by rule and, in fact, cannot be expressed algebraically. Probably the simplest function defined in mathematics by means of a nontrivial algebraic expression, is the function

$$f(x) = mx + b, \qquad (6\text{--}1)$$

where m and b are constants. This function of the first degree in x is called a *linear function*, since the graph of $y = f(x) = mx + b$ is a straight line.* Moreover, any straight line other than $x = k$ (a straight line parallel to the y-axis) can be represented by such an equation with the appropriate m and b. Rather than the geometric properties of such a function, we shall concern ourselves in this discussion with its algebraic properties and its zeros.

Recalling (Article 3–7) that the zero of a function is the x-coordinate or x-value for which y, the value of the function, is zero, by letting $y = 0$, we can find the zeros of a function. This is done by solving the equation $mx + b = 0$ for x. In general, the zeros of a function are the *roots* or *solution* of the equation $f(x) = 0$. Before considering the solving of such a linear equation specifically, let us consider the problem of solving any equation.

Various methods are used in solving an equation. Any device which produces an *equivalent equation*, one which has the same roots, and only those roots, is permitted. Some procedures may introduce new factors, and care should be taken in dividing so that no roots will be lost. It is always wise to check any solution, for the ultimate test is whether a quantity satisfies the equation. The following operations on an equation are permissible, if care is used in checking the results.

1. The same number or algebraic expression may be added to or subtracted from both members of an equation.

---

* The fact that $y = mx + b$ represents the equation of a straight line is proved in analytic geometry and will be assumed here.

2. Both members of an equation may be multiplied or divided by the same nonzero number or algebraic expression.

ILLUSTRATION 1. In solving $4x - 5 = x + 7$, 5 is added, and x is subtracted from both members, giving $3x = 12$. Then, dividing by 3, $x = 4$.

ILLUSTRATION 2. If in the equation $x - 1 = 3$, which has the root $x = 4$, both members are squared (an indirect application of Number 2), $(x - 1)^2 = 9$. This process has introduced the extraneous root -2, which does not satisfy the original equation.*

ILLUSTRATION 3. In solving the equation $\sin^2 \theta - \frac{1}{2} \sin \theta = 0$ for values of θ between 0 and $\pi/2$, we find $\sin \theta = 0$ or $\frac{1}{2}$ and $\theta = 0$ or $\pi/6$. If we were to divide both members by $\sin \theta$, however, we would have only the value $\sin \theta = \frac{1}{2}$. The value $\sin \theta = 0$ would have been "lost."†

Now let us consider the solution of linear equations.

EXAMPLE 1. Solve the equation $\dfrac{2x + 5}{2} - \dfrac{5x}{x - 1} = x$ for all possible values of x.

Solution. We first clear the equation of fractions, by multiplying by the L.C.D., $2(x - 1)$.

$$(2x + 5)(x - 1) - (5x)(2) = (x)(2)(x - 1),$$

or

$$2x^2 + 3x - 5 - 10x = 2x^2 - 2x.$$

Combining similar terms,

$$-7x - 5 = -2x.$$

---

* Any equation which, because of some mathematical process, has acquired an extra root is sometimes called a *redundant equation*.

† Any equation which, because of some mathematical process, has fewer roots, is sometimes called a *defective equation*.

Adding $2x + 5$ to both members,
$$-5x = 5,$$
and dividing by -5,
$$x = -1.$$

This result should be verified by substituting $x = -1$ in the original equation.

EXAMPLE 2. Solve the equation $2 \cos \theta + 3 = 2$ for all values of θ between 0 and 2π.

Solution. Although this type of equation, dealing with the circular functions, is not algebraic and therefore not linear, it is linear in the circular function of θ, and can be solved for $\cos \theta$ as though it were linear. The values of θ can then be found. Problems 7 and 8 of Article 5-5 were simple examples of this type.

$$2 \cos \theta + 3 = 2$$
$$2 \cos \theta = -1$$
$$\cos \theta = -\tfrac{1}{2}$$

Therefore $\theta = 2\pi/3$ or $4\pi/3$, since θ must be in the second or third quadrant when its cosine is negative.

EXAMPLE 3. Solve the equation $s = \dfrac{a - rl}{1 - r}$ for r.

Solution. This type of equation, one in which some or all of the known quantities are represented by letters, is called a *literal* equation. Each step in the solution should be verified.

$$s = \frac{a - rl}{1 - r},$$
$$s(1 - r) = a - rl,$$
$$s - rs = a - rl,$$
$$rl - rs = a - s,$$
$$r(l - s) = a - s,$$
$$r = \frac{a - s}{l - s}.$$

Problems

Find the zeros of the following linear functions:
1. $2x + 4$
2. $-5x + 10$
3. $10 - 12x$
4. $6x - 9$
5. $8x + 24$
6. $5x - 17$

Solve the following linear equations and check:
7. $4x - 2 = 6x + 12$
8. $3x + 7 = 5x - 13$
9. $5 + 2(3 - x) = 4 + 2(x - 2) + 5x$
10. $x^2 - 7x + 10 = x^2 + 5x - 6$
11. $\dfrac{3x + 5}{12} - \dfrac{4 - x}{6} = \dfrac{x - 2}{3}$
12. $\dfrac{3x - 6}{5} = \dfrac{2x - 5}{10} + \dfrac{x - 4}{2}$
13. $\dfrac{3x + 2}{x - 1} - \dfrac{6}{5} = 0$
14. $\dfrac{2}{6x - 7} - \dfrac{5}{3x - 4} = 0$

Using Table I if necessary, find the values of θ between 0 and 2π which satisfy each of the following equations:

15. $2 \sin \theta - \sqrt{3} = 0$
16. $\dfrac{\cos \theta - 2}{3} = 3 - \dfrac{\cos \theta + 9}{3}$
17. $\dfrac{1 + \tan \theta}{1 - \tan \theta} = 2$
18. $\sin 3\theta = 2 - 3 \sin 3\theta$
19. $\dfrac{4 \cos \left(2\theta + \dfrac{\pi}{6}\right)}{5} = \dfrac{4 + \cos \left(2\theta + \dfrac{\pi}{6}\right)}{10}$
20. $\dfrac{1}{2 \tan (\theta + \pi)} = \dfrac{2}{\tan (\theta + \pi)} + 1$

Solve the following equations for the letters indicated:
21. $ax - bx = c$, for x
22. $ay + by = c + dy$, for y
23. $A = \frac{1}{2}bh$, for b
24. $A = \frac{1}{2}(b_1 + b_2)h$, for b_1
25. $l = a + (n - 1)d$, for d
26. $l = a + (n - 1)d$, for n
27. $S = \dfrac{a - rl}{1 - r}$, for l
28. $S = \dfrac{a - rl}{1 - r}$, for a
29. $S = \dfrac{n}{2}(a + l)$, for l
30. $S = v_0 t + \frac{1}{2}gt^2$, for v_0
31. $\dfrac{y}{\tan \theta_1} = \dfrac{a + y}{\tan \theta_2}$, for y
32. $\tan \theta_1 \left(\dfrac{x}{\tan \theta_2} + a\right) = x$, for x

In solving the following problems, *read the problem carefully,* let one of the unknown quantities be x, and express all other unknown quantities as functions of x. Find two expressions or quantities that are equal, equate these, and solve the resulting equation. Check *all* answers.

33. A man 42 years old has a son 12. In how many years will the father be twice as old as his son?

34. The tens digit of a number is 3 less than the units digit. If the number is divided by the sum of the digits, the quotient is 4 and the remainder 3. What is the original number? [*Hint:* If x equals the units digit, $x - 3$ is the tens digit, and the number may be written $10(x - 3) + x$.]

35. A man left $\frac{1}{2}$ of his estate to his wife, $\frac{1}{6}$ to his daughter, and the remainder, an amount of $15,000, to his son. How large was the entire estate?

36. A starts walking along a road at 3 mi/hr. Two hours later B starts in the same direction at 3.5 mi/hr. How far from the starting point will B overtake A? [*Hint:* Rate $\times$ time = Distance.]

37. A can do a certain job in 3 hours, while the same job takes B 4 hours. How long will it take both of them working together? [*Hint:* If x = number of hours for both to complete the work, $1/x$ will be the amount of the work done by both in 1 hour.]

38. If the larger of two integers, whose sum is 88, is divided by the smaller, the quotient is 5 and the remainder is 10. What are the two numbers?

6-2 The quadratic function. The second type of algebraic function of one variable usually considered, is that of the second degree, called the quadratic function,

$$f(x) = ax^2 + bx + c, \qquad (6\text{-}2)$$

where a represents the coefficients of all squared terms, b the coefficient of all first degree terms, and c the constant term. Although either or both b and c may be zero, a is not zero (Why?). The graph of such a function was briefly considered in Article 3–7 (Example 2). Another example seems appropriate.

EXAMPLE 1. Draw the graph of the quadratic function $-2x^2 + 12x - 14$.

Solution. The graph (Fig. 6–1) is drawn by tabulating a sufficient number of points whose coordinates satisfy the equation $y = -2x^2 + 12x - 14$.

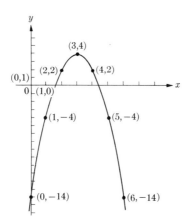

FIGURE 6-1

x	0	1	2	3	4	5	6
y	-14	-4	2	4	2	-4	-14

In comparing this curve with that in Fig. 3–13, we see that both have the same general shape. The graph of any quadratic function in one variable is of this type and is called a parabola. Although we shall not prove it, when the coefficient of x^2 is positive, the curve opens upward, while the curve opens downward if the coefficient of x^2 is negative. Compare the two examples.

The graph of the general quadratic function may be sketched by a more direct process, that of expressing the quadratic function in terms of the square of a linear function. Consider

$$y = ax^2 + bx + c. \tag{6-3}$$

Factoring out a, and grouping the x^2 and x terms together,

$$y = a\left[\left(x^2 + \frac{b}{a}x\right) + \frac{c}{a}\right].$$

Recalling Eq. (1–21), we must add $(b/2a)^2$ to $x^2 + (b/a)x$ in order to form a perfect square. Thus, adding and subtracting $b^2/4a^2$ within the bracket,

$$y = a\left[\left(x^2 + \frac{b}{a}x + \frac{b^2}{4a^2}\right) + \frac{c}{a} - \frac{b^2}{4a^2}\right],$$

and simplifying,

$$y = a\left[\left(x + \frac{b}{2a}\right)^2 + \frac{4ac - b^2}{4a^2}\right]. \tag{6-4}$$

Since the squared quantity $(x + b/2a)^2 \geq 0$, the expression within the bracket has its least value when $x = -b/2a$. If $a > 0$, the function also has its least value at $x = -b/2a$. This least value of the function, $(4ac - b^2)/4a$, is called its *minimum*. If, however, $a < 0$, when $x = -b/2a$, the function has its greatest value, called its *maximum*, and is also equal to $(4ac - b^2)/4a$. In either case, the point $[-b/2a, (4ac - b^2)/4a]$ is called the *vertex* of the parabola. With this point found, and an additional point or two, the graph can be sketched. Rather than actually using these formulas, the square should be completed in each case.

EXAMPLE 2. Find the vertex of the parabola represented by the equation $y = x^2 - x - 6$. Compare with Example 2, Article 3–7.

Solution. Grouping the first two terms on the right and completing the square,

$$y = (x^2 - x) - 6$$
$$= (x^2 - x + \tfrac{1}{4}) - 6 - \tfrac{1}{4}$$
$$= (x - \tfrac{1}{2})^2 - \tfrac{25}{4}.$$

The vertex is $(\tfrac{1}{2}, -\tfrac{25}{4})$. Since $a = 1$, which is positive, the minimum value of the function is $-\tfrac{25}{4}$, and this occurs when $x = \tfrac{1}{2}$. The graph of this function appears in Fig. 3–13.

Problems

For each of the following functions, find the maximum or minimum and draw the graph.

1. $x^2 + 6x + 5$
2. $x^2 + x - 6$
3. $2x^2 + 5x - 12$
4. $-2x^2 + 11x - 15$
5. $6x^2 - 17x + 5$
6. $-2x^2 + 5x + 8$
7. $x^2 + 6x + 11$
8. $-3x^2 + 5x - 4$

9. Find two numbers whose sum is 16 and whose product is a maximum. [*Hint:* Letting x be one number, and $16 - x$ the other, the product y can be expressed as a function of x, namely, $y = x(16 - x) = 16x - x^2$.]

10. Divide 40 into two parts such that the sum of the squares of these parts is a minimum.

11. A man with 160 ft of fencing wishes to fence off an area in the shape of a rectangle. What should be the dimensions of the area, if the enclosed space is to be as large as possible?

12. A man with 160 ft of fencing wishes to fence off an area in the shape of a rectangle. If one side of the area will not require fencing, what should be the dimensions to ensure the largest area possible?

6–3 Solution of the quadratic equation. We are now in a position to find the zeros of any quadratic function. Recalling the definition of the zeros of a function (Article 3–7), we are interested in finding the abscissas of the points where the graph $y = ax^2 + bx + c$ crosses or meets the x-axis. This, of course, may be accomplished graphically by sketching the curve. For example, from Fig. 3–13, the zeros of $x^2 - x - 6$ are -2, and 3. Likewise, from Fig. 6–1, the zeros of $-2x^2 + 12x - 14$ are approximately 1.3 and 4.7. Such values can, of course, be checked by substituting in the original function, set equal to zero.

There are more accurate methods for finding the zeros of the function $ax^2 + bx + c$. In finding its zeros or, equivalently, in solving the equation

$$ax^2 + bx + c = 0, \qquad (a \neq 0), \qquad (6\text{–}5)$$

we may be able to factor the left member. This method, known as the *factoring method*, depends upon the fact that, in order for a product of two factors to be equal to zero, at least one of these factors must equal zero.

EXAMPLE 1. Solve the quadratic equation $x^2 - 7x + 10 = 0$ by factoring.

Solution. Since $x^2 - 7x + 10 \equiv (x - 5)(x - 2)$, (see Article 1–8), the equation may be written

$$(x - 5)(x - 2) = 0.$$

This product on the left will be equal to zero if we can find a value of x for which $x - 5$ will be zero, or if we can find a value of x for which $x - 2$ will be zero. We therefore have the two linear equations

$$x - 5 = 0 \quad \text{and} \quad x - 2 = 0.$$

Solving these, we find $x = 5$ or $x = 2$, either of which is a root of the original equation. All answers should be checked.

EXAMPLE 2. Solve the equation $2\sin^2\theta + \sin\theta - 1 = 0$, for all possible values of θ between 0 and 2π.

Solution. As Example 2 in Article 6–1, this equation is not algebraic, but may be considered as a quadratic equation in $\sin\theta$. Factoring the left-hand member,

$$2\sin^2\theta + \sin\theta - 1 \equiv (2\sin\theta - 1)(\sin\theta + 1).$$

Thus, we have the two equations

$$2\sin\theta - 1 = 0 \quad \text{and} \quad \sin\theta + 1 = 0$$

to solve for $\sin\theta$, which result in

$$\sin\theta = \tfrac{1}{2} \quad \text{or} \quad -1.$$

Since $\sin\theta$ is positive in the first or second quadrant, there are two values of θ between 0 and 2π for which $\sin\theta = \tfrac{1}{2}$. Therefore,

$$\theta = \pi/6, \quad 5\pi/6, \quad \text{or} \quad 3\pi/2.$$

PROBLEMS

1–8. From the graphs of the functions listed in Problems 1–8 of the last article, give the roots of the corresponding equations. Check these by substituting the results into the original equations.

Solve Problems 9–26 by factoring and check by substitution. In the equations involving circular functions, give all roots between 0 and 2π.

SOLUTION OF THE QUADRATIC EQUATION

9. $9x^2 - 16 = 0$
10. $2x^2 - 5x - 12 = 0$
11. $4 \sin^2 \theta = 1$
12. $4 \cos^2 \theta = 3$
13. $6x^2 - 5x = 50$
14. $2x^2 - 2 = x - 4x$
15. $1 - \cos^2 \theta = \cos^2 \theta$
16. $\tan \theta (2 \sin \theta - \sqrt{3}) = 0$
17. $3x^2 - x = 10$
18. $4x^2 - 12x + 9 = 0$
19. $2 \cot \theta \cos \theta - \cot \theta = 0$
20. $2 \tan^2 \theta + \tan \theta = 0$
21. $x^2 + 2ax = b^2 - a^2$
22. $8x^2 + 14ax + 3a^2 = 0$
23. $2 \sin^2 \theta - \sin \theta = 1$
24. $2 \cos^2 \theta + 3 \cos \theta + 1 = 0$
25. $\dfrac{x-2}{x+3} - 3 = \dfrac{4(x+3)}{x-2}$
26. $(x-2)(x+3) = 6$

In general, the quadratic expression involved in an equation may be difficult or in many cases, impossible to factor. As a result, the most useful method of solving any quadratic equation is *by the quadratic formula*. This formula for the solution of any quadratic equation we shall obtain by completing the square, as was done in finding the vertex of the parabola. Let us solve

$$ax^2 + bx + c = 0, \quad a \neq 0. \tag{6-6}$$

Dividing by the coefficient of x^2 and transposing the constant term to the right side of the equation, we have

$$x^2 + \frac{b}{a}x = -\frac{c}{a}.$$

We may complete the square of the left-hand member by adding $b^2/4a^2$ to both sides of the equation,

$$x^2 + \frac{b}{a}x + \frac{b^2}{4a^2} = \frac{b^2}{4a^2} - \frac{c}{a},$$

or

$$\left(x + \frac{b}{2a}\right)^2 = \frac{b^2 - 4ac}{4a^2}.$$

Extracting the square root,

$$x + \frac{b}{2a} = \frac{\pm \sqrt{b^2 - 4ac}}{2a},$$

or
$$x = \frac{-b \pm \sqrt{b^2 - 4ac}}{2a}. \qquad (6\text{--}7)$$

With the use of the plus and then the minus signs, Eq. (6–7) gives the two roots of the quadratic equation $ax^2 + bx + c = 0$. Although any quadratic equation can be solved by the *method of completing the square* used to obtain this formula, direct substitution into Eq. (6–7) is more frequently employed. Consider the examples.

EXAMPLE 3. Solve the quadratic equation $4x^2 + 5x = 21$.

Solution. Transposing all the members to the left-hand side to put the equation in the form $ax^2 + bx + c = 0$, and comparing the two, we have $a = 4$, $b = 5$, and $c = -21$. Substituting these values into the formula (6–7),

$$x = \frac{-5 \pm \sqrt{(5)^2 - 4(4)(-21)}}{2(4)}.$$

Simplifying,

$$x = \frac{-5 \pm \sqrt{25 + 336}}{8}$$

$$= \frac{-5 \pm \sqrt{361}}{8}$$

$$= \frac{-5 \pm 19}{8}.$$

Therefore, choosing the plus and then the minus sign,

$$x = \tfrac{7}{4} \quad \text{or} \quad -3.$$

Both of the answers should be checked by substituting them in the original equation. The fact that the number under the radical sign is a perfect square guarantees that the equation could have been solved by the method of factoring. Specifically, $4x^2 + 5x - 21 \equiv (4x - 7)(x + 3)$.

SOLUTION OF THE QUADRATIC EQUATION

EXAMPLE 4. Solve $\cos^2 \theta + 5 \sin \theta + 2 = 0$ for all values of θ between $0°$ and $360°$.

Solution. Since $\cos^2 \theta$ may be expressed as $1 - \sin^2 \theta$, it is possible to rewrite this equation entirely in terms of $\sin \theta$. Thus,

$$1 - \sin^2 \theta + 5 \sin \theta + 2 = 0,$$
$$\sin^2 \theta - 5 \sin \theta - 3 = 0.$$

Regarding the variable as $\sin \theta$, and using the formula (6–7), $a = 1$, $b = -5$, and $c = -3$. Therefore,

$$\sin \theta = \frac{5 \pm \sqrt{25 + 12}}{2} = \frac{5 \pm 6.0828}{2},$$

or

$$\sin \theta = 5.5414 \quad \text{or} \quad -0.5414.$$

Since $|\sin \theta| \leq 1$, we consider only the value -0.5414, which results in an angle in the third quadrant and one in the fourth quadrant. Using Table I, we find $\sin \alpha = 0.5414$ gives $\alpha = 32°47'$. Therefore,

$$\theta = 180° + 32°47', \quad \text{or} \quad 360° - 32°47'.$$

Thus,

$$\theta = 212°47', \quad \text{or} \quad 327°13'.$$

The values of θ may, of course, be expressed in radian measure. (Recall Example 4, Article 3–5.) Since $32°47' = 0.5721$ radians,

$$\theta = 3.1416 + .5721 \quad \text{or} \quad 6.2832 - .5721$$
$$= 3.7137 \quad \quad \quad \text{or} \quad 5.7111.$$

The general procedure in solving any quadratic equation is:
1. Try to solve the equation by the method of factoring.
2. If this method does not seem possible, use the quadratic formula.

PROBLEMS

Solve the equations of Problems 1–19 by using the quadratic formula. If the equations involve circular functions, find all angles between 0 and 2π. Check all answers.

1. $2x^2 + 5x - 12 = 0$
2. $4x^2 - 2x = 7$
3. $x^2 + x - 1 = 0$
4. $2x^2 + x - 12 = 0$
5. $\tan^2 \theta - 3 \sec \theta + 3 = 0$ [*Hint:* $\tan^2 \theta \equiv \sec^2 \theta - 1$]

6. $2\cos^2\theta + 3\sin\theta = 0$
7. $x^2 - (a+b)x + ab = 0$
8. $6x^2 + 17x + 12 = 0$
9. $(a-b)x^2 + (b-c)x + (c-a) = 0$
10. $4\sin^2\theta - 3\cos\theta - 2 = 0$
11. $3\sec^2\theta + \tan\theta - 5 = 0$
12. $x^2 - 2x + 1 = 4a^2$
13. $\dfrac{x-1}{x^2-9} - \dfrac{3x+5}{x+3} = \dfrac{x+3}{x-3}$
14. $\cot 2\theta = 2 + \tan 2\theta$
15. $4\sin^2 2\theta + 2\cos 2\theta = 3$
16. $6ax^2 - 2bx + 3b = 9ax$
17. $s = v_0 x - gx^2/2$
18. $\tan 2\theta + 5 = 3\sec^2 2\theta$
19. $\dfrac{1}{x^2+3x+2} - \dfrac{1}{1-x} = \dfrac{2}{x^2-1}$

20. The product of two consecutive positive integers is 72. Find the integers.

21. The sum of a number and its reciprocal is 34/15. Find the number.

22. A man traveling 40 miles finds that by traveling one more mi/hr, he would make the journey in 2 hr less time. How many mi/hr did he actually travel?

23. If $(6-x)$, $(13-x)$, and $(14-x)$ are the lengths of the sides of a right triangle, find the value of x.

24. By how much must a radius of 24 inches be reduced in order to decrease the area of a circle by 49π square inches?

25. A rectangular flower bed 30 yards long by 24 yards wide has a walk of uniform width around it. If the area of the path is one-fourth that of the flower bed, find the width of the path.

26. Working together, two men can do a job in 20 days. Working alone, however, it would take one man 9 days longer than it would take the other to complete the job. How long would it take each separately?

6–4 Conditional inequalities. Since most of the inequalities discussed in this book will involve either first or second degree algebraic expressions, it seems appropriate to consider them in this chapter.

We recall the definition from Article 3–1 which stated that for *real numbers*, x is said to be *greater* than y (or y is said to be *less* than x) if the difference $x - y$ is positive. The symbol used to express this condition was $x > y$ (or $y < x$). Any statement that one mathematical expression is greater than or less than another is called an *inequality*. As is true for equations (Article 1–2), there are two general types of inequalities useful in mathematics, the **conditional** inequality, corresponding to the conditional equation, and the absolute inequality, corresponding to the identity. If an inequality is true for *all* permissible values, or if no literal numbers are involved, it is called an *absolute* inequality. The absolute inequality will be considered in more detail in Chapter 14, along with identities. If an inequality, however, is true only for certain permissible values of the letters involved, it is called a *conditional* inequality.

ILLUSTRATION. The inequality $a^2 + b^2 + 1 > 0$ is an absolute inequality. Also, $-4 < 3$ is of this type. But $2x - 6 > 0$ is a conditional inequality, since it is true only for values of x greater than 3. Also, $\sin \theta < 0$ is conditional, for this is true only for values of θ in the third or fourth quadrants.

As is true in dealing with equations, there are certain properties important in working with inequalities. Their proofs follow directly from the definition and only the first will be proved.

1. The sense of an inequality is not changed if both members are increased or decreased by the same number.

Proof. Assume $a > b$. Then $a - b = n$, where $n > 0$. Therefore
$$a + c - b - c = (a + c) - (b + c) = n,$$
or
$$(a + c) - (b + c) \text{ is positive.}$$
Thus
$$a + c > b + c.$$
Similarly,
$$a - c > b - c.$$

The usefulness of this property comes from the fact that a term may be transposed from one member of an equality to the other member by changing the sign of the term without changing the sense of the inequality.

2. The sense of an inequality is not changed if both members are multiplied or divided by the same *positive* number.

3. The sense of an inequality is reversed if both members are multiplied or divided by the same *negative* number.

Since we shall confine our discussion to inequalities of one variable, we are interested in those which may be written $f(x) > 0$ or $f(x) < 0$. To solve such a conditional inequality, we must find the

114   LINEAR AND QUADRATIC FUNCTIONS   [CHAP. 6

domain of values of x for which the inequality is true. When $f(x)$ is linear or quadratic, this domain may be found by either an algebraic or graphical method.

EXAMPLE 1. Solve $\dfrac{x}{3} - 2 < \dfrac{5x + 9}{2}$ both algebraically and graphically.

Algebraic solution. Multiplying both members by 6, we have
$$2x - 12 < 15x + 27.$$
Transposing and collecting terms,
$$-13x < 39,$$
and dividing by -13,
$$x > -3.$$

Graphical solution. If we transpose all terms to the left side of the inequality, we have
$$\frac{x}{3} - 2 - \frac{5x + 9}{2} < 0.$$
Denoting this left-hand member by $f(x)$,
$$f(x) \equiv \frac{2x - 12 - 15x - 27}{6} \equiv \frac{-13x - 39}{6}.$$

From the graph of $y = f(x)$ in Fig. 6–2, which can easily be drawn from the table, it is clear that the inequality is satisfied for $x > -3$. The graphical method is more frequently used for quadratic inequalities.

x	0	-3	-6
y	$-6\frac{1}{2}$	0	$6\frac{1}{2}$

EXAMPLE 2. For what values of x does the inequality $x^2 - x - 6 > 0$ hold?

Solution. Factoring,
$$x^2 - x - 6 \equiv (x - 3)(x + 2).$$

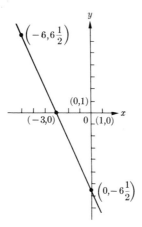

FIGURE 6-2

Since the product of two terms may be positive either if both are positive or both negative, $x - 3$ and $x + 2$ must both be greater than zero, or both less than zero. Thus,

$$\left.\begin{matrix} x - 3 > 0 \\ \text{and} \\ x + 2 > 0 \end{matrix}\right\} \text{ or } \left.\begin{matrix} x > 3 \\ \text{and} \\ x > -2 \end{matrix}\right\} \text{ will both be true if } x > 3.$$

$$\left.\begin{matrix} x - 3 < 0 \\ \text{and} \\ x + 2 < 0 \end{matrix}\right\} \text{ or } \left.\begin{matrix} x < 3 \\ \text{and} \\ x < -2 \end{matrix}\right\} \text{ will both be true if } x < -2.$$

Thus the inequality will be true if $x > 3$ or $x < -2$. This may be checked graphically by Fig. 3-13.

EXAMPLE 3. For what values of x does the inequality $-2x^2 + 12x - 14 > 0$ hold?

Solution. Since the left-hand member is not factorable rationally, we use the graphical method. In Fig. 6-1 we have the graph of the function $y = f(x) = -2x^2 + 12x - 14$, whose zeros we may approximate as 1.6 and 4.4. Thus, the inequality holds for values of x where $1.6 < x < 4.4$.

Should we wish to determine the zeros more accurately, we could use the quadratic formula for the equation $-2x^2 + 12x - 14 = 0$, or the equivalent equation $x^2 - 6x + 7 = 0$.

$$x = \frac{6 \pm \sqrt{36-28}}{2} = \frac{6 \pm 2\sqrt{2}}{2} = 3 \pm \sqrt{2}.$$

Hence,
$$x = 4.414 \text{ or } 1.586,$$
and as a result $1.586 < x < 4.414$.

EXAMPLE 4. Determine the values of θ, such that $0 \leq \theta < 2\pi$, which satisfy the inequality $2\cos^2\theta + \sin\theta < 2$.

Solution. Since $\cos^2\theta \equiv 1 - \sin^2\theta$, we may express the members of the inequality as functions of $\sin\theta$,
$$2(1 - \sin^2\theta) + \sin\theta < 2,$$
and simplifying,
$$2 - 2\sin^2\theta + \sin\theta - 2 < 0,$$
or
$$\sin\theta(1 - 2\sin\theta) < 0.$$

Since the product of two quantities will be negative, if one is negative while the other is positive, we may have

either $\begin{Bmatrix} \sin\theta < 0 \\ \text{and} \\ 1 - 2\sin\theta > 0 \end{Bmatrix}$, which gives $\begin{Bmatrix} \sin\theta < 0 \\ \text{and} \\ \sin\theta < \frac{1}{2} \end{Bmatrix}$,

or $\begin{Bmatrix} \sin\theta > 0 \\ \text{and} \\ 1 - 2\sin\theta < 0 \end{Bmatrix}$, which gives $\begin{Bmatrix} \sin\theta > 0 \\ \text{and} \\ \sin\theta > \frac{1}{2} \end{Bmatrix}$.

The first set is satisfied by $\pi < \theta < 2\pi$, while the second is satisfied by $\pi/6 < \theta < 5\pi/6$.

PROBLEMS

1. Prove that for real numbers a, b, and c, if $a > b$ and $c > 0$, then $a - c > b - c$.

2. Prove the second property for inequalities of real numbers, that is, if $a > b$, and $c > 0$, then $ac > bc$.

3. Prove the third property for inequalities of real numbers, that is, if $a > b$, and $c < 0$, then $ac < bc$.

Find the values of x for which the following inequalities are satisfied. Use either the algebraic or graphical method.

4. $3x - 27 > 0$
5. $2x - 12 < 0$
6. $2x + 5 > 4x - 9$
7. $5x - 3 < 8x - 12$
8. $x^2 + 2x > 99$
9. $2x^2 + 3x < 14$
10. $6x^2 + x < 1$
11. $6x^2 - x > 35$
12. $x^2 + 2x > 12$
13. $x^2 + 2x + 4 > 0$
14. $x^4 + x^2 < 0$
15. $\dfrac{x-2}{x-5} > 0$
16. $\dfrac{1}{x} < \dfrac{1}{5}$
17. $\dfrac{1}{x-2} < \dfrac{1}{3}$
18. $|x - 4| < 1$
19. $|x - 3| > 2$

20. Prove that the sense of an inequality, where both members are positive, is not changed if both members are raised to the same positive power, or if the same positive root (principal value) of both members is taken.

Find the values of θ for which each of the following inequalities holds. Limit the answers to values of θ, such that $0 \leq \theta < 2\pi$:

21. $2 \sin^2 \theta < 1$
22. $4 \cos^2 \theta > 1$
23. $\sin^2 \theta > \sin \theta$
24. $\cos^2 \theta < \cos \theta$
25. $\sin \theta + \sin \dfrac{\theta}{2} < 0$ [*Hint*: Recall Problem 30 (Article 5-3).]
26. $\sin \theta + \cos \dfrac{\theta}{2} > 0$
27. $2 \cos^2 \theta + \sin \theta \geq 1$
28. $2 \sin^2 \theta + \cos \theta > 2$
29. $\tan^2 \theta + \sec \theta + 1 \geq 0$
30. $\sqrt{3} \sin^2 \theta + 2\sqrt{3} > 5 \sin \theta$
31. $\sin \theta > \cos \theta$

6-5 Relation between zeros and coefficients of the quadratic function. In addition to Eq. (6-7), which gives the zeros of the quadratic function, $ax^2 + bx + c$, in terms of the coefficients, there are other relations between the zeros and the coefficients. By letting r_1 and r_2 be the zeros of $ax^2 + bx + c$, where

$$r_1 = \frac{-b + \sqrt{b^2 - 4ac}}{2a}, \qquad r_2 = \frac{-b - \sqrt{b^2 - 4ac}}{2a}, \qquad (6-8)$$

we have the sum of the two zeros, by adding,

$$r_1 + r_2 = -\frac{2b}{2a} = -\frac{b}{a}. \qquad (6-9)$$

Likewise, the product

$$r_1 \cdot r_2 = \frac{(-b)^2 - (\sqrt{b^2 - 4ac})^2}{4a^2}$$

$$= \frac{b^2 - (b^2 - 4ac)}{4a^2}$$

or

$$r_1 \cdot r_2 = \frac{c}{a}. \qquad (6\text{--}10)$$

Thus, for any quadratic function $ax^2 + bx + c$, since the zeros of the function and the roots of the equation are the same, (6–9) and (6–10) are particularly useful in forming such an equation, the sum and product of whose roots are known. By dividing $ax^2 + bx + c = 0$ by a,

$$x^2 + \frac{b}{a}x + \frac{c}{a} = 0. \qquad (6\text{--}11)$$

Thus, comparing (6–11) with (6–9) and (6–10), we have in a quadratic equation in which *the coefficient of x^2 is equal to one:*

1. The sum of the roots is equal to the negative of the coefficient of x.
2. The product of the roots is equal to the constant term.

EXAMPLE 1. Without obtaining the zeros, find the sum and product of the zeros of $3x^2 - 4x + 8$.

Solution. Since $a = 3$, $b = -4$, and $c = 8$, the sum is $-(-4/3) = 4/3$, and the product is $8/3$.

EXAMPLE 2. Write a quadratic equation whose roots are $3 + \sqrt{2}$ and $3 - \sqrt{2}$.

Solution. Since $(3 + \sqrt{2}) + (3 - \sqrt{2}) = 6$, and $(3 + \sqrt{2})(3 - \sqrt{2}) = 9 - 2 = 7$, an equation of this type is

$$x^2 - 6x + 7 = 0.$$

EXAMPLE 3. Without solving, form a quadratic equation whose roots are the squares of the roots of $2x^2 + x - 6 = 0$.

Solution. By letting r_1 and r_2 be the roots of the given equation, we have

$$r_1 + r_2 = -\tfrac{1}{2},$$

$$r_1 r_2 = -3.$$

Since $(r_1 + r_2)^2 \equiv r_1^2 + 2r_1 r_2 + r_2^2$, the sum of the roots for the new equation will be

$$r_1^2 + r_2^2 \equiv (r_1 + r_2)^2 - 2r_1 r_2 = \tfrac{1}{4} + 6 = \tfrac{25}{4},$$

and the product will be

$$(r_1 r_2)^2 = 9.$$

Therefore, $x^2 - \tfrac{25}{4}x + 9 = 0$ or, clearing fractions,

$$4x^2 - 25x + 36 = 0.$$

There is another important expression, $b^2 - 4ac$, in terms of the coefficients from which, without solving, we can discover the nature of the zeros of the function. Specifically, if a, b, and c have values such that $b^2 - 4ac$ is negative, this expression, called the *discriminant* of Eq. (6–7), appearing under the radical in (6–8) results in a square root which has *no* real value. Such zeros are *imaginary* and will be considered in Chapter 15. If $b^2 - 4ac$ is equal to zero, $r_1 = r_2$ and the zeros are equal, while if $b^2 - 4ac$ is positive, the zeros are real, and unequal. Since the zeros of the function $ax^2 + bx + c$ and the roots of the equation $ax^2 + bx + c = 0$ are the same, we have immediately:

3. If $b^2 - 4ac < 0$, the roots are imaginary.
4. If $b^2 - 4ac = 0$, the roots are equal.
5. If $b^2 - 4ac > 0$, the roots are real and unequal.

If we wished to emphasize the graph of $y = f(x) = ax^2 + bx + c$, rather than the roots of $ax^2 + bx + c = 0$, we could state:

3'. If $b^2 - 4ac < 0$, the graph of the parabola does not touch or cross the x-axis.

4′. If $b^2 - 4ac = 0$, the graph of the parabola has its vertex on the x-axis.

5′. If $b^2 - 4ac > 0$, the graph of the parabola intersects the x-axis in two real points.

EXAMPLE 4. Without solving any equation, determine the nature of the zeros of $2x^2 - 9x - 35$.

Solution. Since $a = 2$, $b = -9$, and $c = -35$, $b^2 - 4ac = (-9)^2 - 4(2)(-35) = 81 + 280 = 361$. Since $361 > 0$, the zeros are real and unequal. Moreover, since $b^2 - 4ac = 361$, a perfect square, the zeros will be rational. Is this always true when $b^2 - 4ac$ is a perfect square?

EXAMPLE 5. On the same coordinate axis, sketch graphs of

(a) $y = x^2 - 6x + 5$,
(b) $y = x^2 - 6x + 9$,
(c) $y = x^2 - 6x + 13$,

and check the results by finding $b^2 - 4ac$ in each case.

Solution. The graphs of each are shown in Fig. 6–3. These are consistent with the values for $b^2 - 4ac$, since we have

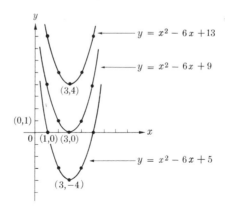

FIGURE 6–3

(a) $b^2 - 4ac = (-6)^2 - 4(1)(5) = 16$
(b) $b^2 - 4ac = (-6)^2 - 4(1)(9) = 0$
(c) $b^2 - 4ac = (-6)^2 - 4(1)(13) = -16$.

PROBLEMS

1. Without solving, find the sum and product of the zeros of the functions listed in Problems 1 through 8, Article 6–2.

Find quadratic equations with integral coefficients having the following given numbers as roots.

2. $2, -3$
3. $-5, 4$
4. $2/3, -2$
5. $-3/4, 2/3$
6. $+\sqrt{3}, -\sqrt{3}$
7. $2 + \sqrt{3}, 2 - \sqrt{3}$
8. $\dfrac{-1 + \sqrt{5}}{2}, \dfrac{-1 - \sqrt{5}}{2}$
9. $\dfrac{-3 + \sqrt{7}}{4}, \dfrac{-3 - \sqrt{7}}{4}$

10. Find a quadratic equation whose roots are the squares of the roots of $4x^2 + 8x - 5 = 0$.

11. Find a quadratic equation whose roots are the reciprocals of the roots of $4x^2 + 8x - 5 = 0$. $\left[Hint: \dfrac{1}{r_1} + \dfrac{1}{r_2} = \dfrac{r_1 + r_2}{r_1 r_2}. \right]$

12. Find a quadratic equation whose roots are the squares of the roots of $3x^2 - 5x - 2 = 0$.

13. Find a quadratic equation whose roots are the reciprocals of the roots of $3x^2 - 5x - 2 = 0$.

14. Find a quadratic equation whose roots are twice the roots of $4x^2 + 8x - 5 = 0$.

15. Find a quadratic equation whose roots are one-third the roots of $4x^2 + 8x - 5 = 0$.

16. Find a quadratic equation whose roots are three times the roots of $3x^2 - 5x - 2 = 0$.

17. Find a quadratic equation whose roots are half the roots of $3x^2 - 5x - 2 = 0$.

Find the value of k so that the equation:

18. $4x^2 + kx + 6 = 0$ has one root $= -2$.
19. $2x^2 + kx - 15 = 0$ has one root $= 3$.
20. $3x^2 + kx - 2 = 0$ has roots whose sum is equal to 6.

21. $5x^2 - 8x + k = 0$ has roots whose product is equal to 5.
22. $4x^2 + 20x + k = 0$ has equal roots.
23. $3x^2 - 7x + k = 0$ has equal roots.
24. $2x^2 + (4 - k)x - 17 = 0$ has roots numerically equal but opposite in sign.
25. $3x^2 - 5x + 8 = kx$ has roots numerically equal but opposite in sign.
26. $3x^2 - 7x + 6 = k$ has one root equal to zero.
27. $4x^2 + 20x + k = 0$ has one root equal to zero.

Find the range of values of k so that the equation:

28. $3x^2 - 4kx + k = 0$ will have real roots.
29. $2x^2 + 3kx - 9 = 0$ will have real roots.
30. $x^2 + (k - 2)x + 4 = 0$ will have real roots.
31. $2x^2 - kx + 8 = 0$ will have imaginary roots.
32. $kx^2 + 4\sqrt{3}x + k = 1$ will have imaginary roots.

Find the range of values of k or value of k so that the graph of the function:

33. $3x^2 - 9x + k$ will touch (have its vertex on) the x-axis.
34. $x^2 + 2kx + \frac{3}{4} - k$ will not intersect the x-axis.
35. $4x^2 + 4\sqrt{2}kx + k + 3$ will intersect the x-axis in two real points.

6–6 Equations in quadratic form. In the last few articles we have discussed the quadratic function, and quadratic equations. In the problems we solved, those involving the circular functions were not specifically quadratic equations, but what might be termed equations in quadratic form. Generally, any equation which can be written in the form

$$av^2 + bv + c = 0, \qquad a \neq 0, \qquad (6\text{--}12)$$

where v is a function of another variable, is called an *equation in quadratic form*. The methods developed in the last articles apply equally to any equation of this type.

EXAMPLE 1. Solve $9x^4 - 37x^2 + 4 = 0$.

Solution. Letting $v = x^2$, and substituting, we have

$$9v^2 - 37v + 4 = 0,$$

or
$$(9v - 1)(v - 4) = 0.$$
Thus,
$$v = x^2 = \tfrac{1}{9} \quad \text{or} \quad 4,$$
and
$$x = \pm\tfrac{1}{3}, \quad \pm 2.$$

EXAMPLE 2. Solve $(x^2 - 5x)^2 + (x^2 - 5x) - 30 = 0$.

Solution. With $v = x^2 - 5x$, we wish to solve
$$v^2 + v - 30 = 0,$$
a quadratic equation in v, whose roots are
$$v = 5 \quad \text{or} \quad -6.$$
Hence, we must solve
$$x^2 - 5x - 5 = 0, \quad \text{and} \quad x^2 - 5x + 6 = 0.$$
Our solution is $x = (5 \pm 3\sqrt{5})/2$, 3, and 2.

PROBLEMS

Solve the following equations for x:

1. $x^4 - 11x^2 + 28 = 0$
2. $9x^4 + 5x^2 - 4 = 0$
3. $x^{-4} - 13x^{-2} + 36 = 0$
4. $x^{-4} - 8x^{-2} + 15 = 0$
5. $x^6 + 7x^3 - 8 = 0$
6. $8x^{-6} + 7x^{-3} = 1$
7. $x^{2/3} + 2x^{1/3} - 8 = 0$
8. $x + x^{1/2} = 20$
9. $(x^2 - 7x)^2 + 9(x^2 - 7x) - 10 = 0$
10. $(x^2 + 2x)^2 + (x^2 + 2x) = 12$
11. $\dfrac{2+x}{2-x} + \dfrac{2-x}{2+x} = 2$
12. $\left(x + \dfrac{1}{x}\right)^2 - 2\left(x + \dfrac{1}{x}\right) + 1 = 0$
13. $3(x+3) + \sqrt{x+3} = 2$
14. $2x - 9\sqrt{x+2} + 14 = 0$
15. $2x^2 - 5x + 10 = 7\sqrt{2x^2 - 5x}$
16. $x^2 - x - 4 = \sqrt{x^2 - x - 2}$

6–7 Equations involving radicals. The last four problems of the last article contained radicals in the equation. Fortunately these were of a special form, and it was possible to solve them by the method described. Many equations involving radicals are not solvable by this method, and the radical or radicals must be eliminated first. After this elimination, former methods may be used. Extreme care must be taken, however, to check all answers in the original equation, since the method of eliminating radicals involves raising both members of an equality to some power, and thus possible introduction of extraneous roots (recall Illustration 2, Article 6–1).

EXAMPLE 1. Solve and check $\sqrt{2x + 5} = 3$.

Solution. In any equation such as this involving one radical, we must first eliminate this radical by squaring both members of the equation,
$$2x + 5 = 9.$$
Solving, we find
$$x = 2.$$
By substituting $x = 2$ in the original equation, we find that it does satisfy:
$$\sqrt{2(2) + 5} = \sqrt{9} = 3.$$

EXAMPLE 2. Solve and check $\sqrt{1 - 5x} + \sqrt{1 - x} = 2$.

Solution. With two radicals, it is simpler to transpose one to the opposite side, before squaring. Thus,
$$\sqrt{1 - 5x} = 2 - \sqrt{1 - x}.$$
Squaring,
$$1 - 5x = 4 - 4\sqrt{1 - x} + 1 - x,$$
and simplifying,
$$-4x - 4 = -4\sqrt{1 - x},$$
or
$$1 + x = \sqrt{1 - x}.$$
Squaring again,
$$1 + 2x + x^2 = 1 - x,$$
$$x^2 = -3x,$$
$$x = 0, -3.$$

By checking, we find 0 satisfies the original equation, but -3 **does not**. Therefore $x = 0$ is the only solution.

EXAMPLE 3. Solve $3 \sin \theta + 4 \cos \theta = 5$ for all values of θ between $0°$ and $360°$.

Solution. Although this is not an algebraic equation, it may be solved by the method discussed in this article. In order to express $\cos \theta$ in terms of $\sin \theta$, we recall $\cos \theta \equiv \pm \sqrt{1 - \sin^2 \theta}$. Thus,

$$\pm 4\sqrt{1 - \sin^2 \theta} = 5 - 3 \sin \theta.$$

Squaring and solving for $\sin \theta$, we have

$$16 - 16 \sin^2 \theta = 25 - 30 \sin \theta + 9 \sin^2 \theta,$$

$$25 \sin^2 \theta - 30 \sin \theta + 9 = 0,$$

$$(5 \sin \theta - 3)^2 = 0,$$

$$\sin \theta = \tfrac{3}{5}.$$

Therefore $\theta = 36°52'$ or $143°8'$. By checking our result, we find $\sin \theta = 36°52'$ is the only solution. This may also be written in radian measure, $\theta = 0.6464$ radians.

PROBLEMS

Solve and check:

1. $\sqrt{2x + 5} = 4$
2. $\sqrt{6x - 3} = 7$
3. $\sqrt{8x - 7} - x = 0$
4. $\sqrt{3x + 1} + 1 = x$
5. $\sqrt{3x + 1} = \sqrt{x} + 3$
6. $2\sqrt{4x + 5} = \sqrt{8 - x} - 1$
7. $\sqrt{11 - x} - \sqrt{x + 6} = 3$
8. $\sqrt{3 - x} - \sqrt{2 + x} = 3$
9. $\sqrt{2x + \sqrt{2x + 4}} = 4$
10. $\sqrt{2x + \sqrt{7 + x}} = 3$

Find the values of θ between 0 and 2π which satisfy the following and check:

11. $4 \sin \theta + 3 \cos \theta = 2$
12. $\cos \theta - 2 = \sqrt{3} \sin \theta$
13. $\sin \theta + \cos \theta = 1$
14. $12 \cos \theta - 5 \sin \theta = 13$
15. $5 \tan \theta - \sqrt{2} \sec \theta + 3 = 0$
16. $\tan \theta + \sec \theta = 1$

6–8 Variation. Many applications from physical and social science make use of a functional dependence known as *proportion* or *variation*. Many of the relations are linear or quadratic, although this is by no means necessary. For example, Ohm's law for an electrical circuit states that the current *varies directly* as the electromotive force, and varies *inversely* as the resistance. Thus the current is a function of both the electromotive force and the resistance. We may write this functional relationship

$$I = k \cdot \frac{E}{R},$$

where k is called the *proportionality constant*. Specifically, we shall define the three different common types of variation.

1. If the two variables x and y are so related (no matter how their values change) that the quotient of y divided by x, called the *ratio* of y to x, is always constant, then y is said to *vary directly* as x. This relationship may be written $y/x = k$, or

$$y = kx. \tag{6-13}$$

2. If the two variables, x and y, are so related (no matter how their values change) that the product of y and x is always constant, then y is said to *vary inversely* as x. This relationship may be written $yx = k$, or

$$y = \frac{k}{x}. \tag{6-14}$$

3. If the variable z varies directly as the product of the variables x and y, then z is said to *vary jointly* as x and y, written

$$z = kxy. \tag{6-15}$$

EXAMPLE 1. Express z as a specific function of x and y, if z varies directly as x and inversely as the square of y, and $z = 18$ when $x = 3$ and $y = 2$.

Solution. The given proportion may be written

$$z = k \frac{x}{y^2}.$$

Substituting the values for x, y, and z, we have

$$18 = k\tfrac{3}{4},$$

and find $k = 24$. Therefore, the function may be written

$$z = \frac{24x}{y^2}.$$

EXAMPLE 2. At constant temperature, the resistance of a wire varies directly as its length and inversely as the square of its diameter. If a piece of wire 0.1 inch in diameter and 50 ft long has a resistance of 0.1 ohm, what is the resistance of a piece of wire of the same material, 2000 ft long, 0.2 inch in diameter?

Solution. Letting R, L, and d represent the resistance, the length and the diameter, respectively, of the wire, we have $R = kL/d^2$. Substituting, $0.1 = k(50)/(0.1)^2$ or $k = (0.1)^3/50$. Thus

$$R = \frac{(0.1)^3}{50} \cdot \frac{L}{d^2}.$$

Therefore,

$$R = \frac{(0.1)^3}{50} \cdot \frac{(2000)}{(0.2)^2} = \frac{(0.001)(2000)}{50(.04)} = 1 \text{ ohm}.$$

Notice that all the measurements do not need to be expressed in the same unit, although we must be consistent for each variable.

EXAMPLE 3. If y varies inversely as the square of x, how is y affected if x is increased 25%?

Solution. We have $y = k/x^2$. If $x = x_1$, then $y = k/x_1^2$. But $x = 5x_1/4$. Thus,

$$y = \frac{k}{\left(\dfrac{5x_1}{4}\right)^2} = \frac{16}{25} \cdot \frac{k}{x_1^2} = 0.64 \frac{k}{x_1^2}.$$

Therefore, y is 0.64 of its original value.

PROBLEMS

In Problems 1–8, express the functional relationship as a single algebraic equation, giving the specific value of k if possible.

1. The variable z varies directly as x and inversely as y.
2. The variable z varies directly as x and inversely as the square of y.
3. The variable z varies jointly as x and y, and $z = 72$, when $x = 4$ and $y = 3$.

4. The variable z varies inversely as x and $z = 8$ when $x = 16$.

5. The circumference of a circle varies directly as the square of the diameter.

6. The area of any triangle varies jointly as the product of the base and altitude.

7. The area of an equilateral triangle varies directly as the square of one side. (Compare Problem 9, Article 3–6.)

8. The distance a falling body travels (neglecting air resistance) varies as the square of the time traveled, and a body, starting from rest, falls 64 ft in 2 sec.

9. Since for any specific angle θ, $\sin \theta$ is a definite value, and is defined as y/r (Article 4–2), we may say that for any constant angle, y varies directly as r. What is the proportionality constant for an angle of 30°, 60°, 38°?

10. For the tangent function, y varies directly as x. What is the proportionality constant for this function if θ is 45°, 60°?

11. If z varies jointly as x^2 and y, and $z = 24$ when $x = 2$ and $y = 3$, find the value of z when $x = 3$ and $y = 5$.

12. If z varies directly as x and inversely as y, and if $z = 5$ when $x = 2$ and $y = 3$, find z when $x = 4$ and $y = 2$.

13. The surface area of a sphere varies directly as the square of the radius. If the surface is 36π in.² when the radius is 3 in., what is the surface area when the radius is 12 in.?

14. With the information in Problem 8, (a) find the distance the body has fallen in 5 sec, (b) find the distance the body fell during the fifth second.

15. The kinetic energy K varies jointly as the mass m and the square of the velocity v. If K is 36 ergs when m is 8 gm, and v is 3 cm/sec, find K if $m = 4$ gm and $v = 6$ cm/sec.

16. The vibrating frequency, or pitch, of a vibrating string varies directly as the square root of the tension of the string. If a string vibrates 216 times/sec, due to a tension of 3 lb, find its rate of vibration caused by a tension of 12 lb.

17. If the intensity of light varies inversely as the square of the distance from its source, how much farther from the light must an object be moved to receive one-fourth the amount of light it now receives, if it is now 2 feet from the light?

18. If z varies directly as the square of x and inversely as y, what effect on z does doubling x and tripling y have?

19. The gravitational attraction F between two bodies varies jointly as their masses m_1 and m_2 and inversely as the square of the distance d between them. What is the effect on the gravitational attraction between

two bodies if the masses are each doubled, and the distance between them is halved?

20. The stiffness of a beam varies jointly as its breadth and depth and inversely as the square of the length. (a) Find the change in stiffness if each of the three dimensions is increased 10%. (b) Find the necessary change in the length in order to increase the stiffness 20% if the breadth and depth are unaltered.

21. Draw the graph of $y = kx$ for $k > 0$. Notice on the graph that y increases as x increases. [*Hint:* Use units of k length on the y-axis.]

22. Draw $y = k/x$ for $k > 0$. Notice on the graph that y decreases as x increases.

6–9 Solution of two linear equations in two unknowns. In Article 6–1 we considered the first degree or linear function in one variable. Any function of the first degree in two variables x and y may be written

$$ax + by + c, \qquad (6\text{–}16)$$

where a, b, and c are constants, and is called a *linear function* in two variables. By setting this function equal to zero, and transposing the constant term to the right side, we have $ax + by = -c$. Since it is convenient to write such an equation with a positive constant term, we denote $-c$ by a new c, so that

$$ax + by = c. \qquad (6\text{–}17)$$

This equation, where a and b are not both zero, is called a *linear equation in two variables*. Since (6–17) may be solved for y in terms of x if $b \neq 0$,

$$by = -ax + c,$$

$$y = -\frac{a}{b}x + \frac{c}{b}, \qquad (6\text{–}18)$$

the graph of Eq. (6–17) is the same as that of Eq. (6–1) where $m = -a/b$, and b* is c/b, and thus Eq. (6–17) represents a straight line in the plane. If $b = 0$, the equation becomes $ax = c$ or $x = c/a$, which represents a straight line, parallel to the y-axis.

---

* The two b's are not the same. The reader should differentiate between them.

Since a solution of any equation must satisfy that equation by definition, any set of values (x,y) which satisfy Eq. (6–17) is called a solution of $ax + by = c$. There are clearly infinitely many solutions, for there are infinitely many points (x,y) whose coordinates satisfy such an equation. This is evident from the equation in the form of Eq. (6–18). Any arbitrary value of x will determine a corresponding value of y, both of which will satisfy the equation. For example, the linear equation $3x - 2y = 5$ has one solution $x = 3$, $y = 2$, another, $x = 1$, $y = -1$, another $x = -1$, $y = -4$, and so on.

We are frequently interested in a solution not of one equation, but of a pair of linear equations in x and y, such as

$$a_1 x + b_1 y = c_1, \qquad a_2 x + b_2 y = c_2. \qquad (6\text{–}19)$$

Thus we are interested in finding a pair of values (x,y) which satisfies both equations. Since the graph of each of these equations is a straight line, interpreted graphically, we are interested in finding the coordinates of a point (x,y) which lies on both lines. Since, in general, two straight lines intersect in one point, there is usually a solution. By drawing a graph of each line, the coordinates of the point of intersection may be approximated.

EXAMPLE 1. Solve graphically the pair of equations

$$x + 2y = 4, \qquad 3x - 2y = -12$$

Solution. The graph of each of these straight lines is shown in Fig. 6–4. It appears that the two lines intersect at the point $(-2,3)$.

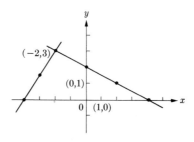

FIGURE 6–4

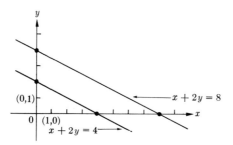

FIGURE 6–5

This is actually the case, as may be verified by direct substitution, so that the solution is $x = -2$, $y = 3$.

We should remark that two parallel lines such as $x + 2y = 4$, and $x + 2y = 8$, when plotted, will not intersect (Fig. 6–5). The fact that the equations of two such lines have no common solution is, of course, apparent from the equations, since there can be no pair of numbers such that the first number plus twice the second is equal to 4 and also 8. Two equations of this type are called *inconsistent*.

Moreover, if two lines such as $x + 2y = 4$ and $3x + 6y = 12$ coincide when their graphs are plotted (Fig. 6–6), any pair of values (x,y) which satisfies one equation will also satisfy the other. Any two equations of this type have an infinite number of solutions, and are said to be *dependent*.

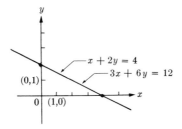

FIGURE 6–6

In summary, the two equations (6–19), either:

1. have a unique solution, when their graphs are two straight lines which intersect in one point;
2. have no solution, when their graphs are two parallel straight lines;
3. have infinitely many solutions, when their graphs are two straight lines which coincide.

Since the graphical method of solution is only approximate, we must also consider some exact method. One of the variables may be eliminated by a proper combination of the two equations (or equations equivalent to them), and the resulting equation may be solved for the other variable. The eliminated variable may then be found by substitution.

EXAMPLE 2. Solve algebraically the two equations of Example 1.

Solution. Since the coefficients of y are equal, but opposite in sign, the members of the first equation may be added to those of the second and thereby the y is eliminated.

$$\begin{array}{rl} x + 2y = & 4 \\ 3x - 2y = & -12 \\ \hline 4x = & -8 \end{array}$$

Solving this equation in one variable, $x = -2$. When this value is substituted in either of the two original equations, we find $y = 3$. Therefore the required solution is $x = -2$, $y = 3$. We might have found y by multiplying each member of the first equation by -3, adding the members of the resulting equation, $-3x - 6y = -12$, to those of the second equation, $3x - 2y = -12$, and finding $-8y = -24$, or $y = 3$.

EXAMPLE 3. Solve the pair of two general equations (6–19) for x and y in terms of the coefficients.

Solution. If we multiply each member of the first equation by b_2 and of the second by b_1, we have

$$a_1 b_2 x + b_1 b_2 y = c_1 b_2,$$
$$a_2 b_1 x + b_1 b_2 y = c_2 b_1.$$

SOLUTION OF TWO LINEAR EQUATIONS

Subtracting each member of the second from those of the first,
$$a_1 b_2 x - a_2 b_1 x = c_1 b_2 - c_2 b_1.$$
Solving this equation for x, we find
$$(a_1 b_2 - a_2 b_1) x = c_1 b_2 - c_2 b_1,$$
$$x = \frac{c_1 b_2 - c_2 b_1}{a_1 b_2 - a_2 b_1}, \quad (a_1 b_2 - a_2 b_1 \neq 0).$$

By multiplying each member of the first equation by a_2 and of the second by a_1, and so on, we find
$$y = \frac{a_1 c_2 - a_2 c_1}{a_1 b_2 - a_2 b_1}, \quad (a_1 b_2 - a_2 b_1 \neq 0).$$

Although not used in this form, these values for x and y can be considered as a *solution by formula*.

Problems

Solve the following pairs of linear equations algebraically. The graphical method might be used as an approximate check. If an angle is involved, find all possible values between 0 and 2π.

1. $2x - y = 5$,
   $x - 3y = 5$.

2. $3x - 2y = -14$,
   $2x + 3y = 8$.

3. $4x + 3y = 27$,
   $2x - 5y + 19 = 0$.

4. $2x - 5y + 43 = 0$,
   $6x - y + 31 = 0$.

5. $3x + 4 = 4y$,
   $9x + 2y = 9$.

6. $6x + 9y = 7$,
   $3x - 6y + 14 = 0$.

7*. $4 \sin \alpha + \cos \beta = 3$,
   $6 \sin \alpha - 2 \cos \beta = 1$.

8. $\sqrt{3} \sin \alpha - \cos \beta = 1$,
   $\sin \alpha - 3\sqrt{3} \cos \beta = -\sqrt{3}$.

9. $2 \sec \alpha - 4 \tan \beta = 0$,
   $\sec \alpha + \tan \beta = \sqrt{3}$.

10. $\dfrac{2}{x} + \dfrac{3}{y} = 2$,
    $\dfrac{4}{x} - \dfrac{9}{y} + 1 = 0$.

11. $\dfrac{15}{x} + \dfrac{4}{y} = 1$,
    $\dfrac{5}{x} - \dfrac{12}{y} = 7$.

12. $\dfrac{15}{2x} - \dfrac{16}{3y} = \dfrac{23}{6}$,
    $\dfrac{4}{3x} + \dfrac{7}{2y} + \dfrac{31}{72} = 0$.

---

* Although Problems 7 through 12 are not linear, they may be solved by the method discussed in this article.

13. $ax + by = a^2 + 2ab + b^2$,
    $bx - ay = b^2 + 2ab - a^2$.

14. $ax - by = a^2 - b^2$,
    $bx + ay = 2ab$.

*15. $\tan k_1 = \dfrac{y}{x}$,

   $\tan k_2 = \dfrac{y+a}{x}$.

*16. $\tan k_1 = \dfrac{a}{y}$,

   $\tan k_2 = \dfrac{a}{y+x}$.

17. The sum of the digits of a two-digit number is 9. If the digits are reversed, the new number is also 9 less than the original number. Find the two numbers.

18. A certain fraction has the value 3/4. If its numerator is decreased by 7, and its denominator increased by 4, the resulting fraction has the value one-half. Find the original fraction.

19. The sum of the two non-right angles in a right triangle is, of course, 90°. If twice the first is 40° more than 3 times the second, find the angles.

20. Find the linear function of x, $f(x) = mx + b$, if its value is 1 when $x = -3$, and 9 when $x = 1$.

21. With the wind, an airplane travels 1120 mi in 7 hr. Against the wind, however, it takes 8 hr. Find the rate of the plane in still air and the velocity of the wind.

22. The sum of the reciprocals of two numbers is 9. Also, 2 times the reciprocal of the first is 12 less than 4 times the reciprocal of the second. Find the numbers.

23. A and B, working together, can do a job in $6\frac{2}{3}$ hr. A became ill after 3 hr of working with B and B finished the job, continuing to work alone in $8\frac{1}{4}$ more hours. How long would it take each working alone to do the job?

24. Find α and β, angles between 0 and $\pi/2$, so that $2\sin(\alpha + \beta) = 2\cos(\alpha - \beta) = \sqrt{3}$.

6–10 Algebraic solution of three linear equations in three unknowns.
Any equation of the form

$$ax + by + cz = d, \qquad (6\text{--}20)$$

where a, b, c, and d are constants, and a, b, and c are not all equal to zero, is called a *linear equation in three variables*. A solution of such an equation is any set of three numbers x, y, and z which satisfies the equation. As was the case for Eq. (6–17), there are an infinite number of solutions for one equation of this type.

Although we shall not prove it in this book, the geometric interpretation of Eq. (6–20) is a plane in a three-dimensional rectangular coordinate system.* Two nonparallel planes intersect in a straight line, and this line intersects a third plane, parallel to neither of the first two, in a point, so that, in general, three planes have one point in common. Algebraically, this may be interpreted by the fact that the system of equations

$$a_1x + b_1y + c_1z = d_1,$$
$$a_2x + b_2y + c_2z = d_2, \qquad (6\text{--}21)$$
$$a_3x + b_3y + c_3z = d_3,$$

has a single solution for x, y, and z.

To solve such a system, we follow a method which is a generalization of that used to solve the pair of equations (6–19). By choosing a pair of these equations, we eliminate one of the variables from the pair, obtaining an equation in two variables. Repeating this procedure for another pair of these equations, we obtain a second equation in the same two variables. We now solve the two resulting equations for the two variables, and by substitution in any of the original equations, find the complete solution. Let us illustrate.

EXAMPLE 1. Solve the system of equations.

$$2x - y + z = 8,$$
$$x + 2y + 3z = 9,$$
$$4x + y - 2z = 1.$$

Solution. Let us eliminate z by combining the first two equations, and then by combining the first and third. Multiply each member of the first equation by -3, and add, member by member, to the second.

$$-6x + 3y - 3z = -24$$
$$\underline{x + 2y + 3z = 9}$$
$$-5x + 5y = -15$$

or

$$x - y = 3.$$

---

* A three-dimensional rectangular coordinate system is a generalization of the one- and two-dimensional systems discussed in Articles 3–1 and 3–2. For example, each point in the space has three coordinates and is denoted (x,y,z), and so on.

Multiply each member of the first by 2, and add, member by member, to the third.

$$4x - 2y + 2z = 16$$
$$\underline{4x + y - 2z = 1}$$
$$8x - y = 17$$

We now solve the two resulting equations. Give the reasons for each step.

$$8x - y = 17$$
$$\underline{x - y = 3}$$
$$7x = 14$$

Therefore, $x = 2$, and $y = -1$. Substituting these values in the first of the original equations,

$$2(2) - (-1) + z = 8,$$
$$z = 3.$$

Therefore, the complete solution is $x = 2$, $y = -1$, $z = 3$. All solutions should be checked.

Problems

Solve each system of equations.

1. $x + 3y - z = 4,$
   $3x - 2y + 4z = 11,$
   $2x + y + 3z = 13.$

2. $3x - y - 2z = -13,$
   $5x + 3y - z = 4,$
   $2x - 7y + 3z = -36.$

3. $2x - y + 3z = 19,$
   $5x - 2y + 4z = 33,$
   $3x + 3y - z = 2.$

4. $6x + 4y - z = 13,$
   $5x - 2y + 7z = 18,$
   $x + y - 8z = -35.$

5. $3x + 5y + 2z = 0,$
   $12x - 15y + 4z = 12,$
   $6x + 25y - 8z = -12.$

6. $7x - 3y + 4z = 18,$
   $13x + 6y + 8z = 30,$
   $11x - 9y - 12z = 16.$

7. $2x + 3y = 28,$
   $3y + 4z = 46,$
   $4z + 5x = 53.$

8. $x - 3y = -11,$
   $2y - 5z = 26,$
   $3z - 7x = 2.$

9. $\dfrac{3}{x} - \dfrac{4}{y} + \dfrac{6}{z} = 1,$
   $\dfrac{9}{x} + \dfrac{8}{y} - \dfrac{12}{z} = 3,$
   $\dfrac{9}{x} - \dfrac{4}{y} + \dfrac{12}{z} = 4.$

10. $x + y + z = a + b + c,$
    $bx - ay + cz = b^2,$
    $ax - ay + cz = ab.$

11. The sum of the digits of a three-digit number is 13. If the tens' and hundreds' digits are interchanged, the new number is 90 less than the original, and if the units' and hundreds' digits are interchanged, the resulting number is 99 less than the original. Find the original number.

12. Twenty-five coins, whose value is $2.75, are made up of nickels, dimes, and quarters. If the nickels were dimes, the dimes were quarters, and the quarters nickels, the total value would be $3.75. How many coins of each type are there?

13. We recall that the sum of the angles of any triangle is 180°. What are the three angles if the sum of two is equal to the third angle, but the difference of these two is only two-thirds of the third angle?

14. Find the specific quadratic function of the form $f(x) = ax^2 + bx + c$, if its value is 1 when $x = 1$, if its value is 2 when $x = 2$, and if its value is 11 when $x = -1$.

15. If the general equation (3-5) of the circle is expanded, it can be written in the form $x^2 + y^2 + Ax + By + C = 0$. Recalling that the coordinate of any point on the circle must satisfy its equation, find the values of A, B, and C, and, thereby, the circle which passes through the points (1,1), (−2,3), and (3,4).

16. If A, B, and C work together on a job, it will take 1-1/3 hr. If only A and B work, it would take 1-5/7 hr, but if B and C work, it would take 2-2/5 hr. How long would it take each man, working alone, to complete the job?

6–11 Solution of one linear and one quadratic equation. In the last two articles, we discussed systems of linear equations. Such systems may be generalized either by considering more variables than two or three, or by considering functions other than the linear ones. The solutions of such systems, if they exist, are often difficult to obtain. We shall consider one simple but useful generalization in two variables, namely, when one equation is linear, and one is of the second degree, or quadratic. Such a system, in general, might be written

$$ax^2 + bxy + cy^2 + dx + ey + f = 0,$$
$$gx + hy + k = 0,$$
(6–22)

where a, b, c, d, e, f, g, h, and k are constants, a, b, and c not all zero, and g and h not both zero.

As in the case of two linear equations, it is possible to solve such a system either graphically or algebraically. The graph of the quadratic equation may best be plotted by solving for one variable in terms of the others, before the table of values is obtained. The points of intersection of the graphs of this quadratic equation and the linear equation may be approximated and their coordinates taken as solutions, since the points lie on both curves.

The simplest algebraic method is carried out by eliminating one of the variables. More specifically, we solve the linear equation for one variable in terms of the other, substitute this value in the quadratic equation, and solve the resulting quadratic equation in one variable. With these results substituted in the original linear equation, we obtain our complete solution.

Since this system reduces to the problem of solving one quadratic equation, we shall have either two real, distinct, solutions, one real solution, or no real solution (recall Article 6–5).

EXAMPLE 1. Solve the system of equations,

$$x^2 - 5x - y + 4 = 0,$$

$$x - 4y = 1.$$

Algebraic solution. Although we are able to eliminate x or y, we choose to solve the linear equation for y in terms of x, and substitute this value in the quadratic, since the work appears to be simpler.

$$4y = x - 1 \quad \text{or} \quad y = \frac{x-1}{4}.$$

Therefore,

$$x^2 - 5x - \left(\frac{x-1}{4}\right) + 4 = 0,$$

$$4x^2 - 20x - x + 1 + 16 = 0,$$

$$4x^2 - 21x + 17 = 0,$$

$$(4x - 17)(x - 1) = 0.$$

Thus, $x = 1$ or $17/4$. Substituting these values in the linear equation gives the corresponding values $y = 0, 13/16$. The two solutions are $x = 1, y = 0$, and $x = 17/4, y = 13/16$.

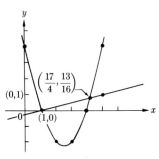

FIGURE 6–7

Graphical solution. The given equations represent a parabola and a straight line, whose graphs are shown in Fig. 6–7. The intersections of the two curves show approximately the same solutions.

EXAMPLE 2. Solve the system of equations:

$$x^2 - 2y^2 + 3x - 4y + 20 = 0,$$
$$2x - y = 1.$$

Solution.* Although we are able to eliminate x or y, we choose to solve the linear equation for y and thus work with x, since the work appears to be simpler. Thus, substituting $y = 2x - 1$ in the first equation for y, we have

$$x^2 - 2(2x - 1)^2 + 3x - 4(2x - 1) + 20 = 0.$$

Simplifying,

$$x^2 - 8x^2 + 8x - 2 + 3x - 8x + 4 + 20 = 0,$$
$$-7x^2 + 3x + 22 = 0,$$
$$7x^2 - 3x - 22 = 0,$$
$$(7x + 11)(x - 2) = 0.$$

Therefore,
$$x = 2, \quad -\tfrac{11}{7}.$$

---

* A graphical solution involves the graphing of $x^2 - 2y^2 + 3x - 4y + 20 = 0$. Such graphing will be discussed in Article 9–1.

Substitution of these values in the linear equation gives the corresponding values $y = 3$, $-29/7$. The two solutions are then $x = 2$, $y = 3$, and $x = -11/7$, $y = -29/7$.

Problems

Solve each of the systems in 1 through 6 both graphically and algebraically.

1. $y = 4x^2$,
   $y = 8x$.
2. $x - 2y = 10$,
   $y = x^2 + 2x - 15$.
3. $y^2 = 4x$,
   $x - y = -4$.
4. $(x - 2)^2 + (y - 3)^2 = 4$,
   $x + y + 3 = 0$.
5. $(x - 2)^2 + (y - 3)^2 = 4$, [*Hint:* Recall Article 3–4 for the graph of
   $x + y = 4$. this quadratic equation.]
6. $xy = 1$,
   $x + y = 2$.

Solve for x and y:

7. $y^2 + 2xy - 3x^2 + 7 = 0$,
   $x - 3y + 1 = 0$.
8. $x^2 + y^2 - 2x = 1$,
   $2x + y - 5 = 0$.
9. $3xy + 6x = 4y$,
   $2y - 3x - 4 = 0$.
10. $4x^2 + 3xy - 2y^2 - x - y + 1 = 0$,
    $x - 2y + 3 = 0$.

11. The circle $x^2 + y^2 = a^2$ and the straight line $y = mx + b$ will intersect in two points, be tangent, or not intersect, depending upon whether the solutions of this system of equations are real and distinct, real and equal, or imaginary. Find the value of b in terms of a and m, so that the straight line will be tangent to the circle.

12. Find the dimensions of a rectangle if its diagonal is 17 in., while its perimeter is 46 in.

13. The product of a two-digit number and the number obtained by reversing its digits is 736. If the difference of the two numbers is 9, find the numbers.

14. A and B, working together, can complete a certain job in $7\frac{1}{2}$ hr. Working separately, it would take A 8 hr longer than it would take B to do the job. How long would it take each working alone?

15. The sum of two numbers is 11, while the sum of their reciprocals is 11/28. Find the numbers.

CHAPTER 7

DETERMINANTS

In mathematics it is frequently convenient to consider square or rectangular arrays of numbers such as

$$\left\| \begin{array}{ccc} a_1 & b_1 & c_1 \\ a_2 & b_2 & c_2 \end{array} \right\|. \tag{7-1}$$

This type of rectangular array of numbers is called a *matrix*. The numbers $a_1, b_1, \cdots c_2$, are called the *elements* of the matrix. The horizontal lines of numbers are called its *rows*, while the vertical lines of numbers are called *columns*. In general, if a rectangular array has m rows and n columns, we call it an m by n matrix. We shall be interested only in square matrices ($m = n$), since any further study is beyond the scope of this book. Such an n by n matrix is said to be of *order n*.

7-1 Determinants of order two and three. With each square matrix, we associate a number called the *determinant* of the matrix. For a matrix of order 2, this number or determinant is defined to be the product of the elements in the upper left and lower right corners, minus the product of the lower left and upper right elements. This determinant is symbolized as

$$D = \begin{vmatrix} c_1 & b_1 \\ a_2 & b_2 \end{vmatrix} = a_1 b_2 - a_2 b_1. \tag{7-2}$$

ILLUSTRATION 1.

(a) $D = \begin{vmatrix} 3 & 2 \\ 5 & 4 \end{vmatrix} = (12) - (10) = 2.$

(b) $D = \begin{vmatrix} -1 & -4 \\ 2 & -3 \end{vmatrix} = (-1)(-3) - 2(-4) = 11.$

Problems

Evaluate the determinants in 1 through 6.

1. $\begin{vmatrix} 3 & 4 \\ 5 & 6 \end{vmatrix}$

2. $\begin{vmatrix} 6 & -1 \\ 3 & -2 \end{vmatrix}$

3. $\begin{vmatrix} -2 & 3 \\ -5 & 6 \end{vmatrix}$

4. $\begin{vmatrix} 8 & 0 \\ 1 & 2 \end{vmatrix}$

5. $\begin{vmatrix} \sin\theta & \cos\theta \\ -\cos\theta & \sin\theta \end{vmatrix}$

6. $\begin{vmatrix} \sec\theta & \tan\theta \\ \tan\theta & \sec\theta \end{vmatrix}$

*7. From Example 3, Article 6–9, show that the solution of the system of equations,

$$a_1 x + b_1 y = c_1,$$
$$a_2 x + b_2 y = c_2,$$

may be written in terms of determinants as

$$x = \frac{\begin{vmatrix} c_1 & b_1 \\ c_2 & b_2 \end{vmatrix}}{\begin{vmatrix} a_1 & b_1 \\ a_2 & b_2 \end{vmatrix}}, \quad y = \frac{\begin{vmatrix} a_1 & c_1 \\ a_2 & c_2 \end{vmatrix}}{\begin{vmatrix} a_1 & b_1 \\ a_2 & b_2 \end{vmatrix}} \tag{7-3}$$

8. Using Eq. (7–3), solve Problems 1, 3, 5 of Article 6–9.

9. Using Eq. (7–3), solve Problems 2, 4, 6 of Article 6–9.

10. Solve for x: $\begin{vmatrix} 3 & 2 \\ x & 4 \end{vmatrix} = 0.$

11. Solve for x: $\begin{vmatrix} 2x & 3 \\ -5 & -4 \end{vmatrix} = 7x.$

12. Prove the identity $\begin{vmatrix} x-2 & 3 \\ x-2 & 5 \end{vmatrix} \equiv 2(x-2).$

13. Prove the identity $\begin{vmatrix} a-b & 0 \\ 1 & c-d \end{vmatrix} \equiv (a-b)(c-d).$

14. For what values of x is $\begin{vmatrix} x & 5 \\ 125 & x \end{vmatrix} > 0?$

15. What can be said about the nature of the roots of $ax^2 + bx + c = 0$ if $\begin{vmatrix} b & 4a \\ c & b \end{vmatrix} > 0?, = 0?, < 0?$

DETERMINANTS OF ORDER TWO AND THREE

Let us now consider the square matrix of order three,

$$M = \begin{Vmatrix} a_1 & b_1 & c_1 \\ a_2 & b_2 & c_2 \\ a_3 & b_3 & c_3 \end{Vmatrix}, \qquad (7\text{–}4)$$

where the elements, rows, and columns were defined at the beginning of this chapter. For each element in such a matrix there exists a matrix of order 2, obtained by deleting the row and column in which this element lies. The second order determinant associated with one such matrix is called the *minor* of the element under consideration. For example, denoting the minor of any element by the corresponding capital letter,

$$A_1 = \begin{vmatrix} b_2 & c_2 \\ b_3 & c_3 \end{vmatrix}, \qquad B_3 = \begin{vmatrix} a_1 & c_1 \\ a_2 & c_2 \end{vmatrix}, \qquad C_2 = \begin{vmatrix} a_1 & b_1 \\ a_3 & b_3 \end{vmatrix}.$$

We now define the determinant of the third order matrix in (7–4) by

$$D = \begin{vmatrix} a_1 & b_1 & c_1 \\ a_2 & b_2 & c_2 \\ a_3 & b_3 & c_3 \end{vmatrix} = a_1 A_1 - b_1 B_1 + c_1 C_1. \qquad (7\text{–}5)$$

EXAMPLE 1. Find the determinant of the matrix $\begin{Vmatrix} 2 & -1 & 3 \\ 3 & -2 & 1 \\ 4 & -3 & 2 \end{Vmatrix}$.

Solution.

$$\begin{vmatrix} 2 & -1 & 3 \\ 3 & -2 & 1 \\ 4 & -3 & 2 \end{vmatrix} = 2 \begin{vmatrix} -2 & 1 \\ -3 & 2 \end{vmatrix} - (-1) \begin{vmatrix} 3 & 1 \\ 4 & 2 \end{vmatrix} + 3 \begin{vmatrix} 3 & -2 \\ 4 & -3 \end{vmatrix}$$

$$= 2(-4 + 3) + 1(6 - 4) + 3(-9 + 8)$$

$$= -2 + 2 - 3 = -3.$$

EXAMPLE 2. Express D [Eq. (7–5)] in terms of the elements only.

Solution.

$$D = \begin{vmatrix} a_1 & b_1 & c_1 \\ a_2 & b_2 & c_2 \\ a_3 & b_3 & c_3 \end{vmatrix} = a_1 \begin{vmatrix} b_2 & c_2 \\ b_3 & c_3 \end{vmatrix} - b_1 \begin{vmatrix} a_2 & c_2 \\ a_3 & c_3 \end{vmatrix} + c_1 \begin{vmatrix} a_2 & b_2 \\ a_3 & b_3 \end{vmatrix}$$

$$= a_1(b_2 c_3 - b_3 c_2) - b_1(a_2 c_3 - a_3 c_2) + c_1(a_2 b_3 - a_3 b_2)$$

$$= a_1 b_2 c_3 + a_2 b_3 c_1 + a_3 b_1 c_2 - a_1 b_3 c_2 - a_2 b_1 c_3 - a_3 b_2 c_1. \qquad (7\text{–}6)$$

PROBLEMS

Evaluate each of the determinants 1 through 4:

1. $\begin{vmatrix} 1 & 2 & 3 \\ -2 & -1 & -2 \\ 3 & 1 & 4 \end{vmatrix}$ 2. $\begin{vmatrix} -1 & 2 & -3 \\ 2 & -1 & -4 \\ 3 & -2 & 1 \end{vmatrix}$

3. $\begin{vmatrix} 2 & 1 & 3 \\ -1 & 4 & 7 \\ 4 & 2 & 6 \end{vmatrix}$ 4. $\begin{vmatrix} 1 & -1 & 1 \\ 4 & 2 & 10 \\ 2 & 2 & 6 \end{vmatrix}$

Verify, using the value in Eq. (7–6):

*5. $D = -a_2 A_2 + b_2 B_2 - c_2 C_2.$ *6. $D = a_3 A_3 - b_3 B_3 + c_3 C_3.$
7. $D = a_1 A_1 - a_2 A_2 + a_3 A_3.$ 8. $D = -b_1 B_1 + b_2 B_2 - b_3 B_3.$
9. $D = c_1 C_1 - c_2 C_2 + c_3 C_3.$

[*Note:* The third order determinant D could have been defined equally well by the equation given in 5, 6, 7, 8, or 9. The definition [Eq (7–5)] gives the expansion of D by minors according to the elements of the first row. Problems 5 and 6 give the expansion by minors according to the elements of the second and third rows, respectively. Problems 7, 8, and 9 give the expansion by minors according to columns.]

10. Using the results of Problems 5 through 9, and Eq. (7–5), prove the statement: The determinant of order three may be expressed as the sum of three products formed by multiplying each element of any row (or column) by its minor, where each such product has assigned to it a plus or minus sign, depending upon whether the sum of the number of the row and the number of the column in which each element is located is even or odd.

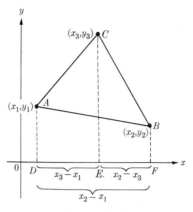

FIGURE 7-1

11. Evaluate Problems 1-4 by expansion by minors according to some row or column other than the first.

*12. Show that
$$\begin{vmatrix} x_1 & y_1 & 1 \\ x_2 & y_2 & 1 \\ x_3 & y_3 & 1 \end{vmatrix} = \begin{vmatrix} x_1 & y_1 \\ x_2 & y_2 \end{vmatrix} - \begin{vmatrix} x_1 & y_1 \\ x_3 & y_3 \end{vmatrix} + \begin{vmatrix} x_2 & y_2 \\ x_3 & y_3 \end{vmatrix}.$$

*13. Using Fig. 7-1 and, recalling that the area of a trapezoid is $\frac{1}{2}(b_1 + b_2)h$, where b_1 and b_2 are its bases and h its altitude, show that the area K of the triangle ABC in terms of the coordinates of its vertices is

$$K = \text{Area of } ADEC + CEFB - ADFB$$
$$= \tfrac{1}{2} \begin{vmatrix} x_1 & y_1 & 1 \\ x_2 & y_2 & 1 \\ x_3 & y_3 & 1 \end{vmatrix}. \quad [\text{Hint: Use Problem 12.}] \quad (7\text{-}7)$$

14. Find the area of the triangle whose vertices are the points (a) (2,3), (5,4), and (4,7); (b) (−4,2), (−2,−3), and (3,−1).

15. Find the area of the triangle the coordinates of whose vertices are (x,y), (2,3) and (5,−1). (See Fig. 7-2.)

16. The condition that the area of the triangle described in Problem 15 is equal to zero is equivalent to the fact that the point (x,y) lies on the line through (2,3) and (5,−1). Using this fact, find the equation of the straight line through (2,3) and (5,−1).

17. Find the equation of the straight line through (−2,−4) and (3,5).

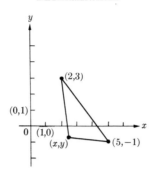

FIGURE 7–2

18. Two nonparallel straight lines in the plane always intersect in one point. The condition that three such lines
$$a_1x + b_1y = c_1,$$
$$a_2x + b_2y = c_2,$$
$$a_3x + b_3y = c_3$$
pass through the same point is that the determinant of the matrix
$$\begin{Vmatrix} a_1 & b_1 & c_1 \\ a_2 & b_2 & c_2 \\ a_3 & b_3 & c_3 \end{Vmatrix}$$
is equal to zero.* Show that the three straight lines
$$x - 2y = -3,$$
$$3x - y = 1,$$
$$5x - 2y = 1$$
pass through the same point.

19. Show that the system of equations
$$3x - 2y + 11 = 0,$$
$$x - y + 5 = 0,$$
$$2x + y = 2$$
has a common solution, and find it.

20. Find the value of x if

(a) $\begin{vmatrix} 1 & 2 & 5 \\ 1 & x & 5 \\ 3 & -1 & 2 \end{vmatrix} = 0,$
(b) $\begin{vmatrix} 1 & 2 & -3 \\ 1 & x & -3 \\ 1 & 4 & -x \end{vmatrix} = 0.$

---
*This fact is proved in **Analytic Geometry**.

7-2 Determinants of order n.

As might be expected, there exist determinants of any finite order. In order to consider them and their general properties. it will be convenient to use a different notation for the elements. Let us denote any element by the symbol a_{ij} (read "a sub i-j"), where i denotes the number of the row and j the number of the column in which the particular element appears. Thus, Eq. (7-6) is written

$$D = \begin{vmatrix} a_{11} & a_{12} & a_{13} \\ a_{21} & a_{22} & a_{23} \\ a_{31} & a_{32} & a_{33} \end{vmatrix}$$

$$= a_{11}(a_{22}a_{33} - a_{23}a_{32}) - a_{12}(a_{21}a_{33} - a_{23}a_{31}) + a_{13}(a_{21}a_{32} - a_{22}a_{31})$$

$$= a_{11}a_{22}a_{33} + a_{12}a_{23}a_{31} + a_{13}a_{21}a_{32} - a_{11}a_{23}a_{32} - a_{12}a_{21}a_{33} - a_{13}a_{22}a_{31}. \quad (7\text{-}8)$$

It should be clear from this expression that each of the six products consists of one, and only one, element from each row and each column. Each of these products has been ordered by placing the first subscripts in natural numerical order. With the exception of the first term, the second subscripts are not in natural order. With regard to these terms, let us consider the number of *inversions* of the second subscripts, that is, the number of times a greater integer precedes a smaller one.

	Second subscripts			Number of inversions
	1	2	3	0
Positive products	2	3	1	2
	3	1	2	2
	1	3	2	1
Negative products	2	1	3	1
	3	2	1	3

From this table, we see that the positive products have an even number of inversions, while the negative products have an odd number. Thus, the correct sign for any term can be expressed $(-1)^k$, where k is the number of inversions. It is by means of a direct generalization of this study that we define the nth order determinant of the n square matrix. The symbol

$$D = \begin{vmatrix} a_{11} & a_{12} & a_{13} & \cdots & a_{1n} \\ a_{21} & a_{22} & a_{23} & \cdots & a_{2n} \\ a_{31} & a_{32} & a_{33} & \cdots & a_{3n} \\ & & \cdots & & \\ a_{n1} & a_{n2} & a_{n3} & \cdots & a_{nn} \end{vmatrix} \qquad (7\text{--}9)$$

is the *nth order determinant*, associated with an n by n matrix. It is equal to the algebraic sum of all possible products formed by taking one and only one element from each row and each column. The sign of each product is chosen as $(-1)^k$, where k is the number of inversions of the second subscripts, when the factors are so arranged that their first subscripts are in natural numerical order.

In considering any possible product of n elements with the first subscripts in natural order, the second subscript in the first element may be any one of n numbers, the subscript for the second element any one of the remaining $n - 1$ numbers, and so on, so that there are $n!^* = n(n - 1)(n - 2) \cdots 2 \cdot 1$ terms in the expansion of D. For example, we recall that there are $3! = 6$ terms in the expansion of the third order determinant. There are $4! = 24$ in the fourth, and $5! = 120$ terms in the expansion of the fifth order determinant. Because of the large number of terms in the expansion of a determinant, our definition is clearly not a convenient way to find its value. Before we consider a practical method of evaluation, we shall prove some of the elementary properties of the determinant function.

PROPERTY 1. *If the corresponding rows and columns of a matrix are interchanged, its determinant remains unchanged.*

For example,

$$\begin{vmatrix} a_{11} & a_{12} & a_{13} \\ a_{21} & a_{22} & a_{23} \\ a_{31} & a_{32} & a_{33} \end{vmatrix} = \begin{vmatrix} a_{11} & a_{21} & a_{31} \\ a_{12} & a_{22} & a_{32} \\ a_{13} & a_{23} & a_{33} \end{vmatrix}.$$

---

* The symbol $n!$ (read "factorial n") denotes the product of the positive integers from 1 to n. Specifically, $1! = 1$, $2! = 2$, $3! = 6$, and so on. This symbol will be used in later chapters.

We recall that for each element our notation indicated the row by the first subscript, and the column by the second subscript. If we now consider the determinant associated with a new matrix M' whose rows and columns are respectively the corresponding columns and rows of M, every term of D' (the determinant associated with M') will again have one element from each row and each column. If the second subscripts are placed in their numerical natural order, the number of inversions in the first subscript will determine the sign of any particular term. Thus, the interchange of rows and columns serves only to change the notation, but the result for each term is the same. Consequently, D' is identical to D.

As a direct result of Property 1, in any further property of determinants, the words "row" and "column" may be interchanged throughout.

PROPERTY 2. *If any two columns (rows) of a determinant are interchanged, the sign of the determinant is changed.*

First consider the interchange of two adjacent columns. In the expansion of the new determinant, the first subscripts will remain the same, but the second subscripts—those representing the columns —will be interchanged, and as a result the number of inversions in each term will be increased or decreased by one. Thus, the sign of every term, and consequently the value of the determinant, will be changed.

Now suppose we wish to interchange the j and k columns (where for convenience $j < k$) and there are m columns between them. This may be done by moving the jth column to the position just to the left of the kth column (m interchanges), interchange these two adjacent columns (1 interchange), and move the kth column back to the original position of the jth column (m interchanges). Since the total number of interchanges, $2m + 1$, is odd, and each requires a change in sign, we again have our required result.

PROPERTY 3. *If two columns (rows) of a matrix are identical, its determinant is zero.*

If D is the value of the determinant, with the two identical columns interchanged, the value becomes $-D$, by Property 2. But since the two columns were identical, the determinant has not changed. Thus, $D = -D$, from which equation we have $D = 0$.

Property 4. *If each element of a column (row) in a determinant is multiplied by the same number m, the value of the determinant is multiplied by m.*

For example,

$$\begin{vmatrix} a_{11} & ma_{12} & a_{13} \\ a_{21} & ma_{22} & a_{23} \\ a_{31} & ma_{32} & a_{33} \end{vmatrix} = m \begin{vmatrix} a_{11} & a_{12} & a_{13} \\ a_{21} & a_{22} & a_{23} \\ a_{31} & a_{32} & a_{33} \end{vmatrix}.$$

This property follows directly from the definition. If each element in any column is multiplied by m, each term in the expansion of the determinant will have m as a factor.

As a direct corollary of Property 4, we list

Property 4'. *Any quantity that is a factor of each element in any column (row) is actually a factor of the expansion of the determinant.*

Property 5. *If three matrices M_1, M_2, and M_3 have corresponding elements equal except for one column (row) in which the elements of M_1 are the sums of the corresponding elements of M_2 and M_3, then $D_1 = D_2 + D_3$, where D_i is the determinant associated with M_i.*

For example,

$$\begin{vmatrix} a_{11} + a'_{11} & a_{12} & a_{13} \\ a_{21} + a'_{21} & a_{22} & a_{23} \\ a_{31} + a'_{31} & a_{32} & a_{33} \end{vmatrix} = \begin{vmatrix} a_{11} & a_{12} & a_{13} \\ a_{21} & a_{22} & a_{23} \\ a_{31} & a_{32} & a_{33} \end{vmatrix} + \begin{vmatrix} a'_{11} & a_{12} & a_{13} \\ a'_{21} & a_{22} & a_{23} \\ a'_{31} & a_{32} & a_{33} \end{vmatrix}.$$

Every term in the expansion of D_1 contains one and only one of these sums, each of which may be expressed as two terms. The expansions of D_2 and D_3 are the direct result.

Property 6. *If each element of any column (row) of a matrix is multiplied by the same number m and added to the corresponding element of another column, the associated determinant remains unchanged.*

For example,

$$\begin{vmatrix} a_{11} & a_{12} & a_{13} \\ a_{21} & a_{22} & a_{23} \\ a_{31} & a_{32} & a_{33} \end{vmatrix} = \begin{vmatrix} a_{11} + ma_{13} & a_{12} & a_{13} \\ a_{21} + ma_{23} & a_{22} & a_{23} \\ a_{31} + ma_{33} & a_{32} & a_{33} \end{vmatrix}.$$

DETERMINANTS OF ORDER n

This follows directly by using Properties 5, 4, and 3 in considering the right-hand determinant.

Many of these properties will be useful in the evaluation of the nth order determinant. Methods for such evaluation will be given in the next article.

PROBLEMS

1. State why the following two determinants are equal. Verify this fact by actual evaluation.

$$\begin{vmatrix} 2 & 4 & 1 \\ 3 & 2 & -1 \\ -1 & 3 & 2 \end{vmatrix} \qquad \begin{vmatrix} 2 & 3 & -1 \\ 4 & 2 & 3 \\ 1 & -1 & 2 \end{vmatrix}$$

2. Show by actual expansion that the sign of the determinant

$$\begin{vmatrix} 3 & -1 & 2 \\ 2 & 4 & 5 \\ -1 & 3 & 2 \end{vmatrix}$$

is changed if the first and third row are interchanged.

Without evaluating, state why each of the determinants in 3 and 4 are zero. Check by evaluating.

3. $\begin{vmatrix} 2 & -3 & 2 \\ 1 & 2 & 1 \\ 6 & 4 & 6 \end{vmatrix}$
4. $\begin{vmatrix} 3 & 1 & -2 \\ 2 & 7 & 3 \\ 6 & 2 & -4 \end{vmatrix}$

5. Evaluate the following determinant by factoring out any common term from any row or column, and then expanding. Check by expanding directly.

$$\begin{vmatrix} 3 & 3 & 5 \\ 6 & 16 & 8 \\ -12 & 6 & 10 \end{vmatrix}$$

6. Use Property 5 to express the sum of the two determinants as a single determinant. Check the work by evaluating all three.

$$\begin{vmatrix} 3 & 1 & -2 \\ 2 & 3 & 5 \\ -1 & 4 & 3 \end{vmatrix} + \begin{vmatrix} -2 & 1 & -2 \\ -1 & 3 & 5 \\ 2 & 4 & 3 \end{vmatrix}$$

7. In the following determinant multiply each of the elements in the second row by three, and form a new determinant by adding these results

to the corresponding elements of the first row. Show by direct expansion that the value of the original determinant is unchanged.

$$\begin{vmatrix} 2 & 5 & -6 \\ -1 & 2 & 4 \\ 3 & 1 & 5 \end{vmatrix}$$

8. In the determinant of Problem 7, multiply each of the elements in the third column by 2 and subtract these from the corresponding elements in the first column, forming a new determinant. Show by direct evaluation that the value of the original determinant is unchanged.

9. Determine the roots of the equation

$$\begin{vmatrix} 1 & 1 & 1 \\ x & a & b \\ x^2 & a^2 & b^2 \end{vmatrix} = 0.$$

[*Hint:* If $x = a$, the first two columns are equal.]

10. Determine the roots of the equation

$$\begin{vmatrix} 1 & a & a^2 & a^3 \\ 1 & b & b^2 & b^3 \\ 1 & c & c^2 & c^3 \\ 1 & x & x^2 & x^3 \end{vmatrix} = 0.$$

11. Find the roots of the equation

$$\begin{vmatrix} 3 & 1 & 9 \\ 2x & 2 & 6 \\ x^2 & 3 & 3 \end{vmatrix} = 0$$

by using Properties 3 and 4, and check by expanding the determinant.

12. Prove that

$$2 \begin{vmatrix} a_1 & b_1 & c_1 \\ a_2 & b_2 & c_2 \\ a_3 & b_3 & c_3 \end{vmatrix} = \begin{vmatrix} b_1 + c_1 & c_1 + a_1 & a_1 + b_1 \\ b_2 + c_2 & c_2 + a_2 & a_2 + b_2 \\ b_3 + c_3 & c_3 + a_3 & a_3 + b_3 \end{vmatrix}.$$

7–3 Expansion of a determinant by minors. In Problem 10 of Article 7–1, the statement may be considered as the expansion of a third order determinant by minors. This same type of property is true in general. The *minor* of an element of a square matrix of order n is the $n - 1$st order determinant obtained by deleting the row and column in which this element lies. The minor of a_{ij} will be denoted by A_{ij}. We wish to show for the nth order determinant given in (7–9),

PROPERTY 7. *The determinant is the algebraic sum of the products obtained by multiplying each element of any column (row) by its minor. The sign of each such product is* $(-1)^{i+j}$, *where the element is in the ith row and jth column.*

For example,

$$\begin{vmatrix} a_{11} & a_{12} & a_{13} & a_{14} \\ a_{21} & a_{22} & a_{23} & a_{24} \\ a_{31} & a_{32} & a_{33} & a_{34} \\ a_{41} & a_{42} & a_{43} & a_{44} \end{vmatrix} = -a_{12}\begin{vmatrix} a_{21} & a_{23} & a_{24} \\ a_{31} & a_{33} & a_{34} \\ a_{41} & a_{43} & a_{44} \end{vmatrix} + a_{22}\begin{vmatrix} a_{11} & a_{13} & a_{14} \\ a_{31} & a_{33} & a_{34} \\ a_{41} & a_{43} & a_{44} \end{vmatrix} \\ -a_{32}\begin{vmatrix} a_{11} & a_{13} & a_{14} \\ a_{21} & a_{23} & a_{24} \\ a_{41} & a_{43} & a_{44} \end{vmatrix} + a_{42}\begin{vmatrix} a_{11} & a_{13} & a_{14} \\ a_{21} & a_{23} & a_{24} \\ a_{31} & a_{33} & a_{34} \end{vmatrix}$$

is the expansion of the 4th order determinant by the second column.

This property is proved by noting two facts. Consider first the product $a_{11}A_{11}$. This product, consisting of all the terms with a_{11} as a factor, has all the signs of each term correct, since the number of inversions in A_{11} is not changed by prefixing a_{11}.

Now consider any element a_{ij}. It may be moved to the original position of a_{11} by first moving the ith row to the first row, requiring $i - 1$ interchanges of rows, and then moving the jth column to the position of the first column, requiring an additional $j - 1$ interchanges of columns. This process will produce $i - 1 + j - 1 = i + j - 2$ changes in sign. If D' is the new determinant, we have its relation to D expressed by

$$D' = (-1)^{i+j-2}D = (-1)^{i+j}D.$$

Thus, the terms of the expansion with a_{ij} as a factor are

$$(-1)^{i+j}a_{ij}A_{ij},$$

where A_{ij} is the original minor of a_{ij} in D. Thus, D may be expanded by means of any column (row) as indicated in the statement of Property 7.

We are now prepared to evaluate any determinant. In the following examples, notice where any of the foregoing properties are used.

EXAMPLE 1. Find the value of the determinant

$$D = \begin{vmatrix} -4 & 2 & 5 & 6 \\ 2 & 1 & 0 & 3 \\ 7 & 2 & -3 & 2 \\ 4 & -1 & 7 & 5 \end{vmatrix}$$

Solution. Since $a_{23} = 0$, it will simplify the work to expand by either the second row or third column. Since the elements in the third column are larger than the second row, we shall choose the second row. Thus,

$$D = -2 \begin{vmatrix} 2 & 5 & 6 \\ 2 & -3 & 2 \\ -1 & 7 & 5 \end{vmatrix} + 1 \begin{vmatrix} -4 & 5 & 6 \\ 7 & -3 & 2 \\ 4 & 7 & 5 \end{vmatrix} + 3 \begin{vmatrix} -4 & 2 & 5 \\ 7 & 2 & -3 \\ 4 & -1 & 7 \end{vmatrix}.$$

Expanding each of the third order determinants,

$$\begin{vmatrix} 2 & 5 & 6 \\ 2 & -3 & 2 \\ -1 & 7 & 5 \end{vmatrix} = 2(-15 - 14) - 5(10 + 2) + 6(14 - 3)$$

$$= -58 - 60 + 66 = -52.$$

$$\begin{vmatrix} -4 & 5 & 6 \\ 7 & -3 & 2 \\ 4 & 7 & 5 \end{vmatrix} = -4(-15 - 14) - 5(35 - 8) + 6(49 + 12)$$

$$= 116 - 135 + 366 = 347.$$

$$\begin{vmatrix} -4 & 2 & 5 \\ 7 & 2 & -3 \\ 4 & -1 & 7 \end{vmatrix} = -4(14 - 3) - 2(49 + 12) + 5(-7 - 8)$$

$$= -44 - 122 - 75 = -241.$$

Using these values, we have
$$D = -2(-52) + 347 + 3(-241) = -272.$$
We notice, in this example, that the work was shortened by the fact that one of the elements was zero. By making use of Property 6, we may introduce other zeros, and shorten the work still further. Let us consider another example.

EXAMPLE 2. Find the value of the determinant
$$D = \begin{vmatrix} 3 & -2 & -1 & 2 \\ 4 & 1 & 2 & -3 \\ -9 & -5 & 7 & -8 \\ 1 & 5 & 3 & -2 \end{vmatrix}.$$

Solution. We look for an element equal to 1 or -1, and work with the row or column containing the ± 1. Let us select the -1 in the first row. By using Property 6, we are now able to introduce zeros in the first row. In turn we (1) multiply the elements of the third column by 3, and add the products to the corresponding elements of the first column; (2) multiply the elements of the third column by -2, and add the products to the corresponding elements of the second column; and (3) multiply the elements of the third column by 2, and add the products to the corresponding elements of the fourth column. Therefore,
$$D = \begin{vmatrix} 0 & 0 & -1 & 0 \\ 10 & -3 & 2 & 1 \\ 12 & -19 & 7 & 6 \\ 10 & -1 & 3 & 4 \end{vmatrix}.$$

Now expanding by the first row as in Example 1, we have
$$D = -1 \begin{vmatrix} 10 & -3 & 1 \\ 12 & -19 & 6 \\ 10 & -1 & 4 \end{vmatrix}.$$

We may again introduce zeros to simplify the work. In this third order determinant, (1) multiply the elements of the first row by -6

and add the products to the corresponding elements of the second row, (2) multiply the elements of the first row by -4 and add the products to the corresponding elements of the third row. This results in

$$D = -1 \begin{vmatrix} 10 & -3 & 1 \\ -48 & -1 & 0 \\ -30 & 11 & 0 \end{vmatrix},$$

which, expanded by the last column, gives

$$D = -1\{1[(-48)(11) - (-30)(-1)]\}$$
$$= -1(-528 - 30) = 558.$$

If there is no element equal to 1 or -1 in any column or row, we are usually able to introduce one by using Property 6, and then continue as above. Although the introduction of a 1 is not necessary, such a process eliminates the possibilities of fractions.

PROBLEMS

Evaluate each of the following determinants:

1. $\begin{vmatrix} 1 & -2 & 2 & -3 \\ 0 & 5 & 0 & -2 \\ 2 & 4 & -6 & 2 \\ 3 & -4 & 1 & -2 \end{vmatrix}$

2. $\begin{vmatrix} 2 & 4 & 3 & -5 \\ 3 & 0 & 2 & -1 \\ -2 & 3 & 1 & 6 \\ 1 & -4 & 2 & 8 \end{vmatrix}$

3. $\begin{vmatrix} 4 & 2 & 4 & 8 \\ 7 & 5 & 2 & 4 \\ -7 & 2 & 3 & -8 \\ 6 & 4 & 5 & 6 \end{vmatrix}$

4. $\begin{vmatrix} 3 & 2 & -1 & 4 \\ 4 & -3 & 5 & -2 \\ 6 & -1 & 4 & 7 \\ 5 & -2 & 8 & 3 \end{vmatrix}$

5. $\begin{vmatrix} 3 & -1 & 2 & 6 \\ 2 & 3 & 5 & 4 \\ 4 & -2 & 2 & 7 \\ -3 & 2 & -1 & -3 \end{vmatrix}$

6. $\begin{vmatrix} 1 & 1 & 1 & 1 \\ 1 & 2 & 3 & 4 \\ 1 & 3 & 6 & 10 \\ 1 & 4 & 10 & 20 \end{vmatrix}$

7-4 Solution of a system of linear equations by determinants.

In Articles 6-9 and 6-10 we discussed the solution of systems of linear equations in two and three unknowns. The solution of such systems, involving any number of unknowns with the proper number of equations, may be expressed in terms of determinants. In considering the general case with n linear equations in n unknowns,

$$\begin{aligned} a_{11}x_1 + a_{12}x_2 + \cdots + a_{1n}x_n &= k_1, \\ a_{21}x_1 + a_{22}x_2 + \cdots + a_{2n}x_n &= k_2, \\ &\vdots \\ a_{n1}x_1 + a_{n2}x_2 + \cdots + a_{nn}x_n &= k_n, \end{aligned} \qquad (7\text{--}10)$$

we are able to prove

THEOREM.* *If D is the determinant of the coefficients of the unknowns, the product of D by any one of the unknowns is equal to the determinant D_i, obtained from D by substituting the constant terms in place of the coefficients of that unknown, and leaving the other elements unchanged.*

Since

$$D = \begin{vmatrix} a_{11} & a_{12} & \cdots & a_{1n} \\ a_{21} & a_{22} & \cdots & a_{2n} \\ \vdots & & & \\ a_{n1} & a_{n2} & \cdots & a_{nn} \end{vmatrix},$$

by Property 4,

$$Dx_1 = \begin{vmatrix} a_{11}x_1 & a_{12} & \cdots & a_{1n} \\ a_{21}x_1 & a_{22} & \cdots & a_{2n} \\ \vdots & & & \\ a_{n1}x_1 & a_{n2} & \cdots & a_{nn} \end{vmatrix}$$

---

* This theorem, known as *Cramer's Rule*, was first stated by the Swiss mathematician Gabriel Cramer (1704–1752).

If we now multiply each element of the second column by x_2, of the third by x_3, and so on, and add all these products to the corresponding elements in the first column, we have, by Property 6,

$$Dx_1 = \begin{vmatrix} a_{11}x_1 + a_{12}x_2 + \cdots a_{1n}x_n, & a_{12} & \cdots & a_{1n} \\ a_{21}x_1 + a_{22}x_2 + \cdots a_{2n}x_n, & a_{22} & \cdots & a_{2n} \\ \cdot & & & \\ \cdot & & & \\ \cdot & & & \\ a_{n1}x_1 + a_{n2}x_2 + \cdots a_{nn}x_n, & a_{n2} & \cdots & a_{nn} \end{vmatrix}$$

$$= \begin{vmatrix} k_1 & a_{12} & \cdots & a_{1n} \\ k_2 & a_{22} & \cdots & a_{2n} \\ \cdot & & & \\ \cdot & & & \\ \cdot & & & \\ k_n & a_{n2} & \cdots & a_{nn} \end{vmatrix} = D_1.$$

In this way, we find

$$Dx_1 = D_1, \qquad Dx_2 = D_2, \cdots, \qquad Dx_n = D_n, \quad (7\text{–}11)$$

where D_i is obtained from D by substituting $k_1, k_2, \cdots k_n$ for the elements $a_{1i}, a_{2i}, \cdots, a_{ni}$ of the ith column of D.

If $D \neq 0$,* the unique solution of (7–10) is obtained from (7–11), by division:

$$x_1 = \frac{D_1}{D},$$

$$x_2 = \frac{D_2}{D}, \cdots, \qquad (7\text{–}12)$$

$$x_n = \frac{D_n}{D}.$$

It is clear that this set of values satisfies the system given in (7–10), and therefore is a solution. For example, the first equation is satisfied, since

$$k_1 D - a_{11} D_1 - a_{12} D_2 - \cdots - a_{1n} D_n$$

---

* If $D = 0$, solutions may or may not exist. A complete discussion is beyond the scope of this book.

is the expansion of

$$\begin{vmatrix} k_1 & a_{11} & a_{12} & \cdots & a_{1n} \\ k_1 & a_{11} & a_{12} & \cdots & a_{1n} \\ k_2 & a_{21} & a_{22} & \cdots & a_{2n} \\ \cdot \\ \cdot \\ \cdot \\ k_n & a_{n1} & a_{n2} & \cdots & a_{nn} \end{vmatrix} \qquad (7\text{--}13)$$

by the first row. But this determinant (7–13) is zero, since the first two rows are identical. The other equations are similarly satisfied.

Problems

1. Solve the systems of equations given in Problems 1, 3, 5, and 7 of Article 6–10 by the method of this article.

2. Solve the systems of equations given in Problems 2, 4, 6, and 8 of Article 6–10 by the method of this article.

Solve by determinants the following systems of equations:

3. $x + y + z + w = -4,$
 $x + 2y + 3z + 4w = 0,$
 $x + 3y + 6z + 10w = 9,$
 $x + 4y + 10z + 20w = 24.$

4. $x + 2y - z = 8,$
 $y + 3z - w = 3,$
 $z + 4w - x = -20,$
 $w + 5x - y = 9.$

CHAPTER 8

FUNCTIONS AND EQUATIONS OF HIGHER DEGREE

In Chapter 6 we considered the first and second degree functions in one variable. We wish to generalize this type of function. Any function of one variable which can be expressed in the form

$$f(x) = a_0 x^n + a_1 x^{n-1} + a_2 x^{n-2} + \cdots + a_{n-1} x + a_n, \quad (8\text{–}1)$$

where $a_0 \neq 0$, n is a positive integer, and a_i ($i = 0, 1, 2, \cdots n$) are constants, is called a *rational integral function* or a *polynomial of the nth degree* in x. (Compare the definition in Article 1–6.) Unless otherwise stated, $f(x)$ will denote such a function in this chapter.

8–1 Certain theorems. There are several theorems which are of considerable importance in the study of the polynomial function which must be established.

THEOREM 1. (*The Remainder Theorem*) *If a polynomial $f(x)$ is divided by $x - r$, where r is any constant, until a constant remainder independent of x is obtained, this remainder is equal to $f(r)$.*

Proof. Let $q(x)$ denote the quotient when $f(x)$ is divided by $x - r$, and let R denote the constant remainder. Then $f(x)$ may be expressed [recall Eqs. (1–16) and (1–17) of Article 1–6] by the identity

$$f(x) \equiv (x - r) \cdot q(x) + R, \quad (8\text{–}2)$$

where clearly $q(x)$ is of degree $n - 1$, since we assume $f(x)$ is of degree n. Since this identity is true for *all* values of x, it is true for $x = r$. Therefore,

$$f(r) = (r - r) \cdot q(r) + R = 0 \cdot q(r) + R,$$

or

$$f(r) = R. \quad (8\text{–}3)$$

ILLUSTRATION 1. Let $f(x) = 5x^3 - 14x + 3$, and $r = 2$. Then, as was shown in Example 3, Article 1–6, $R = 15$. By substituting $r = 2$ for x in $f(x)$, we have $f(2) = 5(2)^3 - 14(2) + 3 = 40 - 28 + 3 = 15$, which is in accord with the Remainder Theorem.

Because of this theorem, the method of synthetic division, described in Article 1–6, is most useful in finding the value of $f(x)$ for different values of x. It has advantages over direct substitution, which are especially evident either when n is large or when r is other than a small integer.

THEOREM 2. (*The Factor Theorem*) *If $f(r) = R$ is zero, that is, r is a zero of $f(x)$, then $(x - r)$ is a factor of $f(x)$.*

Proof. Since r is a zero of $f(x)$, that is, $R = 0$, we have

$$f(x) \equiv (x - r) \cdot q(x) + 0.$$

Thus $(x - r)$ is a factor.

THEOREM 3. (*Converse of the Factor Theorem*) *If $(x - r)$ is a factor of $f(x)$, then $f(r) = R = 0$, and r is a zero of the function $f(x)$.*

Proof. Since $x - r$ is a factor of $f(x)$,

$$f(x) \equiv (x - r) \cdot q(x),$$

where $q(x)$ is the quotient of $f(x)/(x - r)$. Therefore,

$$f(r) = (r - r) \cdot q(r) = 0 \cdot q(r) = 0,$$

which states that r is a zero of $f(x)$.

ILLUSTRATION 2. The quantity $x - 3$ is a factor of $f(x) \equiv x^3 - 27$, since $f(3) = (3)^3 - 27 = 0$.

ILLUSTRATION 3. The function $f(x) = x^3 - 6x^2 + 3x + 10$ is exactly divisible by $x - 2$, since $f(2) = 0$. The fact that $f(2) = 0$ is shown by synthetic division.

$$\begin{array}{rrrr|r} 1 & -6 & 3 & 10 & \underline{2} \\ & 2 & -8 & -10 & \\ \hline 1 & -4 & -5 & 0. & \end{array}$$

PROBLEMS

Using synthetic division, find the remainder, and check by direct substitution, when

1. $3x^2 - 2x - 4$ is divided by $x - 3$.

2. $x^3 + 4x - 7$ is divided by $x - 3$.
3. $x^3 - 2x^2 + 9$ is divided by $x + 2$.
4. $x^4 - 2x^3 - 3x^2 - 4x - 8$ is divided by (a) $x - 2$; (b) $x + 1$.
5. $2x^4 - 3x^3 - 20x^2 - 6$ is divided by (a) $x - 4$; (b) $x + 3$.
6. $x^3 + 3x^2 - 2x - 5$ is divided by (a) $x + 2$; (b) $x + 3$.

By using the Factor Theorem, determine whether the first quantity is a factor of the second in Problems 7–12.

7. $x - 2$, $x^4 + 3x^3 - 5x^2 + 2x - 24$.
8. $x + 3$, $x^3 - 4x^2 - 18x + 9$.
9. $x - 3$, $x^4 - 5x^3 + 8x^2 + 15x - 2$.
10. $x - 5$, $x^3 + 2x^2 - 25x - 50$.
11. $2x + 3$, $2x^4 + 5x^3 + 3x^2 + 8x + 12$.
12. $3x + 1$, $9x^3 + 6x^2 + 4x + 2$.
13. Show that $x - y$ is a factor of $x^5 - y^5$, $x^6 - y^6$, $x^7 - y^7$, and $x^8 - y^8$. Using synthetic division, find the quotient in each case.
14. Show that $x + y$ is a factor of $x^5 + y^5$, and $x^7 + y^7$. Using synthetic division, find the quotient in each case.

One of the more important theorems in connection with the zeros of a polynomial function may be expressed in terms of synthetic division.

THEOREM 4. *In the synthetic division of*

$$f(x) = a_0 x^n + a_1 x^{n-1} + a_2 x^{n-2} + \cdots a_{n-1} x + a_n \text{ by } x - r,$$

where $a_0 > 0$,

(1) *if $r > 0$ and all the numbers in the third row are positive, then r is an upper limit for the positive zeros of $f(x)$;*
(2) *if $r < 0$ and the signs of the numbers in the third row alternate in sign, then r is a lower limit for the negative zeros of $f(x)$.*

Proof. Because of the process of synthetic division, in either (1) or (2) a numerical increase in r will numerically increase all the numbers in the third row except the first. Thus, if r were to be increased numerically, the final number in the third row would be numerically increased, so that it would never be zero, the condition for a zero of the function.

EXAMPLE 1. Find an upper and lower limit to the zeros of $f(x) = x^4 + 3x^3 - 9x^2 + 3x - 10$.

Solution. By using synthetic division,

$$
\begin{array}{rrrrr|r}
1 & 3 & -9 & 3 & -10 & \underline{\,3\,} \\
  & 3 & 18 & 27 & 90 & \\
\hline
1 & 6 & 9 & 30 & 80 &
\end{array}
$$

Since all the numbers in the third row are positive, 3 is an upper limit to the zeros of $f(x)$.

Again,

$$
\begin{array}{rrrrr|r}
1 & 3 & -9 & 3 & -10 & \underline{-6} \\
  & -6 & 18 & -54 & 306 & \\
\hline
1 & -3 & 9 & -51 & 296 &
\end{array}
$$

Since the signs in the third row alternate, there is no zero less than -6.

The next theorem, unfortunately, has no elementary proof but, because of its importance, will be assumed.

THEOREM 5. *(Fundamental theorem of Algebra)* *Every polynomial function,*

$$f(x) = a_0 x^n + a_1 x^{n-1} + a_2 x^{n-2} + \cdots + a_{n-1} x + a_n,$$

$n \geq 1$, $a_0 \neq 0$, *has at least one (real or imaginary) zero.*

Using this theorem, we are able to prove a theorem concerning the number of zeros of a polynomial function.

THEOREM 6. *Every polynomial function,*

$$f(x) = a_0 x^n + a_1 x^{n-1} + a_2 x^{n-2} + \cdots + a_{n-1} x + a_n, \; a_0 \neq 0$$

has exactly n zeros.

Proof. Since, by Theorem 5, $f(x)$ has at least one zero, r_1, Theorem 5 implies that $(x - r_1)$ is a factor of $f(x)$. Thus

$$f(x) \equiv (x - r_1) \cdot q_1(x), \tag{8-4}$$

where $q_1(x)$ is the quotient of $f(x)$ by $(x - r_1)$. Again, $q_1(x)$ has a zero r_2, so that

$$q_1(x) \equiv (x - r_2) \cdot q_2(x),$$

where $q_2(x)$ is the quotient of $q_1(x)$ by $(x - r_2)$. Therefore, we may write
$$f(x) \equiv (x - r_1) \cdot (x - r_2) \cdot q_2(x). \tag{8-5}$$
Since each new quotient is of one degree less than the preceding quotient, we can continue this process, until we finally have
$$f(x) \equiv (x - r_1)(x - r_2) \cdots (x - r_n) \cdot q_n(x), \tag{8-6}$$
where, since there are n factors $(x - r_i)$, $q_n(x)$ must be the constant a_0, so that
$$f(x) \equiv a_0(x - r_1)(x - r_2) \cdots (x - r_n), \tag{8-7}$$
where each r_i is a zero of $f(x)$.

Let r be any number. Since Eq. (8-7) is an identity, it is true for all values of x. Thus
$$f(r) \equiv a_0(r - r_1)(r - r_2) \cdots (r - r_n).$$
If $r \neq r_i$ for all i, none of the factors $(r - r_i)$ is zero. Since $a_0 \neq 0$, $f(r) \neq 0$, and r is not a zero of $f(x)$. Therefore, there are *exactly* n zeros, and the theorem is proved.

ILLUSTRATION 4. The function $(x - 3)^2(x - 1)(x + 2)^3$ is a polynomial of the sixth degree. Its six zeros are 3, 3, 1, -2, -2, -2. Notice that any zero which occurs m times is considered as m zeros.

Problems

Find the zeros of each of the following functions, and give the multiplicity of each.

1. $(x - 2)(x - 3)^2(x + 4)^3$
2. $(x + 1)^4(x - 2)^5$
3. $(x + 7)(2x - 3)^3$
4. $(x^2 - 4x + 4)(x^2 + 3x - 10)$
5. $(3x + 5)(x^2 - 6x + 9)^2$

Find an upper limit and a lower limit for the zeros of the following functions:

6. $x^3 - 3x^2 - 2x + 15$
7. $x^3 + 2x^2 - 7x - 8$
8. $x^4 - 2x^3 - 7x^2 + 10x + 10$
9. $x^4 - x^3 - x^2 - 2x - 6$
10. $x^4 - 4x^3 + x^2 + 6x + 2$
11. $x^4 - 5x^2 + 6x - 9$
12. $x^3 - 8x + 5$
13. $x^3 + 16x - 29$
14. $x^5 + 5x^2 - 7$
15. $x^5 - 3x^3 + 24$

8-2 Graphing of polynomial functions.

One of the methods of finding the real zeros of a function was mentioned in connection with the graphs in Article 3–7. It is quite possible to use this method for any polynomial function. Since, by using synthetic division, the values of the function are easily obtained for any value of the variable, a table of values can be constructed, from which the graph may be sketched. Care must be taken in joining the plotted points in the proper order. We recall that the real zeros of any function $f(x)$ are the abscissas of the points where the graph of $y = f(x)$ crosses or touches the x-axis.

EXAMPLE 1. Draw the graph of $y = f(x) = x^4 - 2x^3 - 7x^2 + 10x + 10$, and verify that it has one real zero between -3 and -2, one between -1 and 0, and two between 2 and 3.

Solution. We first construct the following table of values. For any arbitrary x, the corresponding value of the function is found by synthetic division. Usually it is wise to use all integral values of x between an upper and lower limit (Why?). In addition, fractional values may be necessary to ascertain the shape of the curve. The values from the table have been plotted as points and the graph drawn in Fig. 8–1. Notice that the scales on the two axes are con-

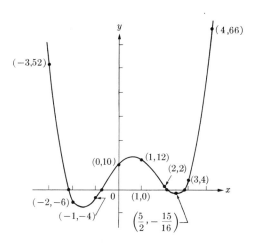

FIGURE 8–1

veniently *not* the same. From the figure it is clear that the zeros are located as suggested in the statement of the example.

x	-3	-2	-1	0	1	2	$2\frac{1}{2}$	3	4
y	52	-6	-4	10	12	2	$-\frac{15}{16}$	4	66

In the last example the fourth degree function has four real zeros. This is not always the case, for some of the zeros may be imaginary. This type of zero cannot be approximated from the graph.

EXAMPLE 2. Draw the graph of the function $y = f(x) = x^4 - x^3 - x^2 - 2x - 6$, and approximate its real zeros.

Solution. Again a table of values is constructed, from which the points are plotted and the graph is drawn (Fig. 8–2).

x	-2	-1	0	1	2	3
y	18	-3	-6	-9	-6	33

Although the function is of the fourth degree, its graph only crosses the x-axis twice. These two real zeros are between -2 and -1, and between 2 and 3, while the other two are imaginary.

There is, unfortunately, no simple method for approximating the imaginary zeros of a general polynomial function.

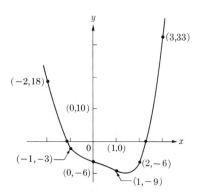

FIGURE 8–2

Problems

Draw the graphs of each of the following functions, and thus verify the statement given in Problems 1–4:

1. $f(x) = x^3 - x^2 - 2x + 1$ has one zero between -2 and -1, 0 and 1, and 1 and 2.
2. $f(x) = x^3 - 3x + 1$ has one zero between -2 and -1, 0 and 1, and 1 and 2.
3. $f(x) = x^4 - 2x^2 + 12x - 17$ has one zero between -3 and -2, and one between 1 and 2.
4. $f(x) = x^4 - 4x^3 + x^2 + 6x + 2$ has two zeros between 3 and 2, and two between 0 and -1.

Draw the graphs of the following functions, showing the location of all the real zeros:

5. $f(x) = x^4 - 20x^2 + 48x - 32$
6. $f(x) = x^3 + x^2 - 3x - 4$
7. $f(x) = 4x^3 + 8x^2 - 11x + 3$
8. $f(x) = x^4 - 2x^3 + 3x^2 + x + 6$
9. $f(x) = x^5 - 3x^3 + 9x^2 - 8x + 11$
10. $f(x) = -x^4 + 2x^2 + 8x + 3$

8–3 General remarks concerning zeros and roots. As we recall, the zeros of any function are identical with the roots of the equation formed by equating the function to zero. Consequently, all our previous remarks with regard to the zeros of the polynomial function apply also to the roots of the associated equation.

In Chapter 6 we found specific methods for solving any polynomial equation,

$$a_0 x^n + a_1 x^{n-1} + a_2 x^{n-2} + \cdots + a_{n-1} x + a_n = 0, \quad a_0 \neq 0,$$

when $n = 1$ (linear equation) and $n = 2$ (quadratic equation). There are also formulas for solving this type of equation when $n = 3$ and 4, although such material is beyond the scope of this book. For equations when $n \geq 5$, it has been proved that no algebraic formulas for the roots in terms of the coefficients are possible. We are, however, able to find all the rational roots and approximate any irrational roots. Imaginary roots will be discussed briefly in Chapter 15.

8–4 Rational roots. With regard to the rational roots of an equation with integral coefficients, we have the following theorem:

THEOREM 7. *If the rational number p/q, a fraction in lowest terms, is a root of the equation*

$$a_0 x^n + a_1 x^{n-1} + a_2 x^{n-2} + \cdots + a_{n-1} x + a_n = 0, \qquad (8\text{–}8)$$

where a_i ($i = 0, 1, 2, \cdots, n$) are integral coefficients, then p is an exact divisor of a_n, and q is an exact divisor of a_0.

Proof. Since p/q is a root of Eq. (8–8), we have

$$a_0 \left(\frac{p}{q}\right)^n + a_1 \left(\frac{p}{q}\right)^{n-1} + a_2 \left(\frac{p}{q}\right)^{n-2} + \cdots + a_{n-1} \left(\frac{p}{q}\right) + a_n = 0.$$

Multiplying each term of this equation by q^n,

$$a_0 p^n + a_1 p^{n-1} q + a_2 p^{n-2} q^2 + \cdots + a_{n-1} p q^{n-1} + a_n q^n = 0. \qquad (8\text{–}9)$$

Transposing $a_n q^n$ to the right side, and dividing both members by p,

$$a_0 p^{n-1} + a_1 p^{n-2} q + a_2 p^{n-3} q^2 + \cdots + a_{n-1} q^{n-1} = \frac{-a_n q^n}{p}.$$

Since each a_i, p, and q is an integer, the left member, and therefore the right member, is an integer. Also, p and q have no common factor, so that p does not divide q^n. Thus, p is an exact divisor of a_n.

If, in Eq. (8–9), we take the term $a_0 p^n$ to the opposite side of the equation, and divide both members by q,

$$a_1 p^{n-1} + a_2 p^{n-2} q + \cdots + a_{n-1} p q^{n-2} + a_n q^{n-1} = \frac{-a_0 p^n}{q}.$$

With the same type of argument, we have the fact that q is an exact divisor of a_0.

A direct corollary of this is clearly the following:

THEOREM 8. *Any rational root of the equation*

$$x^n + a_1 x^{n-1} + a_2 x^{n-2} + \cdots + a_{n-1} x + a_n = 0, \qquad (8\text{–}10)$$

where each a_i is an integral coefficient, must be an integer which is an exact divisor of the constant term a_n.

We are now prepared to find all the rational roots of any polynomial equation of the type given by Eq. (8–8).

EXAMPLE 1. Solve the equation

$$x^4 - x^3 - 7x^2 - 14x - 24 = 0$$

by first finding the rational roots.

Solution. In examining the equation for possible rational roots, we find, due to Theorem 8, that they are ± 1, ± 2, ± 3, ± 4, ± 6, ± 8, ± 12, and ± 24. By synthetic division, 1, 2, and 3 are not roots. For $x = 4$,

```
1   −1   −7   −14   −24  | 4
      4    12    20     24
―――――――――――――――――――――
1    3    5     6     0.
```

Putting this result in algebraic form [Eq. (1–17)],

$$x^4 - x^3 - 7x^2 - 14x - 24 \equiv (x^3 + 3x^2 + 5x + 6)(x - 4).$$

Since $x - 4$ is a factor of the original equation, $x = 4$ is one root and our problem reduces to finding the roots of the *depressed equation* $x^3 + 3x^2 + 5x + 6 = 0$. Since all the signs in this equation are plus, there are no positive roots (Why?). Using synthetic division, -1 is not a root, but $x = -2$ is a root.

```
1    3    5    6   | −2
     −2   −2   −6
―――――――――――――――――
1    1    3    0
```

The new depressed equation is $x^2 + x + 3 = 0$. Solving this by the quadratic formula [Eq. (6–7)], $x = (-1 \pm \sqrt{-11})/2$, so that our complete solution is

$$x = 3, -2, \frac{-1 \pm \sqrt{-11}}{2},$$

with the last two roots imaginary.

EXAMPLE 2. Solve for the exact roots of

$$4x^5 - 16x^4 + 17x^3 - 19x^2 + 13x - 3 = 0.$$

Solution. Since the signs of the terms alternate, this equation has no negative roots (Why?). Its possible rational roots are 1, 3,

$\frac{1}{2}$, $\frac{3}{2}$, $\frac{1}{4}$, and $\frac{3}{4}$. Using synthetic division, 1 is not a root, but $\frac{1}{2}$ is a root.

$$\begin{array}{rrrrr|r} 4 & -16 & 17 & -19 & 13 & -3 \underline{\frac{1}{2}} \\ & 2 & -7 & 5 & -7 & 3 \\ \hline 4 & -14 & 10 & -14 & 6 & 0 \end{array}$$

Since the depressed equation has a common factor of 2 in each term, it may be divided out, reducing to $2x^4 - 7x^3 + 5x^2 - 7x + 3 = 0$. Again we find that $\frac{1}{2}$ is a root, so that it is a double root of the original equation.

$$\begin{array}{rrrrr|r} 2 & -7 & 5 & -7 & 3 \underline{\frac{1}{2}} \\ & 1 & -3 & 1 & -3 \\ \hline 2 & -6 & 2 & -6 & 0 \end{array}$$

Again, factoring out the common 2, the new depressed equation becomes $x^3 - 3x^2 + x - 3 = 0$. The only possible remaining rational root is the integer 3 (Why?).

$$\begin{array}{rrrr|r} 1 & -3 & 1 & -3 & 3 \\ & 3 & 0 & 3 & \\ \hline 1 & 0 & 1 & 0 \end{array}$$

We see that 3 is a root, and the solution of the depressed equation $x^2 + 1 = 0$ gives the final two imaginary roots $\pm \sqrt{-1}$. Thus, the complete solution is $x = \frac{1}{2}, \frac{1}{2}, 3, \pm \sqrt{-1}$.

Problems

Find the exact roots of the following equations:

1. $2x^3 - 3x^2 - 11x + 6 = 0$
2. $x^3 - 6x^2 + 11x - 6 = 0$
3. $x^4 - 16x^3 + 86x^2 - 176x + 105 = 0$
4. $x^3 + x^2 - 24x + 36 = 0$
5. $x^4 - x^3 - 19x^2 + 49x - 30 = 0$
6. $x^4 - 4x^3 + 4x - 1 = 0$
7. $4x^4 + 8x^3 - 7x^2 - 21x - 9 = 0$
8. $2x^4 + 5x^3 - 11x^2 - 20x + 12 = 0$
9. $x^4 - 4x^3 + 6x^2 - 4x + 1 = 0$
10. $10x^4 - 13x^3 + 17x^2 - 26x - 6 = 0$

11. $8x^5 - 12x^4 + 14x^3 - 13x^2 + 6x - 1 = 0$

12. $12x^3 - 52x^2 + 61x - 15 = 0$

Solve each of the following for all positive values of θ less than 2π.

13. $4\sin^4\theta - 12\sin\theta\cos^2\theta - 7\cos^2\theta + 9\sin\theta + 5 = 0$

[*Hint:* Use the identity $\cos^2\theta \equiv 1 - \sin^2\theta$ and then simplify by letting $x = \sin\theta$.]

14. $4\sin^4\theta - 4\sin\theta\cos^2\theta - 11\cos^2\theta + 3\sin\theta + 8 = 0$

15. $3\tan\theta\sec^2\theta + 3\sec^2\theta - 4\tan\theta - 4 = 0$

8–5 Irrational roots. There are several well-known approximation methods* for solving polynomial equations for irrational roots. We shall consider the most elementary method. In Article 8–2, we noticed that any simple root of $f(x) = 0$ could be isolated. If $f(a)$ and $f(b)$ are opposite in sign, there is at least one value of x between a and b where $f(x) = 0$. It is this basic idea that will be used in the following example.

EXAMPLE 1. Find the approximate value of the largest positive root of $x^3 - x^2 - 3x + 1 = 0$.

Solution. By first plotting the graph of the function $y = x^3 - x^2 - 3x + 1$, as shown in Fig. 8–3, we note that the root which we

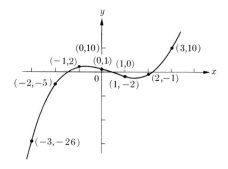

FIGURE 8–3

---

* Methods such as Graeffe's, Horner's, or Newton's appear in most books on Theory of Equations.

wish to approximate is between 2 and 3. If we divide this interval in ten equal parts, and successively use synthetic division for the values $x = 2.1, 2.2, 2.3, \cdots, 2.9, 3$, we find $f(2.1) = -.45$, while $f(2.2) = .21$, so that the root lies between 2.1 and 2.2. If we repeat this process for the values $2.11, 2.12, 2.13, \cdots, 2.19, 2.2$, we find $f(2.17) = -.0006$, but $f(2.18) = .0678$, so that $x = 2.17$ represents the root correct to two decimal places. In fact, $f(2.171) = .00621$, which indicates that $x = 2.170$ is correct to three decimal places. This process may, of course, be continued indefinitely.

The amount of work involved in approximating such a root may be greatly reduced by the method of *linear interpolation*, similar to that used in Article 5–5. We found that the root in question lay between 2.1 and 2.2. This section of the curve appears in Fig. 8–4, with the location of the two points whose abscissas are 2.1 and 2.2. Assuming that the curve approximates a straight line between the two points, by similar triangles

$$\frac{h}{.45} = \frac{.1}{.66},$$

so that $h = .07$ approximately. Therefore, a good estimate for x is $2.1 + .07 = 2.17$. We now evaluate $f(2.17)$ and find it negative. From the way in which the curve is drawn, we realize that the root in question is **greater than 2.17**. Evaluating $f(2.18)$, we isolate this

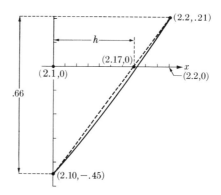

FIGURE 8–4

root without trying all the other eight possible values. This method of linear interpolation can be carried out for each new decimal, until the desired accuracy is obtained.

The beauty of this method is its simplicity. Moreover, it can be used to approximate any real root, provided the graph of the function actually crosses the x-axis, rather than merely being tangent to it.

PROBLEMS

Plot the graphs of each of the functions associated with the following equations and find, accurately to two decimals, the value of the indicated real roots:

1. $x^3 - 3x^2 - x + 2 = 0$ (the least positive).
2. $x^3 + 3x^2 - 6x - 3 = 0$ (the greatest positive).
3. $x^3 - x^2 - 2x + 1 = 0$ (the least positive).
4. $x^3 - 3x + 1 = 0$ (the three real roots).
5. $x^3 - 7x + 7 = 0$ (the two roots between 1 and 2).
6. $x^4 - x^3 + 2x^2 - 3x - 3 = 0$ (all).
7. $x^4 - 2x^3 + x^2 - 1 = 0$ (all).
8. $x^4 - 4x^3 - 4x + 12 = 0$ (all).

Find accurately to three decimals the indicated principal roots:

9. $\sqrt[3]{6}$
10. $\sqrt[3]{15}$
11. $\sqrt[4]{2}$
12. $\sqrt[5]{-9}$

CHAPTER 9

INVERSE FUNCTIONS

9–1 Inverse functions. In all our discussion, we have considered only functions of one independent variable (recall Article 3–6). Actually, we may also define *functions of more than one independent variable.* For example, $x^2 + y^2$, $x + \sin y$, or x/y^2 define functions of two independent variables, and may be represented symbolically as $f(x,y)$, $F(x,y)$, $g(x,y)$, and so on. On many occasions in algebra, trigonometry, or other branches of mathematics, a function of two variables appears as the left member of an equation whose right member is zero. Such was the case in both equations of the system given by Eq. (6–22). One was a first degree function, and the other a second degree function of two variables.

By setting such a function $F(x,y)$ equal to zero, a functional relationship is set up between the variables x and y, although both cease to be independent. If we solve the equation

$$F(x,y) = 0 \qquad (9\text{–}1)$$

for y in terms of x, the relation, implied by Eq. (9–1), is expressed explicitly with y a function of x, or, if we solve for x in terms of y, x is expressed explicitly as a function of y, with the latter variable independent. In either of these cases, $y = f(x)$ or $x = g(y)$, the function is called an *explicit* function, while the same function in the form of Eq. (9–1) is called an *implicit* function.

ILLUSTRATIONS.

	$F(x,y) = 0$	$y = f(x)$	$x = g(y)$
(a)	$3x + 4y - 12 = 0$	$y = \dfrac{12 - 3x}{4}$	$x = \dfrac{12 - 4y}{3}$
(b)	$xy - 1 = 0$	$y = \dfrac{1}{x}$	$x = \dfrac{1}{y}$
(c)	$xy - x^2 + 1 = 0$	$y = \dfrac{x^2 - 1}{x}$	$x = \dfrac{y \pm \sqrt{y^2 + 4}}{2}$
(d)	$x^2 + y^2 - 1 = 0$	$y = \pm\sqrt{1 - x^2}$	$x = \pm\sqrt{1 - y^2}$

It is important to realize that some functions, such as $y - x + \sin x = 0$, cannot readily be solved for both variables.

In this chapter we shall concern ourselves with the functions in the second and third columns of the Illustrations. We say that $f(x)$ and $g(y)$ are *inverse* functions, each of which is obtainable from the other, and both originate in the same equation in two variables. Specifically, each function in the third column is the *inverse* of the corresponding function in the second column, and vice versa.

From the illustrations, we notice several properties of inverse functions. A function may be its own inverse, as in (b), although this is not normally the case. In (a), both the function and its inverse are single-valued (recall the footnote in Article 3–6). In (c), although the function is single-valued, its inverse is *double-valued*, since for each value of y, there are two corresponding values of x. Actually, a function may be single-valued, while its inverse is infinitely many-valued.

A function is frequently more readily plotted by using the expressions listed in both columns.

EXAMPLE 1. Sketch the graph of the function expressed by the equation $x^2 - 2xy + y^2 + 2x - 3y + 2 = 0$.

Solution. By using the quadratic formula to solve for x in terms of y, we find
$$x = y - 1 \pm \sqrt{y - 1}.$$

From this expression we may tabulate a few values, noticing that for each y, there are two corresponding values for x. For example, when $y = 2$, $x = 2$ or 0. When $y = 1$, $x = 0$ is a double root and the curve is tangent to $y = 1$ at this point. Moreover, when $y < 1$, the values of x are imaginary, so that the curve lies only where $y \geq 1$.

Solving for y in terms of x, we find
$$y = \frac{2x + 3 \pm \sqrt{4x + 1}}{2}.$$

Again, we can easily find sets of values which satisfy the equation. In addition, the curve lies only where $x > -\frac{1}{4}$ and is tangent to the

176   INVERSE FUNCTIONS   [CHAP. 9

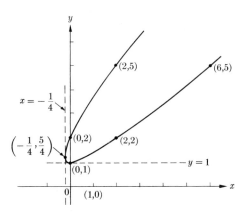

FIGURE 9-1

line $x = -\frac{1}{4}$ at $y = \frac{5}{4}$. (Why?). The resulting curve, shown in Fig. 9-1, is a parabola, although its equation is not as simple as those considered in Article 6-2.

Systems of equations such as that considered in Example 2, Article 6-11, can now be solved graphically.

PROBLEMS

Find the inverse of each of the following functions:

1. $y = 5x - 6$
2. $y = x^2$
3. $y = x^2 - 4x$
4. $y = \dfrac{x}{x - 3}$
5. $y = \dfrac{x^2 - 1}{x^2}$
6. $y = \dfrac{x^2 - 1}{x}$
7. $y = x^n$
8. $y = x^{2n} + 2x^n + 1$
9. $y = x - 1 \pm \sqrt{x - 3}$ [*Hint:* Eliminate the radical.]
10. $y = \dfrac{2x - 1 \pm \sqrt{3x - 4}}{2}$

Sketch the graph of the function expressed by each of the equations 11-20.

11. $16x^2 + 25y^2 - 400 = 0$
12. $x^2 - y^2 - 1 = 0$

13. $xy = 1$ 14. $x - 2 = \dfrac{1}{y-3}$

15. $4x^2 + 4y^2 - 12x - 10y + 5 = 0$

16. $2xy + 4y - 6x = 0$

17. $x^2 + 2xy + y^2 - 2x - 2y = 1$

18. $x^2 + xy + y^2 - 3y - 3 = 0$

19. $2x^2 + 3xy - 2y^2 - 5 = 0$ 20. $x^2 + 2xy + y^2 - 2x - 2y = 0$

21. Solve graphically Example 2, Article 6–11.
22. Solve graphically Problems 7, 9, Article 6–11.
23. Solve graphically Problems 8, 10, Article 6–11.

9–2 Inverse circular functions. The concept of an inverse function is of importance and use in studying the circular functions and their properties. In considering the relation $y = \sin \theta$, we may wish to talk about y, the sine of θ, but we might also wish to consider or emphasize the angle θ, that is, θ whose sine is y. This is done so frequently that "θ an angle whose sine is y," or the inverse function of $y = \sin \theta$, is given a name and a notation. About 1730, Daniel Bernoulli and Leonhard Euler introduced the notation

$$\theta = \arcsin y \qquad (9\text{--}2)$$

to denote an angle whose sine is y, and thus the name for this function, *arcsine of y*. This was the first suitable notation for an inverse circular function. Later, in 1813, John Herschel introduced another notation also in common use today, $\theta = \sin^{-1} y$. When -1 is used in this manner, it must be understood that -1 is not an exponent.

From the above definition, if $y = \arcsin 1/2$, y is an angle whose sine is $1/2$. Thus y may be equal to $\pi/6$, $5\pi/6$, $13\pi/6$, $-7\pi/6$, and so forth, or, in general, $y = \pi/6 \pm 2n\pi$, or $5\pi/6 \pm 2n\pi$, where $n = 0, 1, 2, \cdots$. Moreover, these are the only values which y may have. This will be clear from the graph of the function. From this one observation, we see that although $y = \sin x$ is a single-valued function, the inverse function, $y = \arcsin x$, is infinitely many-valued.

The remaining inverse circular functions are defined in a corresponding way. The function arccos x denotes an angle whose cosine is x; arctan x denotes an angle whose tangent is x; and so forth. Although there are six inverse circular functions, the functions arccot x, arcsec x, and arccsc x are of less importance. They can readily be expressed in terms of arctan x, arccos x, and arcsin x, respectively, by using (4-11, 12, and 13). Thus the expression arccsc x may be considered equivalent to arcsin $1/x$. More specifically,

$$\text{arccsc } 2 = \text{arcsin } \tfrac{1}{2} = \pi/6, \text{ and so on.}$$

The graphs of the inverse circular functions show their behavior quite clearly. In considering the graph of $y = \arcsin x$, we merely think of the equivalent expression $x = \sin y$, and recall its graph. (This graph was shown in Fig. 4-7, with x and y for y and θ.) If we graph $x = \sin y$ on transparent paper with the x-axis vertical and the y-axis horizontal, turn the paper over, and rotate it clockwise through 90°, the result is the graph of $y = \arcsin x$, shown in Fig. 9-2. The graphs of $y = \arccos x$ and $y = \arctan x$, obtained in a similar way, are shown in Fig. 9-3 and Fig. 9-4. Notice that arcsin x and

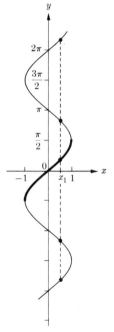

FIGURE 9-2

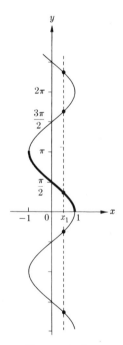

FIGURE 9-3

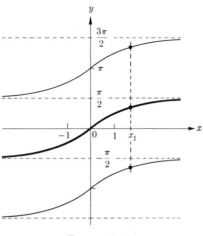

FIGURE 9-4

arccos x are defined only in the domain where x is between -1 and 1, inclusive, although arctan x is defined for all finite values of x. Thus arcsin 2 has no meaning,* since there is no angle whose sine is 2, but arctan 2 does have meaning. We notice again, from the graphs, that these inverse functions have infinitely many values for one value of x. Observe this property in the case of each function.

If a specific value of y, let us say y_1, has its sine equal to x_1, what other values of y have the same sine? We recall that the addition or subtraction of integral multiples of 2π does not change the circular function of an angle. Also, in the case of the sine function, we found that $\sin(\pi - y) = \sin y$. Therefore $\pi - y + 2n\pi$ will have the same sine as y. In other words, an even multiple of π must be added to y, or an even multiple of π must be added to $\pi - y$ for the angle to have the same sine. Put into one equation, this results in

$$\arcsin x_1 = (-1)^n y_1 + n\pi, \ (-1 \leq x_1 \leq 1), \text{ if } \sin y_1 = x_1. \quad (9\text{--}3)$$

This is true for every integral value of n and gives a different value of arcsin x_1 for each value of n. For example, when $n = 0$, arcsin $x_1 = y$; when $n = 1$, arcsin $x_1 = \pi - y$; when $n = -1$, arcsin $x_1 = -(\pi + y_1)$;

---

* This is true in the present context of real values. In more advanced mathematics the circular functions may be treated in the context of complex values (see Chapter 15) and in such context arcsin 2 is a complex number.

and so forth. Also, the formula gives every value of arcsin x_1, which should be clear from Fig. 9–2. These are indicated by the intersection of the dotted vertical line and the curve.

Since $\cos(2\pi + y) = \cos y$, and $\cos(-y) = \cos y$, it is easy to verify that the expression

$$\arccos x_1 = \pm y_1 + 2n\pi, \ (-1 \le x_1 \le 1), \text{ if } \cos y_1 = x_1, \quad (9\text{–}4)$$

gives all the values for arccos x_1.

From the fact that both $\tan(\pi + y)$ and $\tan(2\pi + y) = \tan \theta$, we get the relation

$$\arctan x_1 = y_1 + n\pi, \ (-\infty \le x_1 \le \infty), \text{ if } \tan y_1 = x_1, \quad (9\text{–}5)$$

for all the values of arctan x_1.

As examples of such inverse functions, we notice that

$$\arcsin \frac{1}{2} = (-1)^n \frac{\pi}{6} + n\pi,$$

$$\arccos \frac{1}{2} = \pm \frac{\pi}{3} + 2n\pi,$$

$$\arctan(-1) = -\frac{\pi}{4} + n\pi,$$

where n is any integer.

9–3 Principal values. We noticed in Article 9–2 that the inverse functions are many-valued. Often it is desirable to restrict these functions so that they are single-valued. This single value is called the *principal value* of the function. The reason for the following restrictions for such values will be clear when the study of calculus is undertaken. At this time it merely needs to be stated that these choices are the most convenient for our purpose. The notation for the principal value is the use of the capital A in writing the Arc functions.

The range of the principal values of each of the inverse circular functions is as follows:

$$-\frac{\pi}{2} \le \text{Arcsin } x \le \frac{\pi}{2},$$

$$0 \leq \text{Arccos } x \leq \pi,$$

$$-\frac{\pi}{2} < \text{Arctan } x < \frac{\pi}{2},$$

and the graphs of these functions are indicated in the figures by heavy lines.

In the case of the principal values, the examples listed in Article 9–2 become:

Arcsin $\frac{1}{2} = \pi/6$, Arccos $\frac{1}{2} = \pi/3$, and Arctan $(-1) = -\pi/4$.

Some other examples of principal values are:

Arcsin $(-\frac{1}{2}) = -\pi/6$, Arccot $(-\sqrt{3}) = -\pi/6$,

Arccos $(-\frac{1}{2}) = 2\pi/3$, Arcsec $(-2/\sqrt{3}) = 5\pi/6$.

Problems

Find the values of the following without using tables:

1. arcsin $\sqrt{3}/2$

Hint: We know $\sin \pi/3 = \sqrt{3}/2$. Therefore

$$\text{arcsin } \sqrt{3}/2 = (-1)^n \pi/3 + n\pi.$$

2. arctan 1
3. arccos $(-\frac{1}{2})$
4. arcsin 0
5. arctan 0
6. arccos (-1)
7. arcsec $\sqrt{2}$
8. arccot $-\sqrt{3}$
9. Arccos 0
10. Arctan (-1)
11. Arcsin (-1)
12. Arccsc 2

Find the value of the following with the help of Table I:

13. arcsin 0.4067
14. arctan 1.5399
15. arccos 0.6293
16. Arccos 0.8450
17. Arcsin 0.9951
18. Arctan 0.3281

Solve each of the following for θ:

19. $y = \sin 4\theta$

Hint: We know $4\theta = \arcsin y$. Therefore $\theta = (\arcsin y)/4$.

20. $y = \cos 3\theta$
21. $y = 3 \tan 2\theta$
22. $y = \sin (\theta/2)$
23. $2y = 4 \sec 2\theta$
24. $3y = 2 + \sin 3\theta$

25. Explain why the principal value of $y = \arccos x$ cannot be taken in the interval $-\pi/2 \leq y \leq \pi/2$.

Solve each of the following for x:

26. $y = \arcsin 2x$
27. $y = \frac{1}{3} \arccos (4x - 4)$
28. $y = \arctan (x - 2)$
29. $y = \arctan x - 2$
30. $y = \pi + 2 \arcsin x$
31. $8y = \pi/3 - 4 \arccos (2x + 1)$

9–4 Operations involving inverse circular functions. The most convenient way of considering various operations with the inverse functions is to analyze several examples. It is sometimes clearer, since the inverse circular functions are the same angles for which we have established many formulas, to substitute for an arc function an angle θ or ω. This type of substitution will be illustrated in the examples.

EXAMPLE 1. Find the value of sin (arccos $\frac{3}{5}$).

Solution. This example, similar to many problems in Chapter 4, asks for the sine of an angle whose cosine is $\frac{3}{5}$. Let θ be this angle. Then $\cos \theta = \frac{3}{5}$, and $\sin \theta = \pm\sqrt{1 - (\frac{3}{5})^2} = \pm\frac{4}{5}$.

EXAMPLE 2. Find the value of cos (Arcsin u + Arccos v).

Solution. Letting Arcsin $u = \theta_1$ and Arccos $v = \theta_2$, the example reduces to expressing $\cos (\theta_1 + \theta_2)$ in terms of u and v, where

$$\sin \theta_1 = u, \qquad \cos \theta_2 = v,$$

and therefore,

$$\cos \theta_1 = \sqrt{1 - u^2}, \qquad \sin \theta_2 = \sqrt{1 - v^2}.$$

(Explain why both radicals are positive.) Thus

$$\begin{aligned}\cos (\text{Arcsin } u + \text{Arccos } v) &= \cos (\theta_1 + \theta_2) \\ &= \cos \theta_1 \cos \theta_2 - \sin \theta_1 \sin \theta_2 \\ &= v\sqrt{1 - u^2} - u\sqrt{1 - v^2}.\end{aligned}$$

9-4] OPERATIONS INVOLVING INVERSE CIRCULAR FUNCTIONS

EXAMPLE 3. Prove that Arctan $\frac{1}{2}$ + Arctan $\frac{1}{3}$ = $\pi/4$.

Solution. Since each of the two angles on the left side is less than $\pi/4$, the left side represents an angle between 0 and $\pi/2$, as does the angle $\pi/4$ on the right side. If the tangents of two such angles are equal, the angles themselves are equal. Let us take the tangent of each member of the suspected equality, and if they are equal, we have proved the original relation. It should be emphasized that this will be true only because the tangent function between 0 and $\pi/2$ is single-valued.

$$\tan (\text{Arctan } \tfrac{1}{2} + \text{Arctan } \tfrac{1}{3}) \stackrel{?}{=} \tan \frac{\pi}{4}$$

$$\frac{\tan (\text{Arctan } \tfrac{1}{2}) + \tan (\text{Arctan } \tfrac{1}{3})}{1 - \tan (\text{Arctan } \tfrac{1}{2}) \tan (\text{Arctan } \tfrac{1}{3})} \stackrel{?}{=} 1$$

$$\frac{\tfrac{1}{2} + \tfrac{1}{3}}{1 - \tfrac{1}{2} \cdot \tfrac{1}{3}} = 1.$$

PROBLEMS

Find the values of the following without the use of tables:

1. sin (arctan 3/4)
2. cos (arcsin 7/25)
3. tan (arccos 5/13)
4. sin [arccos $(-24/25)$]
5. cos (Arcsin 5/6)
6. tan [Arcsin $(-3/4)$]
7. sin (arcsin u)
8. cos (Arccos v)
9. tan (arccos u)
10. sin (arctan v)
11. Arcsin (sin $\pi/7$)
12. Arccos [cos $(-\pi/5)$]
13. Arctan (cot $4\pi/9$)
14. Arcsin (cos $\pi/7$)
15. Arccos (sin $\pi/10$)
16. Arccot [tan $(-\pi/5)$]
17. Arcsin (tan π)
18. Arctan (sin $7\pi/2$)
19. sin (arcsin u + arccos v)
20. cos (arccos u + arcsin v)
21. sin (arccos 4/5 + π)
22. cos ($\pi/2$ − arcsin 5/13)
23. sin (arcsin 1/4 + arccos 1/4)
24. cos (arctan 9/40 − arccos 15/17)
25. tan [Arcsin 5/13 + Arctan $(-3/4)$]
26. cos [Arccos $(-1/2)$ + Arcsin $(-1/3)$]
27. sin [2 Arcsin 4/5 + 1/2 Arccos 1/9]

28. $\cos$ (Arcsin $3/5$ + Arccos $5/13$ + Arctan $8/15$)

Verify the following equations without the use of tables:

29. Arctan 3 + Arctan $1/3 = \pi/2$
30. Arcsin $3/5$ + Arccos $12/13$ = Arcsin $56/65$
31. Arctan $1/3$ + Arctan $1/5$ = Arctan $4/7$
32. Arctan $1/7$ + 2 Arctan $1/3 = \pi/4$
33. Arctan $1/8$ + Arctan $1/5$ + Arctan $1/2 = \pi/4$
34. Is the following equation true? Arctan 2 + Arctan $3 = -\pi/4$. Explain your answer.
35. Notice that in Problems 11 through 18 only principal values were involved. Find all possible values of arcsin $(\cos \theta)$.

Hint: Let $y = \arcsin (\cos \theta)$. Then $\sin y = \cos \theta = \sin (\pi/2 - \theta)$. Therefore $y = \pi/2 - \theta + 2n\pi$ or $= \pi - (\pi/2 - \theta) + 2n\pi = \pi/2 + \theta + 2n\pi$. Thus, combining these into a single expression,

$$y = \pi/2 \pm \theta + 2n\pi.$$

Find all possible values of the following:

36. arcsin $(\sin \theta)$
37. arccos $(\sin \theta)$
38. arccos $(\cos \theta)$
39. arctan $(\tan \theta)$

Solve the following equations for all possible values of x without using tables, and check carefully.

40. Arctan x + 2 Arctan $1 = 3\pi/4$
41. Arccos x + 2 Arcsin $1 = \pi$
42. Arcsin x + Arccos $2x = \pi/6$. *Hint:* Let Arcsin $x = \alpha$, Arccos $2x = \beta$, and use $\sin (\alpha + \beta) = \sin \alpha \cos \beta + \cos \alpha \sin \beta$.
43. Arcsin x + Arccos $(1 - x) = 0$
44. Arcsin x + Arccos $(1 - x) = \pi/2$
45. Arctan $(1 + x)$ + Arctan $(1 - x) = \pi/2$

CHAPTER 10

CIRCULAR FUNCTION GRAPHS WITH APPLICATIONS

A large number of the problems with which science deals today are periodic in nature. Such problems are found in astronomy and mechanics, and in dealing with the phenomena of light, sound, and electricity. The analysis of these problems requires, among other things, the frequent use of the circular functions, with special emphasis on certain of their properties and graphs.

10-1 Graphs of the curves $y = a \sin kx$. In drawing the graph of $y = \sin x$ (Fig. 4-7, Article 4-3), we might have plotted it directly by assigning arbitrary values of x, computing the corresponding value of y for each x, plotting the points whose coordinates were thus obtained, and drawing a smooth curve through these points. Although such a method can be used for sketching the graph of any circular function, it is seldom used for more general expressions. It is more convenient to generalize from the simple curves already plotted, although occasionally a few points are used for checking purposes.

The graphs of the function $y = a \sin kx$, for different values of the constants a and k, can be considered as the graphs of the general "sine curves" of which $y = \sin x$ is a special case.

Since the function $y = \sin x$ is periodic with a period of 2π [recall Eq. (4-5)], its graph has its greatest ordinate, one, when $x = \pi/2 \pm 2n\pi$. Thus the general function $y = a \sin kx$ (assuming $a > 0$ and $k > 0$) is also periodic, and repeats itself each time kx varies over a length of 2π, or x over the length $2\pi/k$. Thus its period is $2\pi/k$, and its graph will have its greatest ordinate when $x = \pi/2k \pm 2n\pi/k$. This greatest ordinate, a, is called the *amplitude* of the function, and the length of one complete period, $2\pi/k$, is called its *cycle*.

To sketch the graph of any sine curve of the type $y = a \sin kx$, mark off from the origin on the positive x-axis a distance equal to one period. As in Fig. 4-7, the two end points and the mid-point of this interval are points on the curve. Also, the graph takes on its greatest value at the x-value midway between the first two points, and its greatest negative value midway between the last two points. Knowing that the amplitude is a, these points will suffice to sketch the graph with fair accuracy. It will frequently be convenient to have the subdivisions on the two axes of different length. It may also be clearer to express the horizontal units in terms of π.

EXAMPLE. Give the amplitude and period of the function

$$y = 3 \sin (\pi x/2)$$

and sketch its graph.

Solution. The amplitude is 3 and its period is $2\pi/(\pi/2) = 4$. The distance marked off from the origin for one period of this graph is 4; the end points of this distance are (0,0) and (4,0) and the midpoint is (2,0). Since $a = 3$, the points of maximum value and of minimum value, respectively, are (1,3) and (3,−3). With this one period sketched any number of additional periods can be added to the right or left. The graph of this function appears in Fig. 10–1.

This discussion has assumed that a and k are positive. If $k < 0$, the relationships between functions of positive and negative values of the angles* can be used to change kx to a positive value when x is positive. Thus, to complete the discussion, we need only consider $y = -a \sin kx$, where a and k are positive. This function has

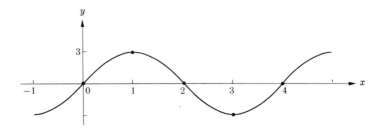

FIGURE 10–1

---

* $-\sin x = \sin(-x)$, $\cos x = \cos(-x)$, and so forth.

precisely the same graph as $y = a \sin kx$, except that for each value of x those values of y which were positive are now negative, and vice versa. In other words, the graph of $y = -a \sin kx$ might be described as a reflection about the x-axis of $y = a \sin kx$.

The same procedure which is used in plotting $y = a \sin kx$ may be used with slight modification in sketching $y = a \cos kx$. Its curve also has an amplitude of a and a period $2\pi/k$. The cosine curve differs from the corresponding sine curve, however, in that it is shifted to the left a distance equal to one-quarter of its period, since $\sin (x + \pi/2) = \cos x$. Compare Figs. 4–7 and 4–8.

Problems

Give the amplitude and period of each of the following and sketch their graphs:

1. $y = 5 \sin 2\pi x$.
2. $y = 2 \sin (x/2)$.
3. $y = 3 \sin (\pi x/3)$.
4. $y = 0.5 \cos 4x$.
5. $y = 1.5 \cos (3x/2)$.
6. $y = \sin 6x$.
7. $y = 2 \sin (-x/4)$; how does this compare with $y = 2 \sin (x/4)$?
8. $y = 2 \cos (-x/4)$; how does this compare with $y = 2 \cos (x/4)$?
9. $y = -0.5 \sin 3t$.
10. $y = 2 \sin 0.001\pi t$.
11. $y = 100 \cos 0.0314t$.
12. $y = \sin 0.0025t$.

10–2 Graphs of the curves $y = a \sin (kx + b)$. In a discussion of the graphing of the circular functions, the more generalized sine function $y = a \sin (kx + b)$ should also be included. We have noticed that $\cos x = \sin (x + \pi/2)$ differed from $\sin x$ merely by a shift to the left a distance $\pi/2$. Similarly, the graph of $y = a \sin(kx + b) = a \sin k(x + b/k)$ differs from $y = a \sin kx$ only in its shift in the horizontal direction a distance b/k. Assuming $k > 0$, this shift, or displacement, is to the right or left according as $b/k < 0$, or $b/k > 0$. We call $|b/k|$ the *phase displacement*, and b the *phase constant* or *phase difference*. The function has the same amplitude a, and period $2\pi/k$.

The graph of $y = a \cos (kx + b)$ can be sketched either as a generalization of $y = a \cos kx$ or by considering it as a "shifted" sine curve.

EXAMPLE 1. Give the amplitude, period, and phase displacement, and sketch the graph of $y = 4 \sin (2x - \pi/3)$.

Solution. This function has an amplitude of 4 and a period of π. Since its phase displacement is $\pi/6$ (Why?), its graph goes through the point $(\pi/6, 0)$ and, continuing to the right, completes one period as it passes through $(7\pi/6, 0)$. The graph is shown in Fig. 10-2.

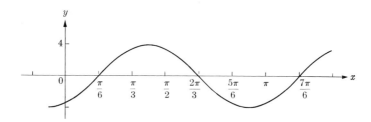

FIGURE 10-2

EXAMPLE 2. Sketch the graph of $y = 3 \sin x + 4 \cos x$.

Solution. In Example 5, Article 5-3, we showed that

$$3 \sin x + 4 \cos x = 5 \sin (x + \theta_1),$$

where $\sin \theta_1 = 4/5$ and $\cos \theta_1 = 3/5$. Thus, using Table I, $\theta_1 = .93$ (approximately), so that we may sketch the graph by considering the expression,

$$y = 5 \sin (x + .93).$$

Be sure to have the location of $(-.93, 0)$ consistent with the length of the period $2\pi = 6.28$, as in Fig. 10-3.

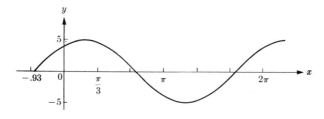

FIGURE 10-3

10–3 Graphing by addition of ordinates.

One other very useful method of sketching is the method known as *composition of ordinates*. An example will best illustrate this method. In Article 10–5 this will be used in connection with certain applications.

EXAMPLE. Sketch the graph of $y = \sin x + \sin 2x$.

Solution. This graph is drawn by first sketching on the same axes the graphs of the two separate functions $y = \sin x$ and $y = \sin 2x$. Then the ordinate for any value of x is the sum of the ordinates for that value of x from each of the graphs, $y = \sin x$ and $y = \sin 2x$. This amounts to finding the height of the curve $y = \sin x + \sin 2x$ by adding the heights of the other two curves. Of course, the sign must be taken into account. The period of one of these functions is 2π, while the other is π. Thus the period of the function $y = \sin x + \sin 2x$ is 2π. The required graph appears in Fig. 10–4.

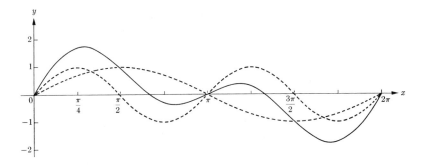

FIGURE 10–4

PROBLEMS

Sketch the graphs of the following relationships, giving their amplitudes, periods, and phase displacements:

1. $y = 2 \sin (x - \pi/6)$
2. $y = 4 \sin (x - 1)$
3. $y = 0.5 \sin (2x + \pi/8)$
4. $y = 3 \cos (t/2 - 1)$
5. $y = \sin (t + 0.25)$
6. $y = 2 \sin [(t - 1)/3]$

7. $13y = 5 \sin x + 12 \cos x$
8. $y = \sin x + \cos x$
9. $y = 15 \sin x + 8 \cos x$
10. $y = 24 \sin x + 7 \cos x$
11. $y = 2 \sin x - 5 \cos x$
12. $y = 0.6 \sin 2x + 0.5 \cos 2x$

Sketch the graphs of the following functions, by the method of addition of ordinates, and give the period of each:

13. $y = \sin x + \cos x$ (Compare this curve with Problem 8.)
14. $y = 2 \sin x + 3 \cos 2x$
15. $y = 4 \cos x/2 - 3 \sin 2x$
16. $y = 0.26 \sin x + 0.47 \cos 3x$
17. $y = \sin x + \sin 2x + \sin 3x$

Sketch the graph of the following functions:

18. $y = 2 (\sin x - \frac{1}{2} \sin 2x + \frac{1}{3} \sin 3x)$
19. $y = \sin x + \frac{1}{3} \sin 3x + \frac{1}{5} \sin 5x$
20. $y = \frac{4}{\pi} \left(\sin x - \frac{1}{3^2} \sin 3x \right)$

10–4 Simple harmonic motion. The notion of simple harmonic motion is fundamental to any discussion of periodic phenomena. Examples of bodies which move approximately according to the laws of such motion are the bob of a simple pendulum, the bobbing up and down of a floating cork, a particle in a vibrating violin string, a point on the prong of a tuning fork, a particle of air during the passage of a sound wave, or a particle of earth during a small earthquake. Actually, any body whose position d on a straight line is given at any instant t by the equation $d = a \sin \omega t$ is said to describe *simple harmonic motion*. Because of this definition, we need only to recall the properties of this function to recognize some of the evident features of simple harmonic motion. The motion is oscillatory in character, repeating itself in definite intervals of time. Thus, it is periodic. The *amplitude* of the motion is the magnitude of the maximum value of $a \sin \omega t$, or a, that is, the magnitude of the displacement from the central point of the motion. The time required for one complete vibration is called the *period* $(2\pi/\omega)$. The term *cycle* is used in place of vibration, in connection with electric current. The number of complete periods per unit time, namely, $1/(2\pi/\omega)$ or $\omega/2\pi$, is called the *frequency* of the simple harmonic motion. The *phase* at any instant is the fractional part of the period which has elapsed since the body passed through its central position in the positive direction. Actually, the position of the body need not be given from the time

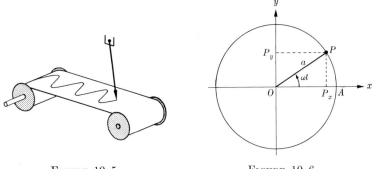

FIGURE 10-5   FIGURE 10-6

when $t = 0$, but may be given from any time t_0. Thus the more general equation may be written $d = a \sin \omega(t - t_0)$. We have learned that this more general equation represents the same curve, except for a shift along the t-axis. This equation also describes simple harmonic motion, for it can be written $d = a \sin \omega t'$, where $t' = t - t_0$. Since $y = a \sin kt$ and $y = a \cos kt$ differ only in phase, $d = a \cos \omega t$ also is used to describe simple harmonic motion.

It is interesting to observe that a graph of a simple harmonic motion can be obtained directly from a demonstration of such a motion. This is illustrated by considering the swinging of a simple pendulum from a fixed point, and the marking of its trace on a sheet of paper moving at a constant velocity. From Fig. 10-5 it should be clear that the displacement of the bob from its central position is a function of the time.

One of the simplest examples of a simple harmonic motion is produced by a point moving at a uniform velocity around the circumference of a circle. Consider the circle with center at O and a radius a in Fig. 10-6. If the point P is thought of as moving with constant or uniform velocity on the circle, the question arises as to the type of motion of the point P_x or P_y, that is, the projection of P on the x-axis or the projection of P on the y-axis. Suppose the constant angular velocity of P is ω radians per second,* and the point

---

* It is left as an exercise to distinguish between angular velocity expressed in radians per second, and linear velocity expressed in linear units per second. *Hint:* Recall $s = r\theta$.

P starts at A. Then the measure of the angle traveled by P in any time t, expressed in seconds, is ωt; thus P_y moves so that its y-coordinate is given by $y = a \sin \omega t$, and the projection of P on the y-axis describes simple harmonic motion. It is interesting to notice that P_y is at the origin when $t = 0$ and, as P_y moves up, its velocity decreases until it is 0 at C. The point P_y then moves down, increasing in velocity until it attains its maximum as it passes through O, diminishing again in velocity until it becomes 0 at C'; it then moves up with increasing velocity to its starting point O. Then the whole motion is repeated. Actually, the velocity of any body describing simple harmonic motion increases and decreases as it does in this case.

Problems

1. What can be said about the motion of the point P_x in Fig. 10–6? Obtain an expression of its position at any time t.

2. The following equations represent simple harmonic motions. Give their amplitudes, their periods, their frequencies, and sketch their graphs.

(a) $y = 8 \sin 2x$
(b) $y = 10 \sin (t/3)$
(c) $y = \frac{1}{4} \sin 3t$
(d) $y = 7 \cos 5x$
(e) $y = 4 \sin 2\pi x$
(f) $y = a \sin \omega t$

3. A particle moves along a line in such a way that its displacement from a fixed point of the line is given by $d = 2 \sin^2 t$. Can this motion be expressed as simple harmonic motion? What is its period? *Hint:* Can $\cos 2\theta$ be expressed in terms of $\sin^2 \theta$?

4. The quantity of electric current is measured in amperes, a unit denoting the number of electrons which pass a fixed point in a wire during one second. With a simple generator this quantity of current is expressed by

$$I = a_1 \sin \omega t.$$

Give the expression for I if the amplitude is 10 and the current is 60-cycle. Sketch the graph of this expression. The 60-cycle current is the frequency in common use at the present time in this country, and represents 60 cycles per second.

5. The flow of electricity also exerts a force called *electromotive force*. The expression for the electromotive force for a simplified generator is given by

$$E = a_2 \sin \omega t,$$

where the unit of measure is the volt. How can E be expressed in terms of t if its amplitude is 8 and the current is 60-cycle?

6. The amount of power of an electrical current to light lamps, generate heat, or operate machinery is dependent on its energy at any time t. This electrical energy or power is usually expressed in kilowatt hours, and is given by the expression, $P = EI$. For the 60-cycle current mentioned in Problems 4 and 5, show that P can be written in the form

$$P = a_1 a_2 \sin^2 \omega t.$$

Find the period of this periodic function. Notice that P, E, and I, in Problems 4, 5, and 6, are all examples of *simple harmonic behavior*.

7. If a flute is played softly in the middle register, the sound which is produced closely approximates a simple sound. A *simple sound* is defined as one which produces on the oscillograph a wave which may be represented by a simple harmonic curve, that is, by $y = a \sin \omega t$. Give the expression and draw the graph of the equation representing such a sound if its frequency is 400 and its amplitude is 0.001 inch.

8. The pressure in a traveling sound wave is given by

$$p = 10 \sin 200\pi \, (t - x/1000) \text{ dynes cm}^2,$$

where t is in seconds and x in centimeters. Sketch p as a function of x at the following definite times: $t = 0, 1/400, 2/400, 3/400,$ and $4/400$ sec.

9. The equation of a traveling transverse wave in a certain chord is given by the expression

$$y = 25 \sin \pi \, (0.20t - 0.01x),$$

where x is in centimeters and t is in seconds. Graph this equation when $x = 10, 25,$ and 100 cm.

10–5 Addition of two general sine functions. In order to present certain applications of graphing, let us consider the combining of two sine functions.

(1) Any two general sine functions of the *same period* are represented by

$$y_1 = a \sin (\omega t + \alpha), \qquad (10\text{–}1)$$

and

$$y_2 = b \sin (\omega t + \beta).$$

Their amplitudes are a and b, their phase difference is $\beta - \alpha$, and their common period is $2\pi/\omega$. We shall show that the sum of these two functions of the same period is itself a sine function of that

period. For
$$y = y_1 + y_2 = a \sin(\omega t + \alpha) + b \sin(\omega t + \beta)$$
$$= a(\sin \omega t \cos \alpha + \cos \omega t \sin \alpha) + b(\sin \omega t \cos \beta + \cos \omega t \sin \beta)$$
$$= A \sin \omega t + B \cos \omega t, \qquad (10\text{--}2)$$

where
$$A = a \cos \alpha + b \cos \beta,$$
and
$$B = a \sin \alpha + b \sin \beta.$$

We recall that $A \sin \omega t + B \cos \omega t = r \sin(\omega t + \delta)$,* where
$$r = \sqrt{A^2 + B^2}$$
$$= \sqrt{(a^2 \cos^2 \alpha + 2ab \cos \alpha \cos \beta + b^2 \cos^2 \beta) + (a^2 \sin^2 \alpha + 2ab \sin \alpha \sin \beta + b^2 \sin^2 \beta)}$$
$$= \sqrt{a^2 + b^2 + 2ab(\cos \alpha \cos \beta + \sin \alpha \sin \beta)}$$
$$= \sqrt{a^2 + b^2 + 2ab \cos(\beta - \alpha)},$$

and where
$$\sin \delta = B/r \quad \text{and} \quad \cos \delta = A/r.$$

Hence, $y = a \sin(\omega t + \alpha) + b \sin(\omega t + \beta) = r \sin(\omega t + \delta)$, where r and δ have the values just found.

We have proved an extremely important theorem:

Theorem. *The sum of any two general sine curves of the same period, regardless of their phase, is a general sine curve with that same period.*

This result may be generalized to include the sum of any finite number of general sine curves of the same period, and it is especially useful in the fields of electricity and sound, as the problems will indicate.

(2) The equations of two general sine functions of *different frequencies*, when they are in phase at $t = 0$, may be given by
$$y_1 = a \sin 2\pi n_1 t, \qquad (10\text{--}3)$$
and
$$y_2 = a \sin 2\pi n_2 t.$$

---

* Compare Example 5, Article 5–3, and Example 2, Article 10–2.

ADDITION OF TWO GENERAL SINE FUNCTIONS

The resulting sine function is then obtained by combining these functions, $y = y_1 + y_2$. Thus,

$$y = y_1 + y_2 = a(\sin 2\pi n_1 t + \sin 2\pi n_2 t),$$

and by Problem 23(a), Article 5–3, with $\alpha + \beta = 2\pi n_1 t$ and $\alpha - \beta = 2\pi n_2 t$,

$$y = 2a \sin 2\pi \frac{n_1 + n_2}{2} t \cos 2\pi \frac{n_1 - n_2}{2} t. \quad (10\text{--}4)$$

This result does not represent an exact sine function, but it does under certain conditions approximate the sine function. If $n_1 - n_2$ is small compared with n_1 and n_2, the resulting equation may be regarded as a "sine function with a slowly varying amplitude." Let

$$n = \frac{n_1 + n_2}{2}, \qquad \gamma = \frac{n_1 - n_2}{2}.$$

Thus, γ is small compared with n and, as a result of this substitution, we have

$$y = 2a \cos 2\pi\gamma t \sin 2\pi n t. \quad (10\text{--}5)$$

Since γ is small in comparison with n, $\cos 2\pi\gamma t$ varies very slowly compared with $\sin 2\pi n t$; that is, the number of oscillations of the frequency n is large during the time it takes the cosine to go through one period. Thus the resultant curve becomes a sine curve of slowly varying amplitude. In the case of sound waves, this periodic variation of amplitude is heard as "beats." The number of beats per second is equal to the number of times the wave, $y = \sin 2\pi n t$, has a value of 1 and -1 per second. The principle of beating is used in connection with radio, television, wireless waves, telephone, and high-frequency electrical current.

EXAMPLE. Sketch the graph of the curve $y = \sin 50\pi t + \sin 60\pi t$.

Solution. By using (10–5), we have,

$$y = \sin 50\pi t + \sin 60\pi t$$
$$= 2 \cos 5\pi t \sin 55\pi t.$$

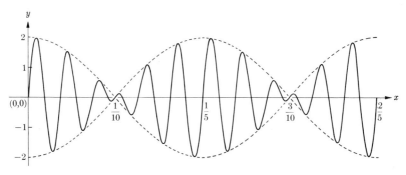

FIGURE 10-7

Since $\sin 55\pi t$ always lies between $+1$ and -1, the curve lies between the two curves $y = \pm 2 \cos 5\pi t$, touching them at the points where $\sin 55\pi t = \pm 1$. Thus it touches when $t = 1/110, 3/110, 5/110$, and so forth. The graph is shown in Fig. 10-7.

We should notice that this discussion has been for two sine functions with the same amplitudes. In the case when the amplitudes are not the same, not only is the amplitude of the resultant a varying amplitude, but also the frequency varies. Thus, any similarity of this behavior to that of a simple sine function is lost, and no discussion will be presented here.

PROBLEMS

1. Two atmospheric waves give rise to pressure variations at a given point in space according to the equations

$$p_1 = a \sin 2\pi nt,$$

and

$$p_2 = a \sin (2\pi nt - 2\pi/3).$$

Calculate the amplitude of the resultant wave at this point in space.

2. Express $y = 4 \sin (\theta + \pi/6) + 5 \sin (\theta - \pi/3)$ in the form $r \sin (\theta + \delta)$.

3. The electromotive force E (the measure of the pressure due to the flow of electricity) in a circuit is given by

$$E = 80 + 8.2 \sin \omega t + 4.8 \cos \omega t - 0.8 \sin 3\omega t + 1.2 \cos 3\omega t.$$

Express E in the form $A_0 + A_1 \sin(\omega t + \alpha_1) + A_2 \sin(3\omega t + \alpha_2)$, finding the values of A_0, A_1, A_2, α_1, and α_2.

4. Sketch the graphs of Problems 2 and 3. The graphs may be completed by the method of Article 10–3, called *composition of ordinates*, or sometimes the *principle of superposition*.

5. Moving or traveling waves are sometimes reflected from the boundaries of the bodies in which they move and thus introduce waves which travel in the opposite direction. These add to the original waves to produce the resulting wave according to the principle of superposition. Consider a column of air in which the equation of the original wave is given by

$$y_1 = a \sin 2\pi n (t - x/V),$$

and the equation of the reflected wave is

$$y_2 = a \sin 2\pi n (t + x/V).$$

The resulting wave is $y = y_1 + y_2$. Prove by using Problem 23 (a), Article 5–3, with $\alpha + \beta = 2\pi n (t - x/V)$ and $\alpha - \beta = 2\pi n (t + x/V)$, that

$$y = 2a \sin 2\pi n t \cos 2\pi n x/V.$$

This is the equation for the so-called *standing wave*. It is thus named since at certain positions in the body in which the waves are moving, y is always 0; these values for x are the values for which $\cos 2\pi n x/V = 0$, that is,

$$\frac{2\pi n x}{V} = \pi/2, 3\pi/2, 5\pi/2, \text{ and so forth.}$$

For what values of x is $y = 0$? It is interesting to note that this resulting wave at any given time t is a simple harmonic expression in x; and at any position x, it represents a simple harmonic expression of t. In all musical instruments the sound is produced by standing waves.

By making use of the method described in (2) of this article, sketch the graphs of:

6. $y = \sin 200t + \sin 210t$
7. $y = \sin 80t + \sin 100t$

10–6 Harmonic analysis and synthesis. A very interesting and extremely important and useful result in mathematics should be mentioned now. It can be demonstrated mathematically that any periodic piecewise continuous* curve may be approximated by a finite

---
* Such a curve, defined on an interval of the x-axis, consists of "pieces of curves" in the intuitive sense. Any technical definition is impossible at this mathematical level.

sum of sine and cosine curves, where the lowest frequency in the sum produces the period of the curve itself, while the remaining terms have frequencies which are integral multiples of the lowest. This method of analyzing a curve or function, known as *harmonic analysis*, was first announced by Baron J. B. J. Fourier in 1807 and first published by him in Paris in 1822. It might be expected that this method was obtained through the study of sound and wave motion but, although the result is extremely useful in such studies, the principle was discovered in the course of Fourier's study of the conduction of heat.

Fourier's Theorem may be stated in mathematical form for the periodic curve y by the approximate equality

$$y \approx a_0 + a_1 \sin \theta + a_2 \sin 2\theta + a_3 \sin 3\theta + \cdots a_n \sin n\theta$$
$$+ b_1 \cos \theta + b_2 \cos 2\theta + b_3 \cos 3\theta + \cdots b_n \cos n\theta.$$

In this expression y is the ordinate of the original curve for any particular value x, and θ, expressed in terms of x, is $2\pi x/L$. Thus, when values of x ranging from 0 to L, where L is the period of the curve, are used, it is equivalent to letting θ range from 0 to 2π. With the exception of a_0, which simply denotes the displacement of the entire curve from the original reference axis, each coefficient a_i and b_i is a factor indicating to what extent each individual sine or cosine curve enters into the composite. In passing, it is interesting to mention that there are machines which will obtain the expressions for all these Fourier coefficients for any curve. In other words, when a curve is given by means of a drawing, it is possible, with such a machine, to obtain the equation of this curve. These machines are called harmonic analyzers.

The converse of the analytic process which we have described is also useful. This converse process, called *harmonic synthesis*, might be described as combining several simple curves in order to obtain their resultant or composite curve. In some cases this may be accomplished by calculations. Always, although only approximately, the result may be accomplished by the graphical method suggested in Article 10–3, that of adding the measured ordinates of the individual curves and plotting the results.

Great advances have been made in the field of sound waves and especially in the study of musical sound waves by making use of harmonic analysis. The tones of the different musical instruments may be recorded, their waves produced on the oscillograph, and the results discussed mathematically. From this study, for example, the electric organ has developed.

The science *seismology*, which deals with the phenomena and origin of earthquakes, employs harmonic analysis because of the nature of its vibratory motion and resulting wave form. In the simplest case, when the vibrations are not large, the motion is essentially a simple harmonic motion; in the more general case, the wave is a composite wave. The waves are recorded by an instrument called the seismograph, and from this record the amplitude, direction of motion, and amount of periodicity of every single vibration may be computed.

Harmonic analysis and synthesis are used in predicting the tides. The United States Coast and Geodetic Survey has for one of its important functions the annual preparation and publication of a tidal calendar. This "Tide Table," issued one to two years in advance, predicts every "high and low" water at most of the principal seaports of the world. The time of the predicted occurrences is correct to the nearest minute, and the height to the nearest tenth of a foot. This is done mechanically by using the circular functions in connection with the periodic movements of the sun and moon. Actually, applications of the circular functions in advanced astronomy are very common.

CHAPTER 11

MATHEMATICAL INDUCTION AND THE BINOMIAL THEOREM

11–1 Mathematical induction. One of the most important methods of proof in mathematics is that of *mathematical induction*. Since it will be a convenient method of proof for several important results which appear in this book, we shall consider this method in some detail. It is usually employed in proving the validity of a statement involving n for all integral values of n.

This method of proof will be illustrated in two examples. One of the methods in finding a square root on the ordinary computing machine is based on the fact that the sum of n odd integers is equal to n^2. Specifically,

$$1 = 1 = 1^2,$$
$$1 + 3 = 4 = 2^2,$$
$$1 + 3 + 5 = 9 = 3^2,$$
$$1 + 3 + 5 + 7 = 16 = 4^2.$$

This property of the integers is stated in the first example.

EXAMPLE 1. Prove by mathematical induction that for all positive integral values,

$$1 + 3 + 5 + \cdots + (2n - 1) \equiv n^2.$$

Solution. Part (a): Verification for a specific value. This, we have done for $n = 1, 2, 3$, and 4, although such verification is necessary only for one value.

Part (b): Induction property. If the statement is true for $n = k$, where k denotes any value of n, then it is true for $n = k + 1$. Thus, we assume

$$1 + 3 + 5 + \cdots + (2k - 1) \equiv k^2.$$

Adding $2k + 1$ to both members, we have

$$1 + 3 + 5 + \cdots + (2k - 1) + (2k + 1) \equiv k^2 + (2k + 1)$$
$$\equiv (k + 1)^2.$$

This is precisely the equation, $1 + 3 + 5 + \cdots + (2n - 1) \equiv n^2$, stated for $n = k + 1$, which completes the proof of Part (b).

Part (c): Conclusion. We know that the equation is true for $n = 1, 2, 3$, and 4. Therefore, since it is true for $n = 4$, due to Part (b), it is also true for $n = 4 + 1 = 5$; since it is true for $n = 5$, it is also true for $n = 5 + 1 = 6$; and so on for *all positive integral values of n*.

We noted in Problem 13, Article 8–1, that $x - y$ was a factor of $x^n - y^n$, for $n = 5, 6, 7,$ or 8. This may be proved in general, and is specifically proved in Example 2.

EXAMPLE 2. Prove by mathematical induction that $x^n - y^n$ is divisible by $x - y$ for all positive integral values of n.

Solution. Part (a): When $n = 1$, $x^n - y^n$ becomes $x - y$, which is clearly divisible by $x - y$. When $n = 2$, $x^2 - y^2 \equiv (x + y)(x - y)$, which again is divisible by $x - y$.

Part (b): We assume that $x^k - y^k$ is divisible by $x - y$. With this assumption, we must show that $x^{k+1} - y^{k+1}$ is also divisible by $x - y$. Add and subtract xy^k to $x^{k+1} - y^{k+1}$. Thus,

$$x^{k+1} - y^{k+1} \equiv x^{k+1} - xy^k + xy^k - y^{k+1}$$
$$\equiv x(x^k - y^k) + y^k(x - y).$$

Since each term of the right member of this identity is divisible by $x - y$, the right, and therefore the left, member is divisible by $x - y$. This completes Part (b).

Part (c): We know that the theorem is true for $n = 1$ and 2. Since it is true for $n = 2$, by Part (b) it is true for $n = 2 + 1 = 3$; and so on for all positive integral values of n.

We observe from these examples that a proof by mathematical induction consists of three parts.

Part (a). *Verification* of the validity of the statement or theorem for the smallest integral value of n for which the theorem is to hold.

Part (b). *Proof of the inductive property:* if the statement or theorem is valid for $n = k$, where k denotes any value of n, then it is valid for $n = k + 1$.

Part (c). *Conclusion.* The statement or theorem is valid for all integral values of n equal to or greater than that for which it was verified in Part (a).

The fact that both Part (a) and (b) are necessary must be stressed. Consider the following examples.

EXAMPLE 3. In Article 1-8, we considered prime numbers. Does the expression $n^2 - n + 11$ represent a prime number for all positive values of n? For $n = 1, 2, 3, 4$, and 5, we have the prime numbers 11, 13, 17, 23, and 31. Certainly Part (a) is satisfied for our statement. To show that the statement is not valid, however, let $n = 11$, and we have $11^2 - 11 + 11 = 11^2$, which is *not* a prime number.

EXAMPLE 4. With regard to even integers, let us assume the *nonvalid* statement that

$$2 + 4 + 6 + \cdots + 2n = n^2 + n + 1$$

is valid when $n = k$. We are able to prove its validity, based on this assumption, for $n = k + 1$. We have

$$2 + 4 + 6 + \cdots + 2k = k^2 + k + 1.$$

Adding $2k + 2$ to both members of the equality,

$$\begin{aligned} 2 + 4 + 6 + \cdots + 2k + 2k + 2 &= k^2 + k + 1 + 2k + 2 \\ &= k^2 + 2k + 1 + k + 1 + 1 \\ &= (k + 1)^2 + (k + 1) + 1. \end{aligned}$$

Thus, if the statement is valid for $n = k$, it is also valid for $n = k + 1$, and Part (b) has been proved. It is clear, however, that Part (a) and thus the statement itself is *not* valid, for consider the case where $n = 1, 2$, and 3. We would have $2 = 3, 6 = 7$, and $12 = 13$, respectively.

PROBLEMS

By the method of mathematical induction, prove that the following are valid for all positive integral values of n:

1. $1 + 2 + 3 + \cdots + n = \dfrac{n(n + 1)}{2}$

2. $2 + 4 + 6 + \cdots + 2n = n(n + 1)$

3. $4 + 8 + 12 + \cdots + 4n = 2n(n + 1)$

4. $1 + 4 + 7 + \cdots + (3n - 2) = \dfrac{n(3n - 1)}{2}$

5. $1 + 3 + 6 + \cdots + \dfrac{n(n + 1)}{2} = \dfrac{n(n + 1)(n + 2)}{6}$

6. $1^2 + 2^2 + 3^2 + \cdots + n^2 = \dfrac{n(n+1)(2n+1)}{6}$

7. $1^2 + 3^2 + 5^2 + \cdots + (2n-1)^2 = \dfrac{n(2n-1)(2n+1)}{3}$

8. $\dfrac{1}{2} + \dfrac{1}{2^2} + \dfrac{1}{2^3} + \cdots + \dfrac{1}{2^n} = 1 - \dfrac{1}{2^n}$

9. $2 + 2^2 + 2^3 + \cdots + 2^n = 2^{n+1} - 2$

10. $1^3 + 2^3 + 3^3 + \cdots + n^3 = \dfrac{n^2(n+1)^2}{4}$

11. $1^3 + 3^3 + 5^3 + \cdots + (2n-1)^3 = n^2(2n^2 - 1)$

12. $\dfrac{1}{1\cdot 2} + \dfrac{1}{2\cdot 3} + \dfrac{1}{3\cdot 4} + \cdots + \dfrac{1}{n(n+1)} = \dfrac{n}{n+1}$

13. $1\cdot 2 + 2\cdot 3 + 3\cdot 4 + \cdots + n(n+1) = \dfrac{n(n+1)(n+2)}{3}$

14. $1\cdot 2 + 3\cdot 4 + 5\cdot 6 + \cdots + (2n-1)(2n) = \dfrac{n(n+1)(4n-1)}{3}$

15. $1\cdot 2\cdot 3 + 2\cdot 3\cdot 4 + 3\cdot 4\cdot 5 + \cdots + n(n+1)(n+2)$
$= \dfrac{n(n+1)(n+2)(n+3)}{4}$

*16. $a + ar + ar^2 + \cdots + ar^{n-1} = \dfrac{a - ar^n}{1 - r}$

*17. $a + (a+d) + (a+2d) + \cdots + a + (n-1)d$
$= \dfrac{n[2a + (n-1)d]}{2}$

18. $x^{2n} - y^{2n}$ is divisible by $x - y$

19. $x^{2n-1} + y^{2n-1}$ is divisible by $x + y$

20. $n^3 + 2n$ is divisible by 3

11–2 The binomial theorem. One of the more important theorems proved by mathematical induction is known as the binomial theorem. We are interested in obtaining a general expansion of the binomial $(a + b)^n$ for any positive integral value of n. By actual multiplication,

$$(a + b)^1 \equiv a + b,$$
$$(a + b)^2 \equiv a^2 + 2ab + b^2,$$
$$(a + b)^3 \equiv a^3 + 3a^2b + 3ab^2 + b^3,$$

$$(a+b)^4 \equiv a^4 + 4a^3b + 6a^2b^2 + 4ab^3 + b^4,$$
$$(a+b)^5 \equiv a^5 + 5a^4b + 10a^3b^2 + 10a^2b^3 + 5ab^4 + b^5.$$

With n denoting the exponent, we notice the following properties for the above identities:

1. The number of terms in any identity is $n+1$.
2. The first term a^n is the first term of the binomial a, raised to the nth power.
3. The exponents of a in each term decrease by 1 and those of b increase by 1, so that the sum of the exponents of a and b in any term is n.
4. If the coefficient in any term is multiplied by the exponent of a, and divided by the exponent of b increased by 1, the result is the coefficient of the next term.
5. For any term counting from the left or right of the identity, the coefficients are the same.

Assuming that these are general properties of the expansion for $(a+b)^n$, we should have the identity

$$(a+b)^n \equiv a^n + \frac{n}{1}a^{n-1}b + \frac{n(n-1)}{1\cdot 2}a^{n-2}b^2 +$$
$$\frac{n(n-1)(n-2)}{1\cdot 2\cdot 3}a^{n-3}b^3 + \cdots + b^n. \quad (11\text{-}1)$$

To prove this identity, we must have the expression for the general or rth term. By examining Eq. (11-1), we see that the rth term contains b^{r-1}, and therefore, by Property 3, $a^{n-(r-1)} = a^{n-r+1}$. Also, the fractional coefficient of the rth term has for its denominator the product of the first $r-1$ integers, $1\cdot 2\cdot 3 \cdots (r-1)$, or factorial $(r-1)$, $(r-1)!$ (see footnote in Article 7-2), while its numerator contains $r-1$ factors, the first being n, and each succeeding factor being one less than the preceding, so that the last factor is $n-r+2$. The rth term, therefore, is

$$\frac{n(n-1)(n-2)\cdots(n-r+2)}{(r-1)!}a^{n-r+1}b^{r-1}. \quad (11\text{-}2)$$

We are now prepared to prove that Eq. (11-1) holds for all positive integral values of n. Since we have verified it for $n = 1, 2, 3, 4$, and

THE BINOMIAL THEOREM

5, we need only to assume that it is valid for $n = k$, that is,

$$(a + b)^k \equiv a^k + ka^{k-1}b + \frac{k(k-1)}{2!} a^{k-2}b^2 + \cdots$$

$$+ \frac{k(k-1)(k-2)\cdots(k-r+2)}{(r-1)!} a^{k-r+1}b^{r-1} + \cdots + b^k, \quad (11\text{-}3)$$

and prove from this that it is valid for $n = k + 1$, or

$$(a + b)^{k+1} \equiv a^{k+1} + (k+1)a^k b + \frac{(k+1)k}{2!} a^{k-1}b^2 + \cdots$$

$$+ \frac{(k+1)(k)(k-1)\cdots(k-r+3)}{(r-1)!} a^{k-r+2}b^{r-1} + \cdots + b^{k+1}. \quad (11\text{-}4)$$

Multiplying each member of the identity (11-3) by $(a + b)$, we have

$$(a+b)^{k+1} \equiv a^{k+1} + ka^k b + \cdots + \frac{k(k-1)\cdots(k-r+2)}{(r-1)!} a^{k-r+2}b^{r-1} + \cdots + ab^k$$

$$+ a^k b + \cdots + \frac{k(k-1)\cdots(k-r+3)}{(r-2)!} a^{k-r+2}b^{r-1} + \cdots + kab^k + b^{k+1}.$$

The first line represents the right member of (11-3) multiplied by a, while the second line represents this same member multiplied by b. To obtain the rth term in the result, we have written the rth term in the first line, and the $(r - 1)$st term in the second, both containing $a^{k-r+2}b^{r-1}$. Simplifying the whole right member, the total coefficient of $a^{k-r+2}b^{r-1}$ will be

$$\frac{k(k-1)\cdots(k-r+2)}{(r-1)!} + \frac{k(k-1)+\cdots(k-r+3)}{(r-2)!}$$

$$= \frac{k(k-1)\cdots(k-r+3)}{(r-1)!}(k-r+2) + \frac{k(k-1)+\cdots(k-r+3)}{(r-1)(r-2)!}(r-1)$$

$$= \frac{k(k-1)\cdots(k-r+3)}{(r-1)!}[(k-r+2)+(r-1)]$$

$$= \frac{(k+1)k(k-1)\cdots(k-r+3)}{(r-1)!}.$$

Thus, the total right side is exactly that of Eq. (11–4). The conclusion [Part (c)] should be clear, and the proof is complete.

EXAMPLE 1. Expand $(a + 2b)^6$ by the binomial theorem and simplify the result.

Solution. With $a = a$ and $b = 2b$, we have

$$(a + 2b)^6 = a^6 + 6a^5(2b) + \frac{6 \cdot 5}{2!} a^4(2b)^2 + \frac{6 \cdot 5 \cdot 4}{3!} a^3(2b)^3$$

$$+ \frac{6 \cdot 5 \cdot 4 \cdot 3}{4!} a^2(2b)^4 + 6a(2b)^5 + (2b)^6$$

$$\equiv a^6 + 12a^5b + 60a^4b^2 + 160a^3b^3 + 240a^2b^4 + 192ab^5 + 64b^6.$$

EXAMPLE 2. Find the first four terms of the expansion of $(x^3 - 3y^2)^{12}$.

Solution. Here $a = x^3$, and $b = -3y^2$. Thus,

$$(x^3 - 3y^2)^{12} \equiv (x^3)^{12} + 12(x^3)^{11}(-3y^2) + \frac{12 \cdot 11}{2}(x^3)^{10}(-3y^2)^2$$

$$+ \frac{12 \cdot 11 \cdot 10}{3!}(x^3)^9(-3y^2)^3 + \cdots$$

$$\equiv x^{36} - 36x^{33}y^2 + 594x^{30}y^4 - 5940x^{27}y^6 + \cdots.$$

We notice from the method of this second example that when the first term of the binomial is plus and the second minus, the signs in the expansion alternate.

The general term (11–2) is of use in certain problems.

EXAMPLE 3. Find the sixth term in the expansion of $\left(\dfrac{1}{2a} - 3\right)^{16}$.

Solution. In this example $a = \dfrac{1}{2a}$, $b = -3$, $n = 16$, and $r = 6$.

Substituting in (11–2), the sixth term is

$$\frac{16 \cdot 15 \cdot 14 \cdot 13 \cdot 12}{2 \cdot 3 \cdot 4 \cdot 5} \left(\frac{1}{2a}\right)^{11}(-3)^5 = -\frac{66339}{128a^{11}}$$

THE BINOMIAL THEOREM

Problems

Expand each of the following by the binomial theorem, and simplify if necessary.

1. $(a + b)^7$
2. $(xy - 2)^4$
3. $(2x + y^2)^5$
4. $\left(a^2 - \dfrac{x}{2}\right)^6$
5. $(5x - y^2)^4$
6. $(x^{-1} + 2y^{-2})^6$
7. $(x^{1/3} + y^{1/3})^6$
8. $(x^{2/3} - y^{2/3})^5$
9. $(x^{2/5} - 3y^{-2})^5$
10. $(ax^{1/2} - by^{1/3})^6$

Write and simplify the first four terms of the expansions for the following:

11. $\left(\dfrac{x^2}{2} + \dfrac{2}{y^2}\right)^{12}$
12. $(x^{1/3} - y^{1/3})^9$
13. $(x^{1/3} - y^{-1/3})^{11}$
14. $(x^{-2/3} + 2y^{2/3})^8$
15. $(1 + x)^k$
*16. $(1 + k)^{1/k}$

Note: Assume that the binomial theorem holds for $n = 1/k$.

Write and simplify the indicated term in the expansions of the following:

17. Seventh term of $(2x - y)^{12}$
18. Ninth term of $\left(2 + \dfrac{x}{4}\right)^{15}$
19. Middle term of $(y^2 - \tfrac{1}{2})^8$
20. Middle term of $\left(2 + \dfrac{3}{x}\right)^{10}$
21. Term involving x^7 of $(2x - 3)^{10}$
22. Term involving x^3 of $(5 - 2x)^{4/3}$

*23. Prove that the $(r + 1)$st term of Eq. (11–1) is
$$\dfrac{n(n-1)(n-2)\cdots(n-r+1)}{r!} a^{n-r} b^r$$

*24. The coefficients in the binomial expansion are called *binomial coefficients*. The coefficient of the $(r + 1)$st term, given in Problem 23, is a function of n and r. If this function is denoted by $C(n,r)$, where

$$C(n,r) = \dfrac{n(n-1)(n-2)\cdots(n-r+1)}{r!},* \qquad (11\text{–}5)$$

---

* This function, $C(n,r)$, also represents the combination of n things taken r at a time, and is well known in the study of permutations and combinations. This will be considered in Chapter 17.

prove that

$$C(n,r) = \frac{n!}{r!(n-r)!}. \qquad (11\text{-}6)$$

[*Hint:* Multiply numerator and denominator of the right member of (11–5) by $(n-r)!$]

25. Prove that $C(n,r) = C(n, n-r)$, and thus establish Property 5 of the binomial expansion.

26. Solve $C(n,4) = 35$ for the positive integer n.

*27. Prove that $2^n = (1+1)^n = 1 + C(n,1) + C(n,2) + C(n,3) + \cdots + C(n,n)$.

28. Check the statement in Problem 27 for $n = 5$.

11–3 The expansion of $(1 + x)^n$. Applying the binomial theorem to $(1 + x)^n$, we have

$$(1+x)^n \equiv 1 + nx + \frac{n(n-1)}{2!}x^2$$

$$+ \frac{n(n-1)(n-2)}{3!}x^3 + \cdots$$

$$+ \frac{n(n-1)\cdots(n-r+2)}{(r-1)!}x^{r-1} + \cdots, \qquad (11\text{-}7)$$

which contains $n + 1$ terms, and is valid for all positive integers n. If n were any real number other than a positive integer, the expansion in (11–7) does not terminate. Under what conditions would this relation still be valid?

Although the proof is beyond the scope of this book, it can be shown that a finite number of terms of the right member of (11–7) approximates $(1 + x)^n$, for any value of n which is not a positive integer, if $|x| < 1$. Of course, the more terms that are considered, the closer the approximation will be.

EXAMPLE 1. Find the value of $(1.02)^{-4}$ correct to four significant figures.

THE EXPANSION OF $(1 + x)^n$

Solution. We expand $(1.02)^{-4}$ by relationship (11–7).

$$(1.02)^{-4} = (1 + .02)^{-4}$$

$$= 1^{-4} + (-4) \cdot 1^{-5}(.02) + \frac{(-4)(-5)}{2!} \cdot 1^{-6}(.02)^2$$

$$+ \frac{(-4)(-5)(-6)}{3!} \cdot 1^{-7}(.02)^3$$

$$+ \frac{(-4)(-5)(-6)(-7)}{4!} \cdot 1^{-8}(.02)^4$$

$$+ \frac{(-4)(-5)(-6)(-7)(-8)}{5!} \cdot 1^{-9}(.02)^5 + \cdots$$

$$= 1 - .08 + .004 - .00016 + .0000056$$

$$= .9238456.$$

We must consider a sufficient number of terms in order to guarantee the required accuracy. Since the last term above has a zero in the fifth decimal place, no additional terms will affect our result, that is,

$$(1.02)^{-4} = .9238,$$

correct to four significant figures.

EXAMPLE 2. Find the value of $\sqrt{15}$ correct to four significant figures.

Solution. Again we use relationship (11–7).

$$\sqrt{15} = \sqrt{16-1} = (16-1)^{\frac{1}{2}} = [16(1 - \tfrac{1}{16})]^{\frac{1}{2}} = 4(1 - \tfrac{1}{16})^{\frac{1}{2}}$$

$$= 4\left[1 + \tfrac{1}{2} \cdot 1^{-\frac{1}{2}}(-\tfrac{1}{16}) + \frac{\tfrac{1}{2}(-\tfrac{1}{2})}{2!} 1^{-\frac{3}{2}}(-\tfrac{1}{16})^2 \right.$$

$$\left. + \frac{\tfrac{1}{2}(-\tfrac{1}{2})(-\tfrac{3}{2})}{3!} 1^{-\frac{5}{2}}(-\tfrac{1}{16})^3 + \cdots \right]$$

$$= 4[1 - \tfrac{1}{32} - \tfrac{1}{2048} - \tfrac{1}{65536} \cdots]$$

$$= 4[1 - .03125 - .000488 - .000015 \cdots]$$

$$= 3.8730, \text{ correct to four significant figures.}$$

Problems

Write out the first four terms of each of the expansions for the following expressions:

1. $\dfrac{1}{1+x} = (1+x)^{-1}$
2. $\sqrt{1+x}$
3. $\dfrac{1}{\sqrt{1-x}}$
*4. $(1+x)^{1/x}$

Compute each of the following to four significant figures.

5. $(1.01)^{-2}$
6. $(1.03)^{-5}$
7. $\sqrt[4]{1.02}$
8. $\sqrt{1.05}$
9. $\sqrt{33}$
10. $\sqrt[4]{17}$
11. $\sqrt[3]{120}$
12. $(1.01)^{10}$

CHAPTER 12

EXPONENTIAL AND LOGARITHMIC FUNCTIONS

We have considered in some detail the simpler algebraic functions (Chapters 6 and 8), and one type of nonalgebraic function,* the circular functions (Chapters 4 and 5). Also, the inverse functions of these were discussed. There are two other important nonalgebraic functions, the exponential function and its inverse function, the logarithmic function. In this chapter we shall examine these two functions, their properties, and simple applications.

12-1 The exponential function $y = a^x$. Before we define any exponential function, we must define a^x, where a is any positive number and x is an irrational number. Since any irrational number may be approximated by rational numbers (Article 1-1) we may define a^x for x irrational, as the limiting value of a^r as r approaches x through rational values. For example, since 1, 1.4, 1.41, 1.414, 1.4142 are rational values which approach $\sqrt{2}$, we define $5^{\sqrt{2}}$ ($= 9.738$, approximately) to be the limiting value of 5 raised to these powers. It is possible to show that such a definition of irrational exponents obeys the laws of exponents set up in Article 2-5, although the proof is beyond the scope of this book.

With this understanding for irrational as well as rational exponents, the simplest exponential function, where a is any positive constant,

$$y = f(x) = a^x, \quad a > 0, \qquad (12\text{-}1)$$

exists for all real values of x and has a single value for each x. From an examination of the function, and the two graphs of the special cases, $y = 2^x$ and $y = 3^x$, sketched by plotting points whose coordinates satisfy the equations (Fig. 12-1), we may state without proof the properties of this function, $y = a^x$.

x	-2	-1	0	1	2	3
y	$\frac{1}{4}$	$\frac{1}{2}$	1	2	4	8

$y = 2^x$

x	-2	-1	0	1	2
y	$\frac{1}{9}$	$\frac{1}{3}$	1	3	9

$y = 3^x$

---
* The class of all nonalgebraic functions is called the class of *transcendental functions*.

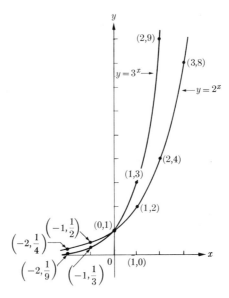

FIGURE 12-1

1. The function is positive (lies above the x-axis) for all values of x.
2. The function is single-valued.
3. If $a > 1$, the function is an increasing function (recall Property 2 of the function $y = \sin \theta$, Article 4-3). As x takes on larger and larger values, so does the function, but as x decreases algebraically, the function approaches but never attains the value zero.
4. For all values of a, the function has the value 1, when $x = 0$.
5. There are no zeros of the function.

PROBLEMS

Sketch the graphs of the functions in 1-6:

1. 4^x
2. 10^x
3. 2^{-x}
4. $(\frac{1}{2})^x$
5. 3^{-x}
6. $(\frac{1}{3})^x$

7. State a property, corresponding to Property 3 if $0 < a < 1$, for the function $y = a^x$.

12-2 The logarithmic function.

Since for any positive number x, and any positive number $a \neq 1$, there is one and only one real value of y which satisfies the equation $a^y = x$,* we can solve this equation for y, and have y equal to the power to which a must be raised to obtain the number x. This power or exponent y is called the *logarithm* of x to the base a, and is written

$$y = \log_a x, \qquad (12\text{-}2)$$

with the conditions that $x > 0$ and $a > 0$, and $\neq 1$. We immediately notice that $y = \log_a x$ and $x = a^y$ are inverse functions of each other, so that many of the logarithm properties should be evident. It is most important to keep in mind that the two expressions,

$$x = a^y \quad \text{and} \quad y = \log_a x, \qquad (12\text{-}3)$$

are equivalent.

ILLUSTRATION 1.

$$\log_3 9 = 2 \quad \leftrightarrow \quad 3^2 = 9.$$

$$\log_2 32 = 5 \quad \leftrightarrow \quad 2^5 = 32.$$

$$\log_6 1 = 0 \quad \leftrightarrow \quad 6^0 = 1.$$

$$\log_2 \tfrac{1}{16} = -4 \quad \leftrightarrow \quad 2^{-4} = \tfrac{1}{16}.$$

$$\log_8 4 = \tfrac{2}{3} \quad \leftrightarrow \quad 8^{2/3} = 4.$$

Let us again sketch the graphs of two logarithmic functions $y = \log_2 x$ and $y = \log_3 x$, and with the help of these graphs (Fig. 12-2) consider some of the fundamental properties of the function $y = \log_a x$.

x	$\tfrac{1}{4}$	$\tfrac{1}{2}$	1	2	4	8
y	-2	-1	0	1	2	3

$$y = \log_2 x \leftrightarrow 2^y = x$$

x	$\tfrac{1}{9}$	$\tfrac{1}{3}$	1	3	9
y	-2	-1	0	1	2

$$y = \log_3 x \leftrightarrow 3^y = x$$

---

* The proof of this statement is beyond the scope of this book.

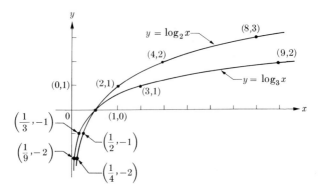

FIGURE 12-2

1. The function is positive for all values of x greater than 1, but negative for all values of x less than 1. It is not defined for negative values of x.
2. The function is single-valued.
3. The function is an increasing function.
4. The logarithm of any number with respect to itself as base is equal to 1. Graphically, all such curves pass through the point $(a,1)$.
5. The function has $x = 1$ for its only zero.

There are several additional properties of logarithms which may be easily derived, recalling that a logarithm is an exponent.

6. The logarithm of the product of two quantities is equal to the logarithm of the first quantity plus the logarithm of the second, that is,

$$\log_a uv \equiv \log_a u + \log_a v. \tag{12-4}$$

Proof. Let u and v be any two positive quantities whose logarithms are x and y, respectively. Then

$$x = \log_a u \quad \text{and} \quad y = \log_a v,$$

or

$$a^x = u \quad \text{and} \quad a^y = v.$$

Multiplying the corresponding members of these equations,

$$uv = a^x a^y = a^{x+y},$$

which, by the definition of a logarithm reduces to

$$\log_a uv = x + y = \log_a u + \log_a v.$$

7. The logarithm of the quotient of two quantities is equal to the logarithm of the first quantity minus the logarithm of the second, that is,

$$\log_a \frac{u}{v} \equiv \log_a u - \log_a v. \tag{12-5}$$

Proof. With the same assumptions as in Property 6,

$$\frac{u}{v} = \frac{a^x}{a^y} = a^{x-y}.$$

Written in terms of logarithms, this becomes

$$\log_a \frac{u}{v} = x - y = \log_a u - \log_a v.$$

8. The logarithm of a power of a quantity is equal to the power multiplied by the logarithm of the quantity, that is,

$$\log_a u^p \equiv p \log_a u. \tag{12-6}$$

Proof. Let u be any quantity, and its logarithm be x. Then $x = \log_a u$ or $a^x = u$. Raising both members to the p power, we have

$$(a^x)^p = a^{xp} = u^p,$$

or, in terms of logarithms,

$$\log_a u^p = xp = p \log_a u.$$

ILLUSTRATION 1. Property 6 implies that

$$\log_{10} (47)(93) = \log_{10} 47 + \log_{10} 93$$

or

$$\log_{10} 4700 = \log_{10} 47 \cdot 100 = \log_{10} 47 + \log_{10} 100.$$

ILLUSTRATION 2. Property 7 implies that

$$\log_{10} \tfrac{82}{53} = \log_{10} 82 - \log_{10} 53.$$

Properties 6 and 7 imply that

$$\log_{10} \frac{(48)(96)}{23} = \log_{10} 48 + \log_{10} 96 - \log_{10} 23.$$

ILLUSTRATION 3. Property 8 is used for both integral powers and roots.

$$\log_{10} 28^5 = 5 \log_{10} 28$$
$$\log_{10} \sqrt[3]{472} = \log_{10} (472)^{1/3} = \tfrac{1}{3} \log_{10} 472.$$

PROBLEMS

Express in logarithmic notation, using (12-3):

1. $3^3 = 27$
2. $5^4 = 625$
3. $4^0 = 1$
4. $10^3 = 1000$
5. $8^{4/3} = 16$
6. $2^{-6} = \frac{1}{64}$
7. $10^{-3} = 0.001$
8. $b^z = w$
9. $4^1 = 4$

Express in exponent notation using (12-3):

10. $\log_2 \frac{1}{8} = -3$
11. $\log_{36} 6 = \frac{1}{2}$
12. $\log_{10} 1 = 0$

Find by inspection the value of x, a, or u in the following expressions:

13. $x = \log_4 16$
14. $x = \log_7 1$
15. $x = \log_9 \frac{1}{9}$
16. $x = \log_{\sqrt{6}} 36$
17. $\log_5 u = 2$
18. $\log_a 16 = 4$
19. $x = \log_2 4^7$
20. $\log_a 32 = -\frac{5}{7}$
21. $\log_a \frac{2}{3} = -\frac{1}{3}$

Sketch the graphs of the following functions:

22. $y = \log_5 x$
23. $y = \log_{\frac{1}{2}} x$
24. $y = \log_{10} x$
25. $y = \log_{\frac{1}{3}} x$

Express as a single logarithm, using Properties 6, 7, and 8:

26. $\log_a x + \log_a y - \log_a z$
27. $\log_b (a + 2) - \log_b (a - 3)$
28. $4 \log_{10} x - 3 \log_{10} y$
29. $\frac{1}{2} \log_a x - \frac{2}{3} \log_a y$
30. $\log_b 2x + 3(\log_b x - \log_b y)$
31. $-\log_a x + 6 \log_a (x - 1) + 3 \log_a x^2$

Write the logarithm of the given expression in terms of the logarithms of its factors:

32. $\log_{10} (895)(1.47)$
33. $\log_{10} (60.3)^4$
34. $\log_{10} (68)^7(\sqrt{147})$
35. $\log_{10} \dfrac{(54.3)^3(67)}{(93.9)(32.5)^2}$

12-3 Common logarithms. Because of the previous article, it should be clear that any positive number, other than 1, can be used as the base for a system of logarithms. For computational purposes, the most convenient is the system with 10 for the base. The advantages of this system, accredited to Henry Briggs, 1560–1631, will become apparent in our discussion. In writing

$$10^3 = 1000$$
$$10^2 = 100$$
$$10^1 = 10$$
$$10^0 = 1$$
$$10^{-1} = 0.1$$
$$10^{-2} = 0.01$$
$$10^{-3} = 0.001$$

and considering this list as extending upward and downward indefinitely, we have a method for representing certain numbers as powers of ten. Although these special numbers are the only ones which can be written as 10 with an integral exponent, all positive numbers can be approximately represented as 10 with some exponent. This exponent, we realize from our definition of a logarithm (12–2), is the logarithm of the number to the base 10, and will be called the *common logarithm* of a number. Throughout the rest of this book, the base 10 will be assumed when no base is indicated. Thus, log 1000 = 3, log 10 = 1, or log .001 = −3.

As was stated, not only do the powers of ten listed above have logarithms, but all positive numbers do. The values of the logarithms of numbers every hundredth of a unit between 1 and 10 have been approximated to four decimal places in Table II. For example, to find log 7.63, we look down the first column N of the table for 7.6, and then move to the right of 7.6 to the number which appears in the column headed by 3. Finding .8825, we have log 7.63 = .8825, which means, of course, $10^{0.8825} = 7.63$ (approximately).

Actually, by making use of the method of *linear interpolation*, similar to that used in Articles 5–5 and 8–5, and Table II, we can find the logarithm of a number with four significant digits. If N lies between x and $x + .01$, then $N = x + .01r$, with r between 0 and 1, so that, assuming the graph of the logarithmic function is a straight line between x and $x + .01$,

$$\log N = \log x + r\,[\log\,(x + .01) - \log x]. \qquad (12\text{--}7)$$

EXAMPLE 1. Find log 3.476.

Solution. Since $3.47 < 3.476 < 3.48$, we have

$$\begin{aligned} \log 3.476 &= \log 3.47 + .6(\log 3.48 - \log 3.47) \\ &= .5403 + .6(.5416 - .5403) \\ &= .5403 + .0008 \\ &= .5411. \end{aligned}$$

This number can have only four significant figures, since the logarithms in the table are only four-figure approximations.

If the logarithm of a number is given to four decimal places, it is also possible to find the number using Table II. If log N appears in the middle of the table, the process used to find N is reversed. If log N lies between log x and log $(x + .01)$, then $N = x + .01r$ with

$$r = \frac{\log N - \log x}{\log (x + .01) - \log x}, \qquad (12\text{–}8)$$

rounded to the nearest tenth. (Why?)

EXAMPLE 2. Find N if log $N = .7281$.

Solution. Looking in the middle of the table, we find $.7275 < .7281 < .7284$, where

$$\log 5.35 = .7284,$$
$$\log 5.34 = .7275.$$

Therefore, we have

$$r = \frac{.7281 - .7275}{.7284 - .7275} = .7 \text{ (rounded off)},$$

and $N = 5.347$.

Problems

Using Table II, find the logarithm of each of the following numbers:

1. 2.57
2. 3.89
3. 6.92
4. 7.65
5. 4.71
6. 9.80
7. 6.875
8. 8.924
9. 3.276

10. 1.892   11. 7.689   12. 5.873

Using Table II, find N in each of the following, if log N is equal to:

13. .3856   14. .8756   15. .6149
16. .6405   17. .9415   18. .7657
19. .2675   20. .5217   21. .9229
22. .7578   23. .5069   24. .6745

In general, the logarithm of N has two parts: a whole number, called the *characteristic*, and a positive decimal (a number n such that $0 \leq n < 1$), called the *mantissa*. If the decimal point in any number is just to the right of the first nonzero digit, the logarithm of that number has 0 for its characteristic. All the numbers between 1 and 10 found in Table II are of this type. Such a number is said to have its decimal point in *standard position*.

If any number is multiplied by 10, the decimal point is moved one place to the right, or if divided by 10, one place to the left. But each time a number is multiplied by 10, since

$$\log 10N = \log N + \log 10$$
$$= \log N + 1,$$

the logarithm of the number is increased by one. Likewise, if a number is divided by 10, its logarithm is decreased by one, for

$$\log \frac{N}{10} = \log N - \log 10$$
$$= \log N - 1.$$

Thus the general rule for obtaining the characteristic can be stated: The characteristic of the logarithm of a number is equal to the number of places the decimal point has been moved from standard position. The *characteristic* is positive if the point has been moved to the right, negative if to the left. The mantissa, that part of the logarithm which appears in the table, is not affected by the position of the decimal point in the number, but depends only on its succession of digits. Thus, using the result of Example 1, log 347.6 = 2.5411, since its characteristic is 2, while log .003476 = -3 + .5411. The characteristic -3 is usually written $7 - 10$ (it is common practice for computational purposes to write any negative characteristic as a positive integer minus a multiple of 10), so that $-3 + .5411 = 7.5411 - 10$.

Problems

Give the characteristic and mantissa of each of the following logarithms:

1. $\log N = 1.3782$
2. $\log N = 3.4729$
3. $\log N = .5728 - 3$
4. $\log N = 9.6847 - 10$
5. $\log N = 5.8723$
6. $\log N = 6.7253 - 10$
7. $\log N = -3.7285$
8. $\log N = -0.6892$

Find the logarithms of each of the following numbers using Table II:

9. 329
10. 0.00874
11. 4728
12. 32.46
13. 0.07284
14. 0.6877

Find N in each of the following, using Table II:

15. $\log N = 3.8228$
16. $\log N = .9643 - 2$
17. $\log N = 2.4268$
18. $\log N = 9.8818 - 10$
19. $\log N = 7.5627 - 10$
20. $\log N = -1.4892$

Much of the numerical work in trigonometry involves computations where a large amount of multiplication and division is employed. To shorten such work the logarithms of the circular functions have been tabulated in Table III. Although the entries are logarithms (with the characteristics included), the form of the table is the same as that of Table I, and it is used in exactly the same manner (see Article 5-5).

EXAMPLE 3. Find $\log \sin 36°17'$.

Solution. Since the angle $36°17'$ lies between $36°10'$ and $36°20'$,

$\log \sin 36°17' = \log \sin 36°10' + \frac{7}{10} [\log \sin 37°20' - \log \sin 37°10']$
$= (9.7710 - 10) + \frac{7}{10} [(9.7727 - 10) - (9.7710 - 10)]$
$= (9.7710 - 10) + \frac{7}{10} (.0017)$
$= (9.7710 - 10) + .0012$
$= 9.7722 - 10$

EXAMPLE 4. Find the angle θ between $0°$ and $90°$ if $\log \tan \theta = .1947$.

Solution. We locate .1947 in the log tangent column between $\log \tan 57°20' = .1930$ and $\log \tan 57°30' = .1958$. Thus

$$\frac{r}{10} = \frac{.1947 - .1930}{.1958 - .1930}$$

or
$$r = 10(\tfrac{17}{28}) = 6$$
so that $\theta = 57°26'$, approximated to the nearest minute.

Problems

1. By using Table III, and interpolation if necessary, find the value of each of the following:
   (a) log sin 13°20'
   (b) log cos 45°30'
   (c) log sin 67°32'
   (d) log cos 38°21'
   (e) log tan 72°47'
   (f) log sin 115°18'
   (g) log cos 68°56'
   (h) log sin 51°49'

2. Find the value of θ between 0° and 90° to the nearest minute by using Table III, and interpolation if necessary, for the following:
   (a) log sin θ = 9.2870 − 10
   (b) log cos θ = 9.8365 − 10
   (c) log tan θ = 9.7353 − 10
   (d) log cos θ = 9.7316 − 10
   (e) log sin θ = 9.2278 − 10
   (f) log tan θ = .4937
   (g) log cos θ = 9.9797 − 10
   (h) log sin θ = 9.8761 − 10

12–4 Computation by the use of logarithms. We are now prepared to show how any computation involving multiplication, division, raising to a power, or extracting roots is greatly simplified through the use of logarithms. The work should be outlined systematically before any actual computation is carried out. An orderly arrangement is most helpful not only in simplifying the work, but also for checking.

EXAMPLE 1. Use logarithms to compute $\dfrac{1280 \cdot 0.849}{62.8}$.

Solution. By letting $N = \dfrac{1280 \cdot 0.849}{62.8}$, we have, using Property 6 and 7, log N = log 1280 + log 0.849 − log 62.8. The work should then be clearly arranged as follows:

$$\begin{aligned}
\log 1280 &= 3.1072 \\
(+)\ \log 0.849 &= \underline{9.9289 - 10} \\
\log \text{numerator} &= 13.0361 - 10 \\
(-)\ \log 62.8 &= \underline{1.7980} \\
\log N &= 11.2381 - 10 \\
N &= 17.3.
\end{aligned}$$

N is given with three significant figures, since the original numbers were of this type.

EXAMPLE 2. Compute by using logarithms: $\sqrt{0.01278}/(0.4825)^3$.

Solution. We know log $N = \frac{1}{2}$ log 0.01278 $-$ 3 log 0.4825. Again arranging our work:

$$\begin{aligned}
\log 0.4825 &= 9.6834 - 10 & \log 0.01278 &= 18.1065 - 20 \\
3 \log 0.4825 &= 29.0502 - 30 & \tfrac{1}{2} \log 0.01278 &= 9.0532 - 10 \\
&= 9.0502 - 10 & (-)\, 3 \log 0.4825 &= \underline{9.0502 - 10} \\
& & \log N &= 0.0030 \\
& & N &= 1.007
\end{aligned}$$

EXAMPLE 3. Find the value of b by using logarithms if

$$b = \frac{32.86 \sin 27°42'}{\sin 54°17'}.$$

Solution. We have

$$\log b = \log 32.86 + \log \sin 27°42' - \log \sin 54°17'.$$

Arranging our work,

$$\begin{aligned}
\log 32.86 &= 1.5167 \\
(+) \log \sin 27°42' &= \underline{9.6673 - 10} \\
\log \text{numerator} &= 11.1840 - 10 \\
(-) \log \sin 54°17' &= \underline{9.9095 - 10} \\
\log b &= 1.2745 \\
b &= 18.81
\end{aligned}$$

EXAMPLE 4. Find the value of $x = 52.8 \log 6.79$.

Solution. We must notice that this example asks for the product of the two factors 52.8 and log 6.79, *not* 52.8 and 6.79. Thus,

$$\log x = \log 52.8 + \log [\log 6.79].$$

$$\begin{aligned}
\log 52.8 &= 1.7226 \\
(+) \log [\log 6.79] = \log 0.832 &= \underline{9.9201 - 10} \\
\log x &= 1.6427 \\
x &= 43.9
\end{aligned}$$

Problems

Using logarithms, compute the value of the following to the correct number of significant figures:

1. $(367)(87.2)$
2. $(47.2)(0.897)$
3. $\dfrac{32.7}{(0.892)^{1/2}}$
4. $\dfrac{(245)(8.62)}{(7.84)^2}$
5. $(32.79)(497.2)(9.738)$
6. $\sqrt{756.9}\ (4.796)$
7*. $(-.8472)^4$
8. $(-3.472)^{-3}$
9. $\dfrac{(6.892)(-0.9245)^{2/3}}{2.475}$
10. $\left(-\dfrac{47.2}{6.783}\right)^3$
11. $\dfrac{\sqrt{738.2}\ (38.74)}{(0.9576)^2(8743)}$
12. $\left[\dfrac{\sqrt{8453}\ (.002477)}{347.9}\right]^{1/2}$
13. $4.72 \log 63.9$
14. $\log(\log 82.4)$
15. $\dfrac{\log 48.5}{\log 67.2}$
16. $\dfrac{\log 0.8924}{\log 5.237}$
17. $48.7 \tan 58°30'$
18. $\dfrac{68.2 \sin 37°20'}{\sin 49°50'}$
19. $897.2 \cos 63°48'$
20. $(189)(256) \cos 27°11'$.

12-5 Compound interest and its generalization. Not only does the study of compound interest make use of logarithmic computation, but it also logically introduces another important system of logarithms.

If an amount of money P is invested at an interest rate of r (expressed in decimals) per year, the amount of interest at the end of one year would be Pr, so that the total amount would be $P + Pr = P(1 + r)$. If this amount were then to draw interest for a second year, at the end of two years the total amount would be $P(1 + r) + P(1 + r)r = P(1 + r)[1 + r] = P(1 + r)^2$. This represents the compound amount of money at the end of two years due to its investment at the interest rate r. Continuing this process, the amount P, invested for n years, compounded annually at the rate r, is given by

$$A = P(1 + r)^n. \qquad (12\text{-}9)$$

---
* Although the logarithm of a negative number is not defined, the calculation may be carried out by considering all factors positive, and prefixing the appropriate sign with the result.

EXAMPLE 1. Find the compound amount at the end of 8 years on an original principal of $500 at 6%, compounded annually.

Solution. Using (12–9), we have $P = 500$, $r = .06$, and $n = 8$. Thus,
$$A = 500(1 + .06)^8 = 500(1.06)^8$$

$\log 1.06 = .0253$ $\log 500 = 2.6990$
$8 \log 1.06 = .2024$ $(+)\ 8 \log 1.06 = \underline{.2024}$
$\log A = 2.9014$
$A = \$796.80$

Since n represents the number of years, and r the rate per year, we can consider the result due to amounts compounded annually, semiannually, quarterly, and so on, by letting s denote the number of conversion periods each year. Thus, in n years, with s conversion periods per year, the number of periods would be ns, and the rate per period r/s, or

$$A = P\left(1 + \frac{r}{s}\right)^{ns}. \tag{12–10}$$

EXAMPLE 2. If the $500 in Example 1 were invested at 6%, compounded quarterly, for the eight years, how large would the result be?

Solution. Since 6% compounded quarterly for eight years is $1\frac{1}{2}$% per period, with 32 periods, we have

$$A = 500(1.015)^{32},$$

which results in $A = \$804.20$.

Both this result and that in the first example are not too accurate, since four-place logarithm tables were used. For accurate results, since the exponent is large, either seven-place logarithmic or compound interest tables should be used.

One of the most important exponential functions is a direct result of a generalization of Eq. (12–10), which is sometimes called the *law of natural growth*. Its applications frequently occur in biology, chemistry, and economics, as well as in mathematics. Let us suppose that the number of conversion periods s increases indefinitely, so that the amount is compounded *continuously*. Letting $r/s = x$, and therefore $s/r = 1/x$, we have, from Eq. (12–10),

$$A = P\left[\left(1 + \frac{r}{s}\right)^{s/r}\right]^{rn} = P[(1 + x)^{1/x}]^{rn}. \quad (12\text{-}11)$$

With s increasing indefinitely, for a fixed r, $r/s = x$ decreases indefinitely through positive values, and approaches zero. If Eq. (12-11) is expanded by the binomial theorem (see Problem 4, Article 11-3),

$$A = P\left[1^{1/x} + \frac{1}{x} \cdot 1^{(1/x)-1} \cdot x + \frac{\frac{1}{x}\left(\frac{1}{x} - 1\right)}{2!} \cdot 1^{(1/x)-2} \cdot x^2 \right.$$
$$\left. + \frac{\frac{1}{x}\left(\frac{1}{x} - 1\right)\left(\frac{1}{x} - 2\right)}{3!} \cdot 1^{(1/x)-3} \cdot x^3 + \cdots\right]^{rn}$$
$$= P\left[1^{1/x} + 1^{(1/x)-1} + \frac{1-x}{2!} \cdot 1^{(1/x)-2} \right.$$
$$\left. + \frac{(1-x)(1-2x)}{3!} \cdot 1^{(1/x)-3} + \cdots\right]^{rn},$$

and evaluating this for $x = 0$,

$$A = P\left(1 + 1 + \frac{1}{2!} + \frac{1}{3!} + \cdots\right)^{rn}. \quad (12\text{-}12)$$

Although we have by no means shown that this infinite sum has a definite value,* it would seem probable, as is the case, that the expression within the brackets of Eq. (12-11) does have a definite value if x is allowed to approach zero. This constant value, denoted by e, is a nonterminating, nonrepeating decimal, and therefore irrational (Article 1-1). To six significant figures

$$e = 2.71828. \quad (12\text{-}13)$$

We therefore have the function

$$A = Pe^{rn}, \quad (12\text{-}14)$$

representing the amount of P compounded continuously for n years at a rate r.

---

* As in Article 11-3, any proof of the existence of a definite value for an infinite sum is beyond the scope of this book.

EXAMPLE 3. The population of a certain locality is 20,000 and is increasing continuously at a rate $r = .037$, according to the law of natural growth [Eq. (12–14)]. Find the approximate population after 25 years.

Solution. With the formula $A = Pe^{rn}$, we have $P = 20,000$, $r = .037$, and $n = 25$. Therefore

$$A = 20{,}000 e^{(.037)(25)}$$

Solving by logarithms,

$$\log A = \log 20{,}000 + (.037)(25) \log e$$

$$\begin{aligned}
\log 20{,}000 &= 4.3010 \\
(+) \ (.037)(25) \log e &= \underline{.4017} \\
\log A &= 4.7027
\end{aligned}$$

$A = 50{,}430$ (approximate) population.

Problems

1. Find the compound amount at the end of 10 years on an original principal of $200 at 4% (a) compounded annually; (b) compounded semi-annually; (c) compounded quarterly; (d) compounded continuously.

2. Find the compound amount at the end of 20 years on an original principal of $3000 at 6% (a) compounded annually; (b) compounded monthly; (c) compounded continuously.

3. What time is required to double a certain amount (a) compounded annually at 6%; (b) compounded continuously at 6%? *Hint:* Let $P = 1$, and $A = 2$.

In Problems 4–10, we shall assume the law of natural growth [Eq. (12–14)].

4. The population of a certain town is 80,000 and has been increasing continuously for the past 20 years at the rate $r = .025$. What was the population 20 years ago?

5. There are originally 1000 bacteria in a culture, and 4 hr later there are 4000. Find the rate of increase per hour of the bacteria.

6. If the growth of a certain bacteria in a culture increases at the rate $r = .24$ per hr, how long will it take 50 bacteria to become 1,000,000?

7. In a certain chemical reaction, the original concentration of .03 is reduced to .01 in 4 min. (a) What is the rate of decrease in the concentration per minute? (b) What will the concentration be in 10 min?

8. If radium decomposes according to the relation $y = y_0 e^{-0.04t}$, where y_0 grams of radium reduce to y grams in t centuries, find how long it will take one gram to reduce to one-half a gram.

12–6 Applications of the exponential functions. In the last article we saw one important application of an exponential function. Use is made of the exponential functions to find the logarithm of a number to any base.

EXAMPLE 1. Find the value of $\log_4 15$.

Solution. Letting $y = \log_4 15$, this reduces to the equivalent equation

$$4^y = 15.$$

Any such equation may be solved by taking the common logarithm of each member, and finding the required value of y. Thus,

$$\log 4^y = \log 15,$$
$$y \log 4 = \log 15.$$

Solving for y,

$$y = \frac{\log 15}{\log 4} = \frac{1.1761}{.6021} = 1.953.$$

If the logarithm of a member to one base is known, it is often desirable to find the logarithm of this number to a different base. Let

$$y = \log_a N.$$

Writing this in the equivalent form

$$a^y = N,$$

and taking the logarithm to the base b of each member, we have

$$\log_b a^y = \log_b N,$$

or

$$y \log_b a = \log_b N.$$

Recalling that $y = \log_a N$, we have

$$\log_a N \cdot \log_b a = \log_b N. \qquad (12\text{--}15)$$

Because of the importance of the function $y = e^{kx}$ (last Article), the system of logarithms with e for the base is frequently used.* In this case, with $a = e$ and $b = 10$, Eq. (12–15) reduces to

$$\log_{10} N = \log_e N \cdot \log_{10} e = 0.4343 \log_e N, \qquad (12\text{–}16)$$

and

$$\log_e N = \frac{1}{\log_{10} e} \cdot \log_{10} N = 2.303 \log_{10} N. \qquad (12\text{–}17)$$

These two expressions are used extensively in analytic work to change from the common logarithmic system to natural logarithms, and vice versa.

EXAMPLE 2. Find $\log_e 3.24$.

Solution. (1) Using (12–17),

$$\begin{aligned}\log_e 3.24 &= 2.303 \log 3.24 \\ &= (2.303)(.5105) \\ &= 1.175.\end{aligned}$$

Solution. (2) This example may, of course, be solved as in Example 1. Letting

$$y = \log_e 3.24,$$
$$e^y = 3.24.$$

Taking the common logarithm of each member,

$$\log e^y = \log 3.24,$$

or

$$y \log e = \log 3.24.$$

Solving for y,

$$y = \frac{\log 3.24}{\log e} = \frac{.5105}{.4343} = 1.175.$$

The method used in Example 1 and Example 2 (2) is similar to that used in solving many exponential as well as logarithmic equations.

---

* The many reasons for the choice of e as a base cannot be discussed in this book. This system, called *natural logarithms*, is used extensively in calculus.

12-6] APPLICATIONS OF THE EXPONENTIAL FUNCTIONS

EXAMPLE 3. Solve $4^{x+3} = 7^{x-1}$ for x.

Solution. Taking the common logarithms of both members,
$$\log 4^{x+3} = \log 7^{x-1}.$$

By using Property 8, we have
$$(x + 3) \log 4 = (x - 1) \log 7.$$

Solving this linear equation for x,
$$x \log 4 + 3 \log 4 = x \log 7 - \log 7,$$
$$x(\log 4 - \log 7) = -\log 7 - 3 \log 4,$$
$$x = \frac{\log 7 + 3 \log 4}{\log 7 - \log 4}$$
$$= \frac{2.6514}{.2430} = 10.92 \text{ (approximately)}.$$

EXAMPLE 4. Solve $\log (x + 3) - \log x = 2$.

Solution. Using Property 7,
$$\log (x + 3) - \log x = \log \frac{x + 3}{x} = 2,$$

so that
$$\frac{x + 3}{x} = 100,$$
$$x + 3 = 100x$$
$$99x = 3, \quad \text{or} \quad x = \tfrac{1}{33}.$$

PROBLEMS

Find the following logarithms:

1. $\log_2 14$
2. $\log_5 27$
3. $\log_7 128$
4. $\log_{27} 15$
5. $\log_e 7$
6. $\log_e 12$
7. $\log_e 1.79$
8. $\log_e 3.78$

Find the value of x in each of the following:

9. $\log_4 x = 23$
10. $\log_{12} x = 17$
11. $\log_e x = 3.28$
12. $\log_e x = 1.72$
13. $\log_e x = .8473$
14. $\log_e x = 2.547$

*15. Prove the relation

$$\log_b a = \frac{1}{\log_a b}.$$

Solve the following exponential equations for x:

16. $3^x = 27$
17. $2^x = 32$
18. $2^x = 27$
19. $3^x = 32$
20. $3^{x+1} = 4^{x-7}$
21. $5(6^x) = 21^{x-2}$
22. $17^{2x-3} = 25^{x-1}$
23. $2.78^x = 7.38^{3x-1}$
24. $e^x + e^{-x} = 2$ *Hint:* This is quadratic in e^x.
25. $e^x - e^{-x} = 2$

Solve the following logarithmic equations for x:

26. $\log x - 2 \log 4 = \log 32$
27. $\log (x + 2) - \log x = \log 12$
28. $\log (3x + 2) = \log (x - 4) + 1$
29. $\log (x + 1) - \log x = 2.4742$

CHAPTER 13

SOLUTION OF TRIANGLES

The entire point of view with regard to solving triangles has changed during the last few years. Many numerical methods, numerous formulas, and detailed study with large numbers of exercises had been considered an important part of trigonometry. More recently, however, the analytic part of trigonometry with its applications in advanced mathematics and science has become more important. Moreover, new developments are constantly being made, resulting in extremely accurate graphical methods as well as the use of high-speed numerical calculators. We shall concentrate our attention on the fundamental theorems, and not emphasize long and detailed processes, although the tables in the back of the book will be used.

13-1 General discussion. If a certain number of the sides and angles of a triangle are known, the *triangle* can be *solved* by finding the remaining parts. We shall derive two of the many formulas used in solving triangles, and consider certain special cases. Other formulas and relations between the sides and angles of a triangle appear in the examples and problems.

While deriving these relations, it is important to recall that a triangle is determined when
 (1) two angles and one side are given,
 (2) two sides and the included angle are given,
 (3) three sides are given, if the longest side is less than the sum of the other two.
Also, there are at most two triangles when
 (4) two sides and an angle opposite one of them is given.
These are the four types of problems we wish to solve. We shall denote the three angles at the vertices of any triangle ABC by α, β, and γ, respectively, and the corresponding opposite sides by a, b, and c.

13-2 The Law of Sines. Let us first derive the *Law of Sines*. By choosing a rectangular coordinate system so that the angle α of the triangle ABC is in standard position (see Fig. 13-1), the co-

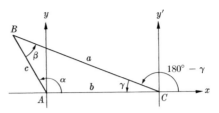

FIGURE 13-1

ordinates of B are $(c \cos \alpha, c \sin \alpha)$. (Recall Problem 53, Article 4-2.) If, however, the origin of the coordinate system is at C, with $(180° - \gamma)$ in standard position, the coordinates of B are

$$[a \cos (180° - \gamma), a \sin (180° - \gamma)].$$

Since in either case the y-coordinates of B are equal (the same distance above the x-axis), we have

$$c \sin \alpha = a \sin(180° - \gamma)$$
$$= a \sin \gamma.$$

Dividing each member of this equation by $\sin \alpha \sin \gamma$, we obtain

$$\frac{c}{\sin \gamma} = \frac{a}{\sin \alpha}.$$

With a different choice of the x-axis, $b/\sin \beta = a/\sin \alpha$, so that

$$\frac{a}{\sin \alpha} = \frac{b}{\sin \beta} = \frac{c}{\sin \gamma}. \tag{13-1}$$

This relationship enables us to solve the problems mentioned in (1) and (4) of Article 13-1. However, before we consider problems solved by (13-1) in general, let us consider an extremely important special case.

13-3 Solution of right triangles. If γ is a right angle equal to 90°, in (13-1), the equations reduce to $\sin \alpha = a/c$, and $\sin \beta = b/c$. But with $\gamma = 90°$, and $\alpha + \beta + \gamma = 180°$, $\beta = 90° - \alpha$, so that $\cos \alpha = b/c$. Since $\tan \alpha = \sin \alpha/\cos \alpha$, we also have $\tan \alpha = a/b$.

SOLUTION OF RIGHT TRIANGLES

Thus, in any right triangle with $\gamma = 90°$,

$$\sin \alpha = \frac{a}{c} = \frac{\text{side opposite } \alpha}{\text{hypotenuse}}, \qquad (13\text{-}2)$$

$$\cos \alpha = \frac{b}{c} = \frac{\text{side adjacent } \alpha}{\text{hypotenuse}}, \qquad (13\text{-}3)$$

$$\tan \alpha = \frac{a}{b} = \frac{\text{side opposite } \alpha}{\text{side adjacent } \alpha}. \qquad (13\text{-}4)$$

Before attempting to solve any triangles, the following remarks are pertinent.

1. Results can be no more accurate than the given sides and angles. We shall agree to set up the following table for accuracy between sides and angles.

Significant figures for sides:	Angles to the nearest:
2	degree
3	ten minutes
4	minute
5	tenth of a minute

2. If the results are required to only two or possibly three significant digits, the slide rule should be used for the computation. Also, the slide rule can be employed as a check upon the work, although the answers may be desired more accurately than the slide rule will allow.

3. If the results are to be correct to several significant digits, the tables should be used. When calculating machines are available, the natural functions and arithmetical methods are usually employed. Since the majority of students do not have access to machines, the logarithmic solution is the logical one. This will be illustrated in most of the examples.

4. In solving any problem it is advisable to draw the triangle, label the known parts, and make a complete systematic outline to follow, before any computation is done.

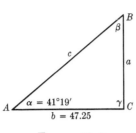

FIGURE 13–2

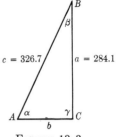

FIGURE 13–3

EXAMPLE 1. In the right triangle ABC, $b = 47.25$, $\alpha = 41°19'$. Find the remaining parts and the area.

Solution. We first draw the triangle and label numerically the parts that are given, as in Fig. 13–2. Then $\beta = 90° - \alpha = 48°41'$. To find a, we use the relation $\tan \alpha = a/b$, and to find c, we use $\cos \alpha = b/c$.

$$a = 47.25 \tan 41°19' \qquad c = \frac{47.25}{\cos 41°19'}$$

$$\log 47.25 = 1.6744 \qquad\qquad \log 47.25 = 11.6744 - 10$$
$$(+) \log \tan 41°19' = 9.9440 - 10 \qquad (-) \log \cos 41°19' = 9.8757 - 10$$
$$\log a = 11.6184 - 10 \qquad\qquad \log c = 1.7987$$
$$a = 41.54 \qquad\qquad c = 62.90$$

The area of the triangle is $K = \tfrac{1}{2}ab$.

$$\log 41.53 = 1.6183 \qquad (-) \log 2 = .3010$$
$$(+) \log 47.25 = 1.6744 \qquad \log K = 2.9917$$
$$3.2927 \qquad K = 981.0$$

This example illustrates a problem with its results correct to four significant figures, as in Table III.

EXAMPLE 2. Solve the right triangle in which $a = 284.1$ and $c = 326.7$.

Solution. Draw the triangle and label numerically the parts that are known, as in Fig. 13–3. Since a and c are given, we use the relation $\sin \alpha = a/c$ to find α, and then $\cos \alpha = b/c$ to find b.

13-3] SOLUTION OF RIGHT TRIANGLES 235

$$\sin \alpha = \frac{284.1}{326.7} \qquad b = 326.7 \cos 60°24'$$

$$\begin{array}{l} \log 284.1 = 12.4534 - 10 \\ (-) \log 326.7 = 2.5141 \\ \hline \log \sin \alpha = 9.9393 - 10 \\ \alpha = 60°24' \\ \beta = 29°36' \end{array} \qquad \begin{array}{l} \log 326.7 = 2.5141 \\ (+) \log \cos 60°24' = 9.6937 - 10 \\ \hline \log b = 12.2078 - 10 \\ b = 161.4 \end{array}$$

Log cos α can be found in the table when log sin α is located. This is one of the advantages in first arranging the work systematically.

In science and engineering, physical entities such as velocity, acceleration, or force not only require magnitude but also direction for their complete determination. They are represented by line segments with an arrowhead on one end to show direction, while the length of the segment, by reference to some scale, denotes magnitude. Such line segments are called *vectors*. For example, with a scale of 40 lb to the unit, the vector in Fig. 13–4 might represent a force of 200 lb acting in the direction of 30° with the positive x-axis. A bold-face letter **v** will denote the vector, while its *length* will be designated by the ordinary v.

Since **v**, with its initial point at O, as in Fig. 13–5, has v for its length and makes an angle θ with the positive x-axis, its terminal point P has coordinates ($v \cos \theta, v \sin \theta$). These coordinates, denoted

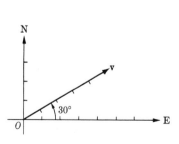

FIGURE 13–4

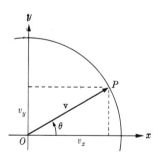

FIGURE 13–5

by v_x and v_y, are called the x- and y- *components* of **v** and satisfy the relations:

$$v_x = v \cos \theta, \qquad v^2 = v_x^2 + v_y^2,$$

(13–5)

$$v_y = v \sin \theta, \qquad \tan \theta = \frac{v_y}{v_x}.$$

EXAMPLE 3. Find v_x and v_y for the vector with $v = 247$ and $\theta = 37°40'$.

Solution. $v_x = 247 \cos 37°40' = (247)(.7916) = 196.$
$v_y = 247 \sin 37°40' = (247)(.6111) = 151.$

By considering $\mathbf{v}_x$ and $\mathbf{v}_y$ as vectors, we can think of **v** as the single vector equivalent to these two. More generally, a vector **v** is the *resultant* or *sum* of two vectors $\mathbf{v}_1$ and $\mathbf{v}_2$ if, as a single force, it will produce the same result as the two forces acting together. If $\mathbf{v}_1$ and $\mathbf{v}_2$ are two nonparallel vectors emanating from the same point, their resultant **v** is the vector from that point, having the length and direction of the diagonal of the parallelogram determined by $\mathbf{v}_1$ and $\mathbf{v}_2$. (What would **v** be if $\mathbf{v}_1$ and $\mathbf{v}_2$ were parallel?) This notion is first used in Problem 22.

PROBLEMS

In solving the problems, a calculating machine should be used for some of the computations if possible. Logarithms should be used for others. All solutions should be checked.

1. Find the unknown sides and angles of each of the following triangles. In each $\gamma = 90°$.

(a) $\alpha = 37°20', a = 243$
(b) $\alpha = 62°40', b = 796$
(c) $a = 3.28, b = 5.74$
(d) $b = 68.4, c = 96.2$
(e) $\beta = 51°10', c = 0.832$
(f) $\alpha = 37°40', a = 54.8$
(g) $a = 37.9, b = 57.3$
(h) $\beta = 41°25', c = 3265$
(i) $a = 5429, c = 6294$
(j) $a = 3.273, b = 7.647$
(k) $\beta = 62°57', a = 0.8263$
(l) $\beta = 47°23', b = 72.55$
(m) $b = 3572, c = 4846$
(n) $\alpha = 24°47', b = 318.4$

SOLUTION OF RIGHT TRIANGLES

2. In a circle of radius 96.4 inches, what is the central angle that subtends a chord of 40.3 inches?

3. A rectangular lot is 102 ft by 296 ft. Find the length of the diagonal and the angles it makes with the longest side.

4. A telegraph pole is held to the ground by wires 18.6 ft up the pole. Find the length of one of the wires that makes an angle of 26°20′ with the vertical.

5. Find the area of a parallelogram whose sides are 33.7 and 15.2 inches if the angle between them is 67°40′.

6. One of the equal sides of an isosceles triangle is 6.73 inches, and one of the base angles is 27°10′. Find the base and altitude.

7. A 36-ft ladder is used to reach the top of a 28-ft wall. If the ladder extends 2 ft past the top of the wall, find its inclination to the horizontal.

8. A hemispherical bowl of inside radius 8.00 inches is standing level and is filled with water to a depth of 2.00 inches. Through what angle may it be tilted before the water spills?

9. A piece of wire 24.78 inches long is bent so as to form an isosceles triangle with one angle 97°26′. Find the length of each side of the triangle formed.

*10. From a point on the ground 152.3 ft away from the foot of a flag pole, the angle of elevation of the top of the pole is 31°46′. How high is the flag pole? *Hint:* The *angle of elevation* of an object is the angle between the line of sight from the eye to the object and the horizontal, when the object observed is above the horizontal plane. When the object observed is below this horizontal plane, the angle is called the *angle of depression*.

11. From a lighthouse 75.3 ft above the level of the water, the angle of depression of a boat is 23°40′. How far is the boat from a point at water level directly under the point of observation?

12. Find the height of a balloon directly above a town A, if the angle of depression of town B, 6.23 miles from A, is 15°20′.

13. From a lookout tower 80.0 ft high a man observes from a position 6.5 ft below the top of the tower, that the angle of elevation of the top of a certain tree is 12°40′, and the angle of depression of its base is 72°20′. If the base of the tower and tree are at the same level, what is the height of the tree?

14. From a point 20.75 ft above the surface of the water, the angle of elevation of a building at the edge of the water is 38°16′, while the angle of depression of its image in the water is 56°28′. Find the height of the building, and the horizontal distance from the point of observation.

15. From a mountain 1780 ft high the angle of depression of a point on the nearer shore of a river is 48°40′, and of a point directly across on the

opposite side, the angle of depression is 22°20′. What is the width of the river between the two points?

16. At a certain point the angle of elevation of a mountain peak is 40°20′. At a point 9560 ft farther away in the same horizontal plane, its angle of elevation is 29°50′. Find the distance of the peak above the horizontal plane.

*17. Lighthouse B is 6.56 miles directly east of lighthouse A. A ship at O observes that A is due north, and that the bearing of OB is N 46°10′ E. How far is the ship from A? From B? *Hint:* The *bearing* of a line in a horizontal plane is the acute angle made by this line with a north-south line. In giving the bearing of a line, the letter N or S is written, followed by the letter E or W. Thus, in Fig. 13-6, the bearing of OB is read north, 46°10′ east.

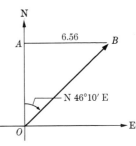

FIGURE 13-6

18. An airplane is 115 miles due east of a radio station A; a second radio station is 136 miles due north of A. What are the distance and bearing of the second radio station from the airplane?

19. Because of a certain wind, a boat sails 4728 ft in the direction S 47°29′ W. How far south has it gone? How far west has it gone?

20. Find v_x and v_y for each of the vectors with v and θ given:
   (a) $v = 75$, $\theta = 60°$ (b) $v = 48$, $\theta = 136°$
   (c) $v = 4.72$, $\theta = 217°10′$ (d) $v = 58.47$, $\theta = 47°18′$

21. Find the length and direction for each of the vectors if
   (a) $v_x = 3$, $v_y = 4$ (b) $v_x = 23$, $v_y = 45$
   (c) $v_x = -16.2$, $v_y = 28.7$ (d) $v_x = 382.4$, $v_y = -768.3$

22. If a force of 658.4 lb is acting east, and another of 316.2 lb is acting north, what are the magnitude and direction of their resultant?

23. A balloon is rising at the rate of 12 ft/sec and at the same time is being blown horizontally by a wind traveling 18 ft/sec. Find the angle its path makes with the vertical, and determine its actual velocity.

24. A river runs directly east at 1.28 mi/hr. If a swimmer can swim at the rate of 1.75 mi/hr in still water, and he starts swimming so that he is headed north directly across the river, in what direction is he actually swimming? Where does he hit the opposite bank if the river is one mile wide?

25. In what direction should the swimmer in Problem 24 head in order to reach a point directly across the river?

26. An airplane travels with a speed of 145 mi/hr in calm air. The wind is blowing with a velocity of 23 mi/hr from N 27° E. (a) If the plane is headed in a direction N 63° W, find the magnitude of the speed, and the direction of the airplane with reference to the ground. (b) In what direction must the airplane be headed in order to fly in the direction N 63° W, and what would be its actual speed in the air?

27. Find the resultant of the following sets of forces, where f is the magnitude and θ the angle each force makes with the positive x-axis:

(a) $f_1 = 15$, $\theta_1 = 65°$, $f_2 = 37$, $\theta_2 = 142°$
(b) $f_1 = 6280$, $\theta_1 = 37°10'$, $f_2 = 2840$, $\theta_2 = -16°40'$
(c) $f_1 = 6800$, $\theta_1 = 210°$, $f_2 = 7200$, $\theta_2 = 315°$, $f_3 = 5600$, $\theta_3 = 90°$

Note: Although we could find the resultant as the diagonal of the parallelogram, it is simpler in such problems to work with components. Obtain the sum of the x-components as one force, the sum of the y-components as another, and then find the resultant of these two.

28. The circle with center at O in Fig. 13-7 is circumscribed about the triangle and OD is drawn perpendicular to BC. What is the relation of $\angle BOD$ to α? Using the right triangle BDO, prove $\sin \alpha = a/2R$. With the necessary additional construction, prove the Law of Sines.

29. In the right triangle ABC, with $\alpha = 30°$, $AB = 2$. D is chosen on AC, so that $DC = BC$, and from D a line is drawn perpendicular to AB, meeting AB at K. Show $\angle DBK = 15°$, and by finding the lengths of the various lines, prove

$$\sin 15° = \frac{\sqrt{6} - \sqrt{2}}{4}.$$

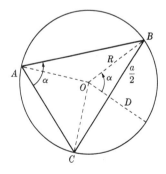

Figure 13-7

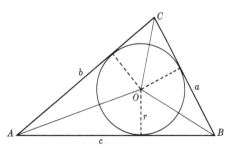

Figure 13-8

*30. Using Fig. 13–8, where the inscribed circle in triangle ABC has r for its radius, prove the area $K = rs$, where $s = (a + b + c)/2$.

13–4 The Law of Cosines. Before considering examples in which we use the Law of Sines to solve general triangles, we shall prove another important relationship, the *Law of Cosines*.

Refer to Fig. 13–1, where the angle α of the triangle ABC is in standard position, with the vertex C at $(b,0)$. Recalling that the coordinates of B are $(c \cos \alpha, c \sin \alpha)$, we have, using (3–4), the distance between $B(c \cos \alpha, c \sin \alpha)$ and $C(b,0)$:

$$\sqrt{(c \cos \alpha - b)^2 + (c \sin \alpha - 0)^2}.$$

But this length is the length of a, so that

$$a = \sqrt{c^2 \cos^2 \alpha - 2bc \cos \alpha + b^2 + c^2 \sin^2 \alpha},$$

or

$$a^2 = b^2 + c^2 - 2bc \cos \alpha. \tag{13–6}$$

Similarly, we obtain

$$b^2 = a^2 + c^2 - 2ac \cos \beta, \tag{13–7}$$

$$c^2 = a^2 + b^2 - 2ab \cos \gamma. \tag{13–8}$$

These three relations, called the Law of Cosines, hold for all triangles, and are used to solve problems of the type (2) and (3) of Article 13–1. If $\gamma = 90°$ in (13–8), $c^2 = a^2 + b^2$. For this reason, this law is sometimes called *the generalization of the Pythagorean Theorem*.

13–5 Applications involving oblique triangles. To clarify the use of the Laws of Sines and Cosines, let us consider the following examples.

EXAMPLE 1. Solve the triangle ABC of Fig. 13–9 if $a = 524.7$, $\beta = 46°24'$, and $\gamma = 98°41'$.

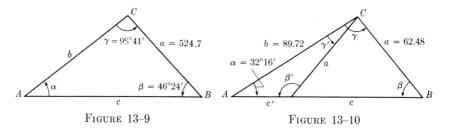

FIGURE 13-9 FIGURE 13-10

Solution. First, $\alpha = 180° - (46°24' + 98°41') = 34°55'$. With the value of α known, we can use the Law of Sines to find b and c:

$$b = \frac{a}{\sin \alpha} \sin \beta \qquad\qquad c = \frac{a}{\sin \alpha} \sin \gamma$$

$\log 524.7 =$	$12.7199 - 10$	$\log \dfrac{a}{\sin \alpha} = 2.9622$
$(-) \log \sin 34°55' =$	$9.7577 - 10$	$(+) \log \sin 98°41' = 9.9950 - 10$
$\log \dfrac{a}{\sin \alpha} =$	2.9622	$\log c = 12.9572 - 10$
$(+) \log \sin 46°24' =$	$9.8599 - 10$	$c = 906.2$
$\log b =$	$12.8221 - 10$	
$b =$	663.9	

This example illustrates (1) in Article 13-1.

EXAMPLE 2. Find all the triangles with $a = 62.48$, $b = 89.72$, and $\alpha = 32°16'$.

Solution. The *ambiguous type* (4) is illustrated here. We have the possibility of two solutions, since $\sin(180° - \beta) = \sin \beta$. Using the Law of Sines, we solve for β, and find two values, for each of which there is a corresponding γ and c. Thus we have two solutions, as is suggested in Fig. 13-10.

$$\sin \beta = b \frac{\sin \alpha}{a}$$

$$\log \sin 32°16' = 9.7274 - 10$$
$$(-) \log 62.48 = 1.7958$$
$$\log \frac{\sin \alpha}{a} = 7.9316 - 10$$
$$(+) \log 89.72 = 1.9529$$
$$\log \sin \beta = 9.8845 - 10$$

Thus
$$\beta_1 = 50°2' \quad \text{or} \quad \beta_2 = 129°58'$$
$$\gamma_1 = 180° - (\alpha + \beta_1) \qquad \gamma_2 = 180° - (\alpha + \beta_2)$$
$$= 97°42' \qquad\qquad\qquad = 17°46'$$

Then
$$c_1 = \frac{a}{\sin \alpha} \sin \gamma_1 \qquad\qquad c_2 = \frac{a}{\sin \alpha} \sin \gamma_2$$

$$\log \frac{a}{\sin \alpha} = 2.0684 \qquad\qquad \log \frac{a}{\sin \alpha} = 2.0684$$
$$(+) \log \sin 97°41' = 9.9961 - 10 \qquad (+) \log \sin 17°47' = 9.4849 - 10$$
$$\log c_1 = 12.0645 - 10 \qquad\qquad \log c_2 = 11.5533 - 10$$
$$c_1 = 116.0 \qquad\qquad\qquad c_2 = 35.75$$

EXAMPLE 3. Find the third side of the triangle with $b = 47$, $c = 58$, and $\alpha = 63°$.

Solution. For this problem, type (2) of Article 13–1, we use the Law of Cosines. It should be emphasized that using four-place tables, this law produces no more than two significant figures for the third side, since a square root is involved.

$$a^2 = (47)^2 + (58)^2 - 2 \cdot 47 \cdot 58 \cos 63°$$
$$= 2209 + 3364 - 2475.2$$
$$= 3097.8$$

Thus
$$a = 56 \text{ (approximately)}.$$

The other angles could be found by using the Law of Sines.

We notice in Example 3 that the Law of Cosines does not lend itself to these problems when more than two significant figures are desired. The First Law of Tangents can be used in such cases. From the relationship $a/\sin \alpha = b/\sin \beta$, it is not difficult to show that

$$\frac{\sin \alpha - \sin \beta}{\sin \alpha + \sin \beta} = \frac{a - b}{a + b}.$$

Using this and Problem 25, Article 5-3, we have

$$\frac{\tan \frac{1}{2}(\alpha - \beta)}{\tan \frac{1}{2}(\alpha + \beta)} = \frac{a - b}{a + b}, \qquad (13\text{-}9)$$

which, together with the five similar formulas obtained by interchanging and rotating letters, are called the *First Law of Tangents*. (See Problem 7.)

EXAMPLE 4. Solve the triangle with $a = 16.47$, $b = 25.49$, and $c = 33.77$.

Solution. This example of type (3) again uses the Law of Cosines. We find

$$a^2 = 271.26 \qquad 2ab = 839.64$$
$$b^2 = 649.74 \qquad 2ac = 1112.4$$
$$c^2 = 1140.4 \qquad 2bc = 1721.6$$

so that

$$\cos \alpha = \frac{b^2 + c^2 - a^2}{2bc} = .8823, \quad \text{or} \quad \alpha = 28°5',$$

$$\cos \beta = \frac{a^2 + c^2 - b^2}{2ac} = .6850, \quad \text{or} \quad \beta = 46°46',$$

$$\cos \gamma = \frac{a^2 + b^2 - c^2}{2ab} = -.2613, \quad \text{or} \quad \beta = 105°9'.$$

This example is easily checked, since $\alpha + \beta + \gamma = 180°$.

Example 4 could be solved using five-place tables of logarithms, but the process is long, due to the form of the Law of Cosines. The

Second Law of Tangents lends itself to large figures as well as logarithmic computation. By adding one to both members of the expression for cos α in Example 4, dividing each side by 2, and factoring the right side, we get

$$\frac{1 + \cos \alpha}{2} = \frac{(b + c + a)(b + c - a)}{4bc}.$$

Also,

$$\frac{1 - \cos \alpha}{2} = \frac{(a + b - c)(a - b + c)}{4bc}.$$

Now, letting $\dfrac{a + b + c}{2} = s$ in these expressions, and recalling Problems 31 and 32, Article 5–3, we have

$$\sin \frac{\alpha}{2} = \sqrt{\frac{(s - b)(s - c)}{bc}},$$

and

$$\cos \frac{\alpha}{2} = \sqrt{\frac{s(s - a)}{bc}}.$$

Therefore, dividing,

$$\tan \frac{\alpha}{2} = \frac{r}{s - a}, \qquad (13\text{--}10)$$

where $r = \sqrt{\dfrac{(s - a)(s - b)(s - c)}{s}}$.*

Formula (13–10) and

$$\tan \frac{\beta}{2} = \frac{r}{s - b}, \qquad \tan \frac{\gamma}{2} = \frac{r}{s - c}$$

are considered the Second Law of Tangents, and give the angles in terms of the three sides of any triangle. (See Problem 10.)

---

* With the results of Problem 30, Article 13–3, and Problem 12 of this article, r is found to be the expression for the radius of the inscribed circle of any triangle.

Problems

1. Solve the following triangles ABC, given
   - (a) $\alpha = 62°40'$, $\beta = 79°20'$, $a = 147$
   - (b) $\beta = 81°43'$, $\gamma = 57°51'$, $c = 47.35$
   - (c) $\alpha = 47°57'$, $\gamma = 118°11'$, $b = 87270$
   - (d) $\beta = 14°36'$, $\gamma = 53°8'$, $b = 8.367$

2. Find all the triangles with $a = 62.48$, $b = 43.17$, and $\alpha = 32°16'$ and draw the figure. Compare Example 2. *Hint:* It is impossible to have two solutions, since $\beta_2 + \alpha > 180°$.

3. Is there a triangle with $a = 62.48$, $b = 143.4$, and $\alpha = 32°16'$? Compare Example 2, and draw the figure. *Hint:* Can $\sin \beta$ be greater than one?

4. In the type of problem where two sides and an angle opposite one of these sides are given, we noticed various possibilities when the angle is less than 90°, by considering Example 2, and Problems 2 and 3. Assuming a, b, and α are given, explain the following with the aid of Fig. 13–11:

$\alpha < 90°$ $\qquad\qquad\qquad\qquad$ $\alpha > 90°$

$a < b \sin \alpha$ gives no solution. $\qquad$ $a \leqq b$ gives no solution.
$a = b \sin \alpha$ gives a special right triangle. $\qquad$ $b < a$ gives one solution.
$b \sin \alpha < a < b$ gives two solutions.
$a \geqq b$ gives one solution.

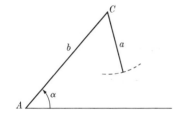
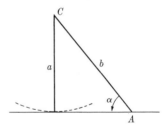
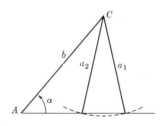
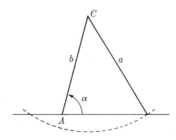

Figure 13–11

5. In each of the following either show that there is no solution or find all solutions:

(a) $b = 59.4$, $c = 72.3$, $\beta = 38°40'$
(b) $a = 49.3$, $c = 8.72$, $\alpha = 45°10'$
(c) $a = 14.72$, $b = 25.64$, $\beta = 147°47'$
(d) $b = 4.927$, $c = 5.764$, $\gamma = 57°18'$

6. Solve each of the following triangles for the side opposite the given angle:

(a) $a = 4$, $b = 7$, $\gamma = 30°$ (b) $a = 7.6$, $c = 9.2$, $\beta = 47°$
(c) $b = 8.3$, $c = 4.4$, $\alpha = 138°$

*7. Solve the following triangles with

(a) $b = 18.62$, $c = 35.61$, $\alpha = 52°18'$

Hint: Use the First Law of Tangents with $c - b = 16.99$, $c + b = 54.23$, $\frac{1}{2}(\gamma + \beta) = 63°51'$, and find $\frac{1}{2}(\gamma - \beta)$. Then $\frac{1}{2}(\gamma + \beta) + \frac{1}{2}(\gamma - \beta) = \gamma$, and so on.

(b) $a = 463$, $b = 628$, $\gamma = 57°40'$

8. Solve the following triangles with

(a) $a = 356.8$, $c = 551.4$, $\beta = 87°48'$
(b) $b = 321.0$, $c = 672$, $\alpha = 124°16'$

9. Solve the following triangles by the Law of Cosines:

(a) $a = 4$, $b = 5$, $c = 6$
(b) $a = 32$, $b = 56$, $c = 63$

*10. Solve the following triangles by the most convenient method:

(a) $a = 18.76$, $b = 25.31$, $c = 29.65$

Hint: If using logarithms and the Second Law of Tangents, arrange your work systematically:

$2s = 73.72$
$s = 36.86$
$s - a = 18.10$
$s - b =$
$s - c =$

$\log (s - a) =$
$\log (s - b) =$
$(+) \log (s - c) = \underline{\hspace{2cm}}$
$\log \text{Numerator} = 3.1782$
$(-) \log s = \underline{\hspace{2cm}}$
$\log r^2 =$
$\log r = .8058$

$\log r =$
$(-) \log (s - a) = \underline{\hspace{2cm}}$
$\log \tan \dfrac{\alpha}{2} =$
$\dfrac{\alpha}{2} =$

$\log r =$
$(-) \log (s - b) = \underline{\hspace{2cm}}$
$\log \tan \dfrac{\beta}{2} =$
$\dfrac{\beta}{2} =$

$\log r =$
$(-) \log (s - c) = \underline{\hspace{2cm}}$
$\log \tan \dfrac{\gamma}{2} =$
$\dfrac{\gamma}{2} =$

(b) $a = 523$, $b = 576$, $c = 615$
(c) $a = .8147$, $b = .6834$, $c = .3449$
(d) $a = 4.32$, $b = 5.78$, $c = 13.44$

*11. Considering the triangle in Fig. 13-1, prove that its area is given by the formula
$$K = \tfrac{1}{2}bc \sin \alpha.$$
Rotating the letters will also give
$$K = \tfrac{1}{2}ac \sin \beta = \tfrac{1}{2}ab \sin \gamma.$$

*12. Since $\tfrac{1}{2} \sin \alpha = \sin \alpha/2 \cos \alpha/2$ (Problem 30, Article 5-3), using the values for $\sin \theta/2$ and $\cos \theta/2$, given in the discussion which follows Example 4, prove
$$K = \sqrt{s(s-a)(s-b)(s-c)}.$$
It is from this expression, and the result, $K = rs$, of Problem 30, Article 13-3, that we establish the value $r = \sqrt{\dfrac{(s-a)(s-b)(s-c)}{s}}$, where r is the radius of the inscribed circle.

13. Using the results of Problem 11, and the Law of Sines, show
$$K = \frac{a^2 \sin \beta \sin \gamma}{2 \sin \alpha} = \frac{b^2 \sin \gamma \sin \alpha}{2 \sin \beta} = \frac{c^2 \sin \alpha \sin \beta}{2 \sin \gamma}.$$

14. Find the areas of the triangles listed in previous problems designated by the instructor.

15. Find the lengths of the sides of a parallelogram if its diagonal, which is 72.83 inches long, makes angles with the sides of 27°52′ and 16°41′, respectively.

16. What angle does the slope of a hill make with the horizontal, if a tree 74.3 ft tall, growing on the slope of the hill, is found to subtend an angle of 19°30′ from a point 147 ft from the foot of the tree, measured along the slope straight down the hill?

17. Two airways cross each other at an angle of 49°. At a certain instant airplane A is 32 miles from the crossing, while B is 76 miles from the crossing. What is the distance between them at this instant? (Two solutions.)

18. Two sides of a parallelogram are 68 and 83 inches, and one of the diagonals is 42 inches. Find the angles of the parallelogram.

19. From a point in the same horizontal plane with the base of a building, the angles of elevation of the top and the bottom of a flagpole standing on top of the building are 64°40′ and 59°50′ respectively. If the building is 112 ft high, how tall is the flagpole?

20. As a train is traveling due north on a certain track, the engineer observes a column of smoke in a direction N 20°20′ E. After traveling 475 ft, he observes the same smoke in a direction S 71°40′ E. How far was the smoke from the first point of observation? From the second? How far was it away from the track?

21. A battleship is moving along the shoreline in a direction N 18°40′ E, at a constant rate of 36.5 mi/hr. If a squadron of airplanes which travels 186 mi/hr is due east of the battleship, in what direction should they fly in order to reach the battleship as quickly as possible?

*22. If **R** is the resultant of two forces $\mathbf{f}_1$ and $\mathbf{f}_2$, and R, f_1, and f_2 represent their respective magnitudes, explain the formula $R^2 = f_1^2 + f_2^2 + 2f_1f_2 \cos \theta$, where θ is the angle between the two forces. Recall Problem 22, Article 13-3. Find the magnitude and direction of the resultant of two forces, one of 75 lb acting due north, and the other of 93 lb acting N 63° W.

23. An airplane travels with a speed of 153 mi/hr in calm air. If the wind is blowing with a velocity of 27 mi/hr from S 21° W, and the plane is headed in a direction S 53° E, find the magnitude of the speed of the airplane and its direction with reference to the ground.

24. The resultant of two forces of 61.3 lb and 34.9 lb, respectively, is a force of 73.7 lb. What angle does the resultant make with each of the two forces?

CHAPTER 14

IDENTITIES AND RELATED SUBJECTS

In Article 1–2, we defined the two different types of equalities or equations mentioned in this book. We have discussed the conditional equation in some detail, primarily in Chapters 6 and 8. In this chapter we shall consider the identity, and in so doing, emphasize the distinction between it and the conditional equation. We have had occasion to use identities in several instances: in factoring (Article 1–8), in statements proved by mathematical induction (Chapter 11), and in establishing the relationships between the circular functions (Chapters 4 and 5).

14–1 The fundamental circular function identities. The *fundamental circular function identities* which are given in Chapter 4 as relationships have been defined or proved but are listed again as identities for emphasis.

The reciprocal relations,

$$\sin \theta \csc \theta \equiv 1, \qquad (14\text{–}1)$$

$$\cos \theta \sec \theta \equiv 1, \qquad (14\text{–}2)$$

$$\tan \theta \cot \theta \equiv 1; \qquad (14\text{–}3)$$

the tangent and cotangent relations,

$$\tan \theta \equiv \frac{\sin \theta}{\cos \theta}, \qquad (14\text{–}4)$$

$$\cot \theta \equiv \frac{\cos \theta}{\sin \theta}; \qquad (14\text{–}5)$$

and the Pythagorean relations,

$$\sin^2 \theta + \cos^2 \theta \equiv 1, \qquad (14\text{–}6)$$

$$\tan^2 \theta + 1 \equiv \sec^2 \theta, \qquad (14\text{–}7)$$

$$1 + \cot^2 \theta \equiv \csc^2 \theta, \qquad (14\text{–}8)$$

are called the eight *fundamental* identities in trigonometry.

With the use of these, any circular function may be expressed in terms of any other function. More specifically, for example, the identities (14–1), (14–2), (14–4), and (14–5) imply that any expression involving the circular functions of θ can be expressed in terms of $\sin \theta$ and $\cos \theta$ only. With these and (14–6) in the form $\cos \theta \equiv \pm \sqrt{1 - \sin^2 \theta}$, where the choice of the sign will depend on the quadrant of θ, any expression can be written completely in terms of $\sin \theta$.

It is, of course, possible to prove other circular function identities than the ones just given, although there is no general rule for a procedure.

Factoring and the addition of fractions or other algebraic simplifications are often advantageous, but the introduction of radicals should be avoided whenever possible. If in doubt, it may be helpful to express all the functions in terms of sines and cosines and simplify. The establishing of an identity is accomplished by reducing (1) the left member into the exact form of the right, (2) the right into the exact form of the left, or (3) each side separately into the same form. In the following examples, give the reasons for each step.

EXAMPLE 1. By reducing the left member to the form of the right, prove that $\dfrac{1}{\sin \theta} - \sin \theta \equiv \cot \theta \cos \theta$.

Solution.

$$\frac{1}{\sin \theta} - \sin \theta \equiv \frac{1 - \sin^2 \theta}{\sin \theta}$$

$$\equiv \frac{\cos^2 \theta}{\sin \theta}$$

$$\equiv \frac{\cos \theta}{\sin \theta} \cos \theta$$

$$\equiv \cot \theta \cos \theta.$$

EXAMPLE 2. By reducing each side separately to the same form, prove that $\tan \theta \sin \theta \equiv \sec \theta - \cos \theta$.

14-1] THE FUNDAMENTAL CIRCULAR FUNCTION IDENTITIES

Solution.

$$\tan \theta \sin \theta \qquad \sec \theta - \cos \theta$$

$$\frac{\sin \theta}{\cos \theta} \sin \theta \qquad \frac{1}{\cos \theta} - \cos \theta$$

$$\frac{\sin^2 \theta}{\cos \theta} \qquad \frac{1 - \cos^2 \theta}{\cos \theta}$$

$$\frac{\sin^2 \theta}{\cos \theta} \equiv \frac{\sin^2 \theta}{\cos \theta}.$$

EXAMPLE 3. Prove the identity $\dfrac{1 - \sin \theta}{\cos \theta} \equiv \dfrac{\cos \theta}{1 + \sin \theta}$.

Solution.

$$\frac{1 - \sin \theta}{\cos \theta} \equiv \frac{(1 - \sin \theta)(1 + \sin \theta)}{\cos \theta \, (1 + \sin \theta)}$$

$$\equiv \frac{1 - \sin^2 \theta}{\cos \theta \, (1 + \sin \theta)}$$

$$\equiv \frac{\cos^2 \theta}{\cos \theta \, (1 + \sin \theta)}$$

$$\equiv \frac{\cos \theta}{1 + \sin \theta}.$$

There are three different reasons for learning to prove identities such as those in the following set of problems. By working with and proving such relationships, one more easily masters the formulas and definitions of the circular functions. One also matures mathematically through the acquisition or review of certain algebraic manipulations. Most important, however, many such identities are often employed in more advanced mathematics.

PROBLEMS

By means of the fundamental identities, express each of the following in terms of $\sin \theta$ only:

1. $\csc \theta$
2. $\cos^2 \theta$
3. $\cos \theta$
4. $\sec \theta$
5. $\tan \theta$
6. $\cot \theta$

By means of the fundamental identities, express each of the following in terms of $\cos\theta$ only:

7. $\sec\theta$
8. $\sin^2\theta$
9. $\sin\theta$
10. $\csc\theta$
11. $\tan\theta$
12. $\cot\theta$

13. Express all six functions of θ in terms of $\tan\theta$. *Hint:* Use the identity $\sec^2\theta \equiv 1 + \tan^2\theta$.

14. Express $\dfrac{\sin\theta + \tan\theta}{\sec\theta + 1}$ in terms of $\sin\theta$ only.

15. Express $\dfrac{\tan\theta + \cot\theta}{\sec\theta \sin\theta}$ in terms of $\cos\theta$ only.

Prove the following identities:

16. $\tan\theta + \cot\theta \equiv \sec\theta \csc\theta$
17. $1 - 2\sin^2\theta \equiv 2\cos^2\theta - 1$
18. $\sin\theta \cos\theta \sec\theta \csc\theta \equiv 1$
19. $\dfrac{1}{1 + \sin\theta} + \dfrac{1}{1 - \sin\theta} \equiv 2\sec^2\theta$
20. $\cos\theta + \tan\theta \sin\theta \equiv \sec\theta$
21. $\cos^4\theta - \sin^4\theta \equiv \cos^2\theta - \sin^2\theta$
22. $\dfrac{\tan\theta - \cot\theta}{\tan\theta + \cot\theta} \equiv 2\sin^2\theta - 1$
23. $\dfrac{\sin\theta}{1 + \cos\theta} + \dfrac{1 + \cos\theta}{\sin\theta} \equiv 2\csc\theta$
24. $\dfrac{1 - \cos\theta}{\sin\theta} \equiv \dfrac{\sin\theta}{1 + \cos\theta}$
25. $\cos^2\theta - \sin^2\theta \equiv \dfrac{1 - \tan^2\theta}{1 + \tan^2\theta}$
26. $\cot\theta + \tan\theta \equiv \cot\theta \sec^2\theta$
27. $\dfrac{1 + \tan^2\theta}{\tan^2\theta} \equiv \csc^2\theta$
28. $(\csc\theta - \cot\theta)^2 \equiv \dfrac{1 - \cos\theta}{1 + \cos\theta}$
29. $(\sec\theta - \tan\theta)^2 \equiv \dfrac{1 - \sin\theta}{1 + \sin\theta}$
30. $(\cos\theta - \sin\theta)^2 + 2\sin\theta \cos\theta \equiv 1$
31. $\dfrac{\cot^2\theta - 1}{1 + \cot^2\theta} \equiv 2\cos^2\theta - 1$

32. $\dfrac{1 + \csc\theta}{\csc\theta - 1} \equiv \dfrac{1 + \sin\theta}{1 - \sin\theta}$

33. $\dfrac{\tan\theta}{1 - \cot\theta} + \dfrac{\cot\theta}{1 - \tan\theta} \equiv 1 + \tan\theta + \cot\theta$

34. $\dfrac{1 - \tan^2\theta}{1 + \tan^2\theta} \equiv 1 - 2\sin^2\theta$

35. $\dfrac{2\sin^2\theta - 1}{\sin\theta \cos\theta} \equiv \tan\theta - \cot\theta$

36. $\sec\theta \csc\theta - 2\cos\theta \csc\theta \equiv \tan\theta - \cot\theta$

37. $\dfrac{\cos\theta - \sin\theta}{\cos\theta + \sin\theta} \equiv \dfrac{\cot\theta - 1}{\cot\theta + 1}$

38. $\dfrac{\sin\theta}{\csc\theta - \cot\theta} \equiv 1 + \cos\theta$

39. $\dfrac{(\cos^2\theta - \sin^2\theta)^2}{\cos^4\theta - \sin^4\theta} \equiv 1 - 2\sin^2\theta$

40. $\dfrac{\sec^2\theta}{1 + \sin\theta} \equiv \dfrac{\sec^2\theta - \sec\theta \tan\theta}{\cos^2\theta}$

41. $1 + \cot\theta \equiv \dfrac{(1 - \cot^2\theta)\sin\theta}{\sin\theta - \cos\theta}$

42. $\dfrac{\sec\theta + \tan\theta}{\cos\theta - \tan\theta - \sec\theta} \equiv -\csc\theta$

43. $\sin\theta + \cos\theta + \dfrac{\sin\theta}{\cot\theta} \equiv \sec\theta + \csc\theta - \dfrac{\cos\theta}{\tan\theta}$

44. $\dfrac{\sin\theta \cos\theta + \cos\theta \sin\theta}{\cos\theta \cos\theta - \sin\theta \sin\theta} \equiv \dfrac{\tan\theta + \tan\theta}{1 - \tan\theta \tan\theta}$

14–2 General identities. We recall that not only the reduction formulas but the formulas for the functions of the sum or difference of two angles (Chapter 5) were proved for any angles and, thus, holding for all angles, were identities. The general addition identities should be reviewed, and are listed below:

$$\sin(\alpha \pm \beta) \equiv \sin\alpha \cos\beta \pm \cos\alpha \sin\beta, \qquad (14\text{–}9)$$

$$\cos(\alpha \pm \beta) \equiv \cos\alpha \cos\beta \mp \sin\alpha \sin\beta, \qquad (14\text{–}10)$$

$$\tan(\alpha \pm \beta) \equiv \dfrac{\tan\alpha \pm \tan\beta}{1 \mp \tan\alpha \tan\beta}. \qquad (14\text{–}11)$$

These can be used to prove other identities, several of which appear in the next list of problems. Also, they are used to obtain the important double- and half-angle identities. Since they are true for any α and β, letting $\alpha = \beta$, we immediately have

$$\sin 2\alpha \equiv \sin(\alpha + \alpha)$$
$$\equiv \sin \alpha \cos \alpha + \cos \alpha \sin \alpha,$$

or

$$\sin 2\alpha \equiv 2 \sin \alpha \cos \alpha. \qquad (14\text{--}12)$$

Also,

$$\cos(\alpha + \alpha) \equiv \cos \alpha \cos \alpha - \sin \alpha \sin \alpha,$$

or

$$\cos 2\alpha \equiv \cos^2 \alpha - \sin^2 \alpha \qquad (14\text{--}13)$$
$$\equiv 1 - 2 \sin^2 \alpha \quad (\text{Why?}) \qquad (14\text{--}14)$$
$$\equiv 2 \cos^2 \alpha - 1. \quad (\text{Why?}) \qquad (14\text{--}15)$$

Moreover,

$$\tan(\alpha + \alpha) \equiv \frac{\tan \alpha + \tan \alpha}{1 - \tan \alpha \tan \alpha},$$

or

$$\tan 2\alpha \equiv \frac{2 \tan \alpha}{1 - \tan^2 \alpha}. \qquad (14\text{--}16)$$

The half-angle identities are also readily established. By using (14–14) with $2\alpha = \theta$, or $\alpha = \theta/2$,

$$\cos \theta \equiv 1 - 2 \sin^2 \frac{\theta}{2}.$$

Solving for $\sin \theta/2$, we have,

$$2 \sin^2 \frac{\theta}{2} \equiv 1 - \cos \theta, \qquad (14\text{--}17)$$

so that

$$\sin \frac{\theta}{2} \equiv \pm \sqrt{\frac{1 - \cos \theta}{2}}, \qquad (14\text{--}18)$$

where the choice of the sign before the radical is determined by the quadrant in which $\theta/2$ lies. Similarly, using (14–15) with the same substitution $\alpha = \theta/2$,

$$\cos \theta \equiv 2 \cos^2 \frac{\theta}{2} - 1,$$

and solving for $\cos \theta/2$, we get

$$2 \cos^2 \frac{\theta}{2} \equiv 1 + \cos \theta, \qquad (14\text{–}19)$$

or

$$\cos \frac{\theta}{2} \equiv \pm \sqrt{\frac{1 + \cos \theta}{2}}, \qquad (14\text{–}20)$$

with the choice of the sign again depending on the location of $\theta/2$.

There are two identities for $\tan \theta/2$, obtained by using (14–17) and (14–19). In the identity

$$\tan \frac{\theta}{2} \equiv \frac{\sin \theta/2}{\cos \theta/2},$$

by multiplying the numerator and denominator of the right member by $2 \sin \theta/2$, we have

$$\tan \frac{\theta}{2} \equiv \frac{2 \sin^2 \theta/2}{2 \sin \theta/2 \cos \theta/2},$$

or

$$\tan \frac{\theta}{2} \equiv \frac{1 - \cos \theta}{\sin \theta}. \qquad (14\text{–}21)$$

In the same expression, multiplying by $2 \cos \theta/2$, we have

$$\tan \frac{\theta}{2} \equiv \frac{2 \sin \theta/2 \cos \theta/2}{2 \cos^2 \theta/2},$$

or

$$\tan \frac{\theta}{2} \equiv \frac{\sin \theta}{1 + \cos \theta}. \qquad (14\text{–}22)$$

Some of the uses of the identities of this article will be more clearly understood by considering the following examples.

EXAMPLE 1. Compute the value of sin $\pi/12$ and cos $\pi/12$ from the functions of $\pi/6$.

Solution. Using (14–18),

$$\sin \frac{\pi}{12} = \sin \frac{1}{2} \cdot \frac{\pi}{6} = \sqrt{\frac{1 - \cos \pi/6}{2}} = \sqrt{\frac{1 - \sqrt{3}/2}{2}} = \frac{\sqrt{2 - \sqrt{3}}}{2}.$$

Also, by (14–20),

$$\cos \frac{\pi}{12} = \cos \frac{1}{2} \cdot \frac{\pi}{6} = \sqrt{\frac{1 + \cos \pi/6}{2}} = \sqrt{\frac{1 + \sqrt{3}/2}{2}} = \frac{\sqrt{2 + \sqrt{3}}}{2}.$$

EXAMPLE 2. Express sin 2θ, cos 2θ, and tan 2θ in terms of x, if $x = \tan \theta$.

Solution. First let us find sin θ and cos θ in terms of x. Since

$$\sec \theta \equiv \pm \sqrt{1 + \tan^2 \theta} \equiv \pm \sqrt{1 + x^2},$$

$$\cos \theta \equiv \pm \frac{1}{\sqrt{1 + x^2}}.$$

Also,

$$\sin \theta \equiv \tan \theta \cos \theta \equiv \pm \frac{x}{\sqrt{1 + x^2}}.$$

Therefore,

$$\sin 2\theta \equiv 2 \sin \theta \cos \theta \equiv \frac{2x}{1 + x^2}.$$

A complete analysis is necessary to show that the sign in the above example is correct, by considering θ in the first, second, third, and fourth quadrants. Also

$$\cos 2\theta \equiv \cos^2 \theta - \sin^2 \theta \equiv \frac{1 - x^2}{1 + x^2},$$

and

$$\tan 2\theta \equiv \frac{2 \tan \theta}{1 - \tan^2 \theta} \equiv \frac{2x}{1 - x^2}.$$

EXAMPLE 3. Reduce $\sin^4 \theta$ to an expression involving only functions of θ raised to the first power.

Solution. By (14–17), we have

$$\sin^2 \theta \equiv \frac{1 - \cos 2\theta}{2}.$$

Thus we notice that by doubling the angle we have changed the exponent of the circular function from 2 to 1. Thus

$$\sin^4 \theta \equiv (\sin^2 \theta)^2 \equiv \frac{(1 - \cos 2\theta)^2}{4} \equiv \frac{1 - 2\cos 2\theta + \cos^2 2\theta}{4},$$

and replacing $\cos^2 2\theta$ by $\dfrac{1 + \cos 4\theta}{2}$ (why is this possible?), we obtain

$$\sin^4 \theta \equiv \frac{3 - 4\cos 2\theta + \cos 4\theta}{8}.$$

This type of transformation is extremely useful in calculus.

PROBLEMS

1. Verify the identities for $\sin 2\theta$ and $\tan 2\theta$ for the value $\theta = \pi/3$.

2. Use the double-angle identities to compute $\sin 4\pi/3$, $\cos 4\pi/3$, and $\tan 4\pi/3$ from the functions of $2\pi/3$.

3. Compute $\sin 7\pi/12$, $\cos 7\pi/12$, and $\tan 7\pi/12$ from the values of the functions of $7\pi/6$.

4. If $\sin \theta = 3/5$, and θ is in the first quadrant, find the exact value of
   (a) $\sin 2\theta$ (b) $\cos 2\theta$ (c) $\tan 2\theta$
   (d) $\sin \theta/2$ (e) $\cos \theta/2$ (f) $\tan \theta/2$

5. If $\cos \theta = -5/13$, and θ is in the second quadrant, find the exact value of
   (a) $\sin 2\theta$ (b) $\cos 2\theta$ (c) $\tan 2\theta$
   (d) $\sin \theta/2$ (e) $\cos \theta/2$ (f) $\tan \theta/2$

6. Reduce $\sin^2 \theta$ to an expression involving only circular functions of θ, raised to the first power.

7. Reduce $\cos^4 \theta$ to an expression involving only circular functions of θ, raised to the first power.

8. Derive identity (14–22) by using the identity $\sin \omega/2 \equiv \sin (\omega - \omega/2)$.
Hint: $\sin \omega/2 \equiv \sin \omega \cos \omega/2 - \cos \omega \sin \omega/2$ or $(1 + \cos \omega) \sin \omega/2 \equiv \sin \omega \cos \omega/2$.

9. Derive identity (14-21) by using the identity
$$\cos \frac{\omega}{2} \equiv \cos\left(\omega - \frac{\omega}{2}\right).$$

10. Derive the identity $\tan \dfrac{\omega}{2} \equiv \pm \sqrt{\dfrac{1 - \cos \omega}{1 + \cos \omega}}$.

11. Derive the identity $\tan \omega/2 \equiv \csc \omega - \cot \omega$.

Prove the following identities:

12. $\sin 3\theta \equiv 3 \sin \theta - 4 \sin^3 \theta$

13. $\cos 3\theta \equiv 4 \cos^3 \theta - 3 \cos \theta$

14. $\sin \dfrac{\theta}{2} \cos \dfrac{\theta}{2} \equiv \dfrac{\sin \theta}{2}$

15. $\dfrac{1 - \cos 2\theta}{\sin 2\theta} \equiv \tan \theta$

16. $\left(\cos \dfrac{\theta}{2} - \sin \dfrac{\theta}{2}\right)^2 \equiv 1 - \sin \theta$

17. $\csc 2\theta - \cot 2\theta \equiv \tan \theta$

18. $\csc 2\theta + \cot 2\theta \equiv \cot \theta$

19. $\tan 3\theta \equiv \dfrac{3 \tan \theta - \tan^3 \theta}{1 - 3 \tan^2 \theta}$

20. $\dfrac{\tan \theta/2 + \cot \theta/2}{\cot \theta/2 - \tan \theta/2} \equiv \sec \theta$

21. $\dfrac{\sin 2\theta}{\sin \theta} - \dfrac{\cos 2\theta}{\cos \theta} \equiv \sec \theta$

22. $\dfrac{\sin 3\theta}{\sin \theta} - \dfrac{\cos 3\theta}{\cos \theta} \equiv 2$

23. $\dfrac{\sin 3\theta}{\cos \theta} + \dfrac{\cos 3\theta}{\sin \theta} \equiv 2 \cot 2\theta$

24. $\dfrac{2 \tan \theta}{1 + \tan^2 \theta} \equiv \sin 2\theta$

25. $\dfrac{\cot^2 \theta - 1}{\csc^2 \theta} \equiv \cos 2\theta$

26. In Fig. 14-1, A is the mid-point of an arc of the unit circle subtended by a central angle θ. Using this figure, where $\theta < \pi$, and (14-18), show

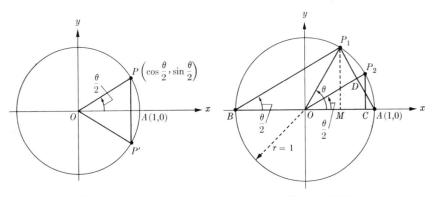

FIGURE 14-1   FIGURE 14-2

$$\sin\frac{\theta}{2} \equiv \sqrt{\frac{1-\cos\theta}{2}}.$$

27. In the unit circle in Fig. 14–2, P_2 is the mid-point of the arc $\widehat{AP_1}$, cut by the central angle θ. BP_1 is drawn parallel to OP_2, so that $\angle P_1BA = \angle P_2OA = \theta/2$. (Why?)

(a) Verify that $BP_1 \equiv 2\cos\theta/2$ and $P_1A = 2\sin\theta/2$. *Hint:* $CP_2 = AD = P_1A/2$, and $OC = OD = BP_1/2$.

(b) Verify from the figure that $\sin\theta = 2\sin\theta/2\cos\theta/2$.

(c) Verify from the figure that $\sin\dfrac{\theta}{2} = \sqrt{\dfrac{1-\cos\theta}{2}}$.

(d) Verify from the figure that $\cos\dfrac{\theta}{2} = \sqrt{\dfrac{1+\cos\theta}{2}}$.

Prove by the method of mathematical induction:

28. $\sin\theta + \sin 2\theta + \cdots + \sin n\theta \equiv \dfrac{\sin\frac{1}{2}(n+1)\theta \sin\frac{1}{2}n\theta}{\sin\dfrac{\theta}{2}}$.

29. $\cos\theta + \cos 2\theta + \cdots + \cos n\theta \equiv \dfrac{\cos\frac{1}{2}(n+1)\theta \sin\frac{1}{2}n\theta}{\sin\dfrac{\theta}{2}}$.

30. $\sin\theta + \sin 3\theta + \cdots + \sin(2n-1)\theta \equiv \dfrac{\sin^2 n\theta}{\sin\theta}$.

31. $\cos\theta + \cos 3\theta + \cdots + \cos(2n-1)\theta \equiv \dfrac{\sin 2n\theta}{2\sin\theta}$.

By using any necessary identities, solve the following equations for all values of θ. Recall Eqs. (9–3,4,5).

32. $2\sin 3\theta\cos 3\theta = -1$
33. $\sin 2\theta = \cos\theta$
34. $\cos 2\theta = \sin\theta$
35. $\sin\theta\cos 2\theta + \cos\theta\sin 2\theta = 1$
36. $\cos 3\theta\cos\theta + \sin 3\theta\sin\theta = -1$
37. $\sin(\theta + \frac{1}{6}\pi) = \cos(\theta + \frac{1}{6}\pi)$
38. $\tan(\theta + \frac{1}{4}\pi) - \tan(\theta - \frac{1}{4}\pi) = 4$
39. $3\sin\theta + 4\cos\theta = 5$. *Hint:* Recall Example 5, Article 5–3.
40. $\cos\theta - 2 = \sqrt{3}\sin\theta$
41. $12\cos\theta - 5\sin\theta = 13$
42. $\sin\theta + \cos\theta = 1$
43. $\sin\theta\cos\theta - 2\cos\theta - 3\sin\theta + 6 = 0$

44. Using the identity established in Problem 12, find sin 10°, correct to 3 significant figures, by the method of linear interpolation, discussed in Article 8–5.

45. Using the identity established in Problem 13, find cos 20°, correct to 3 significant figures.

46. With the exact value of sin 3°, or cos 3° given by Eqs. 5–20, 21, explain a procedure for approximating values of sin 1° or cos 1°. *Hint:* See Problem 44 or 45.

14–3 Conversions of sums and products. It is quite often necessary to convert a product of two circular functions into a sum of two functions, and vice versa. This is possible by employing the identities (14–9, 10). The following identities are not as important as the previous ones in this chapter. By adding the members of the two identities given by (14–9),

$$\sin(\alpha + \beta) + \sin(\alpha - \beta) \equiv 2\sin\alpha\cos\beta,$$

or

$$\sin\alpha\cos\beta \equiv \tfrac{1}{2}[\sin(\alpha + \beta) + \sin(\alpha - \beta)]. \quad (14\text{–}23)$$

By subtracting, we obtain

$$\cos\alpha\sin\beta \equiv \tfrac{1}{2}[\sin(\alpha + \beta) - \sin(\alpha - \beta)]. \quad (14\text{–}24)$$

Similarly, using (14–10), by adding and then subtracting, we obtain

$$\cos\alpha\cos\beta \equiv \tfrac{1}{2}[\cos(\alpha + \beta) + \cos(\alpha - \beta)], \quad (14\text{–}25)$$

$$\sin\alpha\sin\beta \equiv \tfrac{1}{2}[\cos(\alpha - \beta) - \cos(\alpha + \beta)]. \quad (14\text{–}26)$$

As an example, the product $\cos 6\theta \cos 3\theta$ may be represented as a sum. Using (14–25) with $\alpha = 6\theta$ and $\beta = 3\theta$, we have

$$\cos 6\theta \cos 3\theta \equiv \tfrac{1}{2}(\cos 9\theta + \cos 3\theta).$$

The same four identities can also be used for transforming a sum into a product. Such a transformation is useful when logarithms are to be used, for products are more readily calculated by logarithms than are sums. For convenience, we change the notation by letting $\alpha + \beta = \theta$, and $\alpha - \beta = \omega$. Solving these equations for α and β, we obtain $\alpha = (\theta + \omega)/2$, and $\beta = (\theta - \omega)/2$. Substituting these values in the above identities and simplifying, we obtain

$$\sin\theta + \sin\omega \equiv 2\sin\frac{\theta+\omega}{2}\cos\frac{\theta-\omega}{2}, \quad (14\text{-}27)$$

$$\sin\theta - \sin\omega \equiv 2\cos\frac{\theta+\omega}{2}\sin\frac{\theta-\omega}{2}, \quad (14\text{-}28)$$

$$\cos\theta + \cos\omega \equiv 2\cos\frac{\theta+\omega}{2}\cos\frac{\theta-\omega}{2}, \quad (14\text{-}29)$$

$$\cos\theta - \cos\omega \equiv -2\sin\frac{\theta+\omega}{2}\sin\frac{\theta-\omega}{2}. \quad (14\text{-}30)$$

EXAMPLE. Express $\sin\theta + \sin 3\theta + \sin 5\theta + \sin 7\theta$ as a product.

Solution. By grouping the first two and last two terms, and using (14-27),

$$\sin\theta + \sin 3\theta + \sin 5\theta + \sin 7\theta \equiv 2\sin\frac{\theta+3\theta}{2}\cos\frac{\theta-3\theta}{2} +$$
$$2\sin\frac{5\theta+7\theta}{2}\cos\frac{5\theta-7\theta}{2}$$
$$\equiv 2\sin 2\theta\cos(-\theta) + 2\sin 6\theta\cos(-\theta)$$
$$\equiv 2\cos\theta\,(\sin 2\theta + \sin 6\theta)\ (\text{Why?})$$
$$\equiv 2\cos\theta\,(2\sin 4\theta\cos 2\theta)\ (\text{Why?})$$
$$\equiv 4\cos\theta\cos 2\theta\sin 4\theta.$$

PROBLEMS

Express each of the following products as a sum.

1. $\sin 3\theta \cos 5\theta$
2. $4\cos\pi/3 \cos\pi/6$
3. $\cos 7\theta \sin 5\theta$
4. $6\sin 2\pi/3 \sin\pi/3$
5. $\sin 5\theta \cos 2\theta$
6. $\cos 2\theta \cos 6\theta$
7. $\cos\theta \sin\theta/2$
8. $2\sin 7\theta \sin 2\theta$

Express each of the following sums as products.

9. $\sin\pi/9 + \sin 2\pi/9$
10. $\cos 2\theta - \cos\theta$
11. $\cos 7\pi/9 + \cos 2\pi/9$
12. $\sin\pi/4 - \sin\pi/5$
13. $\sin 8\theta + \sin 4\theta$
14. $\cos 4\theta - \cos 8\theta$
15. $\sin 5\theta/3 - \sin 5\theta/6$
16. $\cos 6\theta + \cos 7\theta$

Prove the following identities:

17. $\dfrac{\sin 5\theta - \sin 3\theta}{\cos 5\theta + \cos 3\theta} \equiv \tan \theta$

18. $\dfrac{\cos 6\theta + \cos 4\theta}{\sin 6\theta - \sin 4\theta} \equiv \cot \theta$

19. $\dfrac{\sin 8\theta + \sin 2\theta}{\cos 8\theta + \cos 2\theta} \equiv \tan 5\theta$

20. $\dfrac{\sin 3\theta - \sin \theta}{\cos^2 \theta - \sin^2 \theta} \equiv 2 \sin \theta$

21. $\dfrac{\sin \alpha - \sin \beta}{\cos \alpha + \cos \beta} \equiv \tan \dfrac{\alpha - \beta}{2}$

22. $\dfrac{\sin \alpha - \sin \beta}{\sin \alpha + \sin \beta} \equiv \dfrac{\tan \frac{1}{2}(\alpha - \beta)}{\tan \frac{1}{2}(\alpha + \beta)}$

23. $\dfrac{\sin \theta + \sin 3\theta + \sin 5\theta}{\cos \theta + \cos 3\theta + \cos 5\theta} \equiv \tan 3\theta$

24. $\sin\left(\dfrac{\pi}{4} - \theta\right) \sin\left(\dfrac{\pi}{4} + \theta\right) \equiv \tfrac{1}{2} \cos 2\theta$

14–4 Absolute inequalities. In Article 6–4, both conditional and absolute inequalities were defined. Having considered conditional inequalities in that article, we are now prepared to discuss that type of inequality associated with identities, namely, absolute inequalities. Such inequalities also satisfy the properties stated in Article 6–4.

Since we must start with a valid inequality in any proof, it is often convenient to tentatively assume the proposition we are proving, reduce it to a simpler inequality, which is known to be valid, and then the actual proof will consist of retracing the steps. This method will be illustrated.

EXAMPLE 1. Prove that the sum of any positive number and its reciprocal is greater than or equal to 2.

Solution. Letting x be any positive number, we wish to prove

$$x + \frac{1}{x} \geq 2. \tag{14-31}$$

Tentatively assuming this to be true, and multiplying both members by x,

$$x^2 + 1 \geq 2x. \tag{14-32}$$

By subtracting $2x$ from both members, we have

$$x^2 - 2x + 1 \geq 0, \tag{14-33}$$

which is known to be true, since $x^2 - 2x + 1 \equiv (x - 1)^2$. Our proof therefore, starts with inequality (14–33). Since

$$x^2 - 2x + 1 \geq 0,$$

for all positive x,

$$x^2 + 1 \geq 2x$$

and dividing by x, we have our result,

$$x + \frac{1}{x} \geq 2.$$

We may also be interested in proving that certain absolute inequalities hold, when they involve the circular functions.

EXAMPLE 2. Prove the absolute inequality $3 \sin \theta + 4 \cos \theta \leq 5$.

Solution. Recalling Example 5, Article 5–3, we have

$$3 \sin \theta + 4 \cos \theta = 5 \sin (\theta + \theta_1) \leq 5,$$

since $\sin \alpha \leq 1$.

PROBLEMS

Prove the truth of each of the following inequalities. The letters represent unequal positive numbers, while θ represents any angle.

1. $a^2 + b^2 > 2ab$
2. $\dfrac{a}{b} + \dfrac{b}{a} > 2$
3. $\dfrac{x + y}{2} > \sqrt{xy}$
4. $\dfrac{a + b}{2} > \dfrac{2ab}{a + b}$
5. $\dfrac{x^2}{y} + \dfrac{y^2}{x} > x + y$
6. $x^3 + y^3 > x^2y + xy^2$
7. $x^2 + y^2 + z^2 < (x + y + z)^2$
8. $a^2 + 2a + 2 > 0$
9. $|a + b| \leq |a| + |b|$
10. $|a - b| \geq |a| - |b|$
11. $\sin \theta + \csc \theta \geq 2$
12. $\cos \theta + \sec \theta \geq 2$
13. $|\cos^4 \theta - \sin^4 \theta| \leq 1$
14. $|\sin (\alpha + \beta) \cos (\alpha - \beta)| \leq 1$
15. $\sin \theta + \cos \theta \leq \sqrt{2}$
16. $\sin^2 \theta + \cot^2 \theta \geq 2 \cos \theta$
17. $12 \sin \theta - 5 \cos \theta \leq 13$
18. $4 \sin \theta + 3 \cos \theta \leq 6$

14–5 Approximations of functional values for small angles. One of the uses of inequalities can best be illustrated by finding the approximate value of the circular functions for small values of θ. By recalling the definition for $\sin \theta$ and $\tan \theta$, and restricting θ to a small positive value, we would expect the values of $\sin \theta$, $\tan \theta$, and θ to be approximately the same. This is actually the case, for consider Fig. 14–3 involving a unit circle with θ in standard position. Let A and P be the points where the initial and terminal sides of θ intersect the circle. The line from P is drawn perpendicular to the x-axis, intersecting the circle at P'. The x-axis bisects the vertical chord $P'P$ at C and the arc $\overparen{P'P}$ at A. The tangent to the circle at A is drawn, intersecting the two lines OP and OP' (extended) at R and R' respectively. The tangents to the circle at P and P' are drawn intersecting at D, a point on the x-axis. (Why?) Consider the chord $P'CP$, the arc $\overparen{P'AP}$, and the segments of the tangents to the circle, $P'D$ and DP. From a proposition in plane geometry,

$$P'CP < \overparen{P'AP} < P'DP.$$

Dividing these lengths by 2,

$$CP < \overparen{AP} < DP.$$

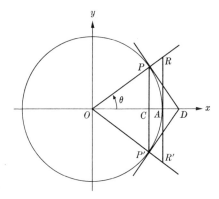

Figure 14–3

But $DP = AR$. (Why?) Therefore,

$$CP < \widehat{AP} < AR,$$

or

$$\sin \theta < \theta < \tan \theta. \qquad (14\text{--}34)$$

This inequality will enable us to find an approximation to the values of the sine or tangent of small values of θ.

By dividing each member of (14–34) by $\sin \theta$, we have

$$1 < \frac{\theta}{\sin \theta} < \sec \theta.$$

Considering the reciprocals, we get

$$\cos \theta < \frac{\sin \theta}{\theta} < 1, \qquad (14\text{--}35)^*$$

and since $\cos \theta = \sqrt{1 - \sin^2 \theta} > \sqrt{1 - \theta^2}$ (Why?), we have

$$\sqrt{1 - \theta^2} < \frac{\sin \theta}{\theta} < 1,$$

or

$$\theta\sqrt{1 - \theta^2} < \sin \theta < \theta. \qquad (14\text{--}36)$$

This inequality can be used in connection with the sine function. Again, by dividing each member of (14–34) by $\tan \theta$, we have

$$\cos \theta < \frac{\theta}{\tan \theta} < 1, \qquad (14\text{--}37)$$

or

$$\sqrt{1 - \theta^2} < \frac{\theta}{\tan \theta} < 1.$$

---

* Consider inequality (14–35). As θ becomes smaller and smaller (approaches 0), $\cos \theta$ approaches 1. Thus $\sin \theta/\theta$, as θ approaches 0, is always between 1 and a number approaching 1. Therefore, the ratio must approach 1. This is written $\lim_{\theta \to 0} \sin \theta/\theta = 1$, and is read: the limit of $\sin \theta/\theta$, as θ approaches zero, is one. This is an extremely important result in more advanced mathematics.

Using the reciprocal relation, this becomes

$$1 < \frac{\tan \theta}{\theta} < \frac{1}{\sqrt{1 - \theta^2}}, \qquad (14\text{--}38)$$

or

$$\theta < \tan \theta < \frac{\theta}{\sqrt{1 - \theta^2}}. \qquad (14\text{--}39)$$

Relations (14–36) and (14–39) give us bounds for $\sin \theta$ and $\tan \theta$.

EXAMPLE. Find an approximate value of $\sin \pi/60$.

Solution. Substituting in (14–36) for $\theta = \pi/60$, where $\pi = 3.1416$, we find

$$.05229 < \sin \pi/60 < .05236.$$

Thus, correct to four decimal places, $\sin \pi/60 = .0523$. Notice the value listed in Table I.

It is also possible to find the value of $\cos \theta$, for small positive values of θ. Since $\cos \theta \equiv 1 - 2 \sin^2(\theta/2)$, we have

$$\cos \theta > 1 - 2\frac{\theta^2}{4},$$

and thus,

$$1 > \cos \theta > 1 - \frac{\theta^2}{2}. \qquad (14\text{--}40)$$

PROBLEMS

1. Using $\pi = 3.1416$, find the best possible approximation for $\sin \pi/30$, $\cos \pi/30$, and $\tan \pi/30$, using the above inequalities.

2. Find approximations for $\cos \pi/60$ and $\tan \pi/60$. Compare these values with those given in Table I.

3. Using formula (14–38), show that $\lim\limits_{\theta \to 0} \tan \theta/\theta = 1$.

4. Using (14–35) or (14–37), prove the inequality

$$\cos \theta = \frac{\sin \theta}{\tan \theta} < 1.$$

5. Using the result of Problem 4, prove $\lim\limits_{\theta \to 0} \dfrac{\sin \theta}{\tan \theta} = 1$.

6. With the use of (14–35) and (14–40), obtain the bounds for $\sin \theta$:

$$\theta(1 - \tfrac{1}{2}\theta^2) < \sin \theta < \theta.$$

This inequality does not involve a square root, and thus has advantages in certain cases.

7. Recall from plane geometry the expression for the perimeter P of a regular n-gon inscribed in a circle of radius r, $P = 2nr \sin(\pi/n)$, and $\lim\limits_{n \to \infty} P = 2\pi r$ (the limit of the perimeter of a regular n-gon as the number of sides increases indefinitely is the circumference, $2\pi r$). Set up a limit by eliminating P from these two expressions and, by substituting $\theta = \pi/n$, prove $\lim\limits_{\theta \to 0} \sin \theta/\theta = 1$.

CHAPTER 15

COMPLEX NUMBERS

The possibility of extensions or generalizations of the real number system was mentioned in Chapter 1. One of these is a most useful algebra of all ordered pairs of real numbers, which we shall discuss briefly. The most common interpretation of this algebra will then be considered in some detail.

15–1 Algebra of ordered pairs. With the set of all ordered pairs of numbers (x,y) as elements, where x and y are real numbers, we can set up an algebra. We shall define two ordered pairs (x_1,y_1) and (x_2,y_2) to be *equal** if and only if $x_1 = x_2$ and $y_1 = y_2$. Thus, $(x + 2y, 2x - y) = (4,3)$ if and only if $x + 2y = 4$ and $2x - y = 3$ or, specifically, $x = 2$ and $y = 1$.

We next define the fundamental operations of addition, subtraction, multiplication, and division. The *sum* of two ordered pairs (x_1,y_1) and (x_2,y_2) is here defined as the ordered pair $(x_1 + x_2, y_1 + y_2)$ and is written

$$(x_1,y_1) + (x_2,y_2) = (x_1 + x_2, y_1 + y_2). \tag{15-1}$$

As in the case of real numbers, where the solution for x of the equation $a + x = b$, written $b - a$, is called the difference, the solution for (x,y) in the equation

$$(x_1,y_1) + (x,y) = (x_2,y_2) \dagger \tag{15-2}$$

is called the *difference* of the two ordered pairs (x_2,y_2) and (x_1,y_1). Since (15–2) implies

$$x_1 + x = x_2, \qquad y_1 + y = y_2,$$

or

$$x = x_2 - x_1, \qquad y = y_2 - y_1,$$

---

* These definitions of equality, addition, and multiplication are by no means the only possible ones for ordered pairs, but have been chosen specifically for the introduction of the complex number system.

† Equations (15–2) and (15–5) have unique solutions for (x,y).

the difference (x,y) is written as the ordered pair, $(x_2 - x_1, y_2 - y_1)$, or

$$(x_2,y_2) - (x_1,y_1) = (x_2 - x_1, y_2 - y_1). \tag{15-3}$$

The definition of multiplication is not as obvious as that of addition. We define the *product* of two ordered pairs by the equation

$$(x_1,y_1) \cdot (x_2,y_2) = (x_1x_2 - y_1y_2, x_1y_2 + x_2y_1). \tag{15-4}$$

For example,

$$(2,1) \cdot (3,2) = (2 \cdot 3 - 1 \cdot 2, 2 \cdot 2 + 3 \cdot 1) = (4,7).$$

Again, as in the case of real numbers, where division is defined in terms of multiplication, the definition for the quotient of two ordered pairs follows the same pattern. Since the real number $x = a/b$ is defined by the equation $bx = a$, we define as the *quotient* of two ordered pairs, the ordered pair, (x,y), which is the solution of the equation

$$(x_2,y_2) \cdot (x,y) = (x_1,y_1);^* \qquad (x_2,y_2) \neq (0,0). \tag{15-5}$$

Using (15–4), this becomes

$$(xx_2 - yy_2, xy_2 + x_2y) = (x_1,y_1),$$

and, therefore,

$$x_2x - y_2y = x_1, \qquad y_2x + x_2y = y_1.$$

Solving for x and y, we get the solution

$$x = \frac{x_1x_2 + y_1y_2}{x_2^2 + y_2^2},$$

$$y = \frac{x_2y_1 - x_1y_2}{x_2^2 + y_2^2}.$$

Thus the quotient of two ordered pairs may be written

$$\frac{(x_1,y_1)}{(x_2,y_2)} = \left(\frac{x_1x_2 + y_1y_2}{x_2^2 + y_2^2}, \frac{x_2y_1 - x_1y_2}{x_2^2 + y_2^2}\right). \tag{15-6}$$

---

* Equations (15–2) and (15–5) have unique solutions for (x,y).

For example,

$$\frac{(4,8)}{(3,1)} = (2,2).$$

It can easily be verified that the elements of this algebra of ordered pairs satisfy the first five properties stated in Article 1–3.

15–2 Complex numbers. One of the most common interpretations of the algebra of ordered pairs, discussed in Article 15–1, is the algebra of complex numbers. The symbol $x + yi$, where x and y are real numbers and i has the property that $i^2 = -1$, is called a *complex number*. If $y = 0$, of course the complex number is real; if $y \neq 0$, the complex number is said to be *imaginary*. With $y \neq 0$, and $x = 0$, $x + yi$ is a *pure imaginary*.

We notice immediately, with the numbers x and y obeying the ordinary laws for real numbers, and the new element i obeying the law

$$i^2 = -1,$$

that the complex numbers $x + yi$ may be thought of as an abbreviation for the ordered pairs, since $x + yi$ satisfies the same operations as the ordered pairs (x,y). Specifically, in considering the sum of two complex numbers,

$$(x_1 + y_1 i) + (x_2 + y_2 i) = (x_1 + x_2) + (y_1 + y_2)i. \quad (15\text{–}7)$$

With regard to the product of two complex numbers,

$$(x_1 + y_1 i) \cdot (x_2 + y_2 i) = x_1 x_2 + x_1 y_2 i + x_2 y_1 i + y_1 y_2 i^2$$
$$= (x_1 x_2 - y_1 y_2) + (x_1 y_2 + x_2 y_1)i. \quad (15\text{–}8)$$

We see in this equation the real reason for defining the product of two ordered pairs as given by (15–4). The quotient of two complex numbers is given by

$$\frac{x_1 + y_1 i}{x_2 + y_2 i} = \frac{(x_1 + y_1 i)(x_2 - y_2 i)}{(x_2 + y_2 i)(x_2 - y_2 i)}$$
$$= \frac{x_1 x_2 - x_1 y_2 i + x_2 y_1 i - y_1 y_2 i^2}{x_2^2 - x_2 y_2 i + x_2 y_2 i - y_2^2 i^2}$$
$$= \frac{x_1 x_2 + y_1 y_2}{x_2^2 + y_2^2} + \frac{x_2 y_1 - x_1 y_2}{x_2^2 + y_2^2} i. \quad (15\text{–}9)$$

Thus, the ordered pair (x,y) may be considered as representing the complex number $x + yi$.

Two complex numbers which differ only in the sign of their imaginary parts are called *conjugates* of each other. Thus, $3 + 2i$ and $3 - 2i$ or $5i$ and $-5i$ are conjugate complex numbers. In general, $x + yi$ and $x - yi$ are numbers of this type. Note Problems 24 and 25.

It is unfortunate that the word imaginary has been applied to these numbers. Although they were originally introduced to solve quadratic equations, they have been extremely useful in physics and engineering, especially in the description of certain electrical phenomena. In such situations, the "imaginary numbers" have significance which is quite as real as that of the "real" numbers.

A few quadratic equations in Chapter 6 and other equations in Chapter 8 had imaginary numbers as solutions. These were listed as square roots of negative numbers with no additional explanation at that time. Graphically, we noticed that the functions involved had real zeros only where the corresponding curves crossed or touched the horizontal axis. We are now able to write such results in terms of the imaginary symbol i.

EXAMPLE 1. Solve the equation $x^2 - 4x + 6 = 0$.

Solution. Using the quadratic formula [Eq. (6–7)], with $a = 1$, $b = -4$, and $c = 6$,

$$x = \frac{-(-4) \pm \sqrt{(-4)^2 - 4(1)(6)}}{2(1)}$$

$$= \frac{4 \pm \sqrt{16 - 24}}{2} = \frac{4 \pm \sqrt{-8}}{2}$$

$$= \frac{4 \pm 2\sqrt{-2}}{2} = 2 \pm \sqrt{2}i.$$

PROBLEMS

In this algebra defined for ordered pairs, show

1. $(x,y) \cdot (0,0) = (0,0)$
2. $(x,y) + (0,0) = (x,y)$
3. $(x,y) \cdot (1,0) = (x,y)$

4. $(x,y) = (y,x)$ if and only if $x = y$
5. $(0,1)^2 = (0,1) \cdot (0,1) = (-1,0)$
6. What ordered pair in this algebra of ordered pairs takes the place of zero in our ordinary number system? What ordered pair takes the place of one?

Find the value of (x,y) in Problems 7 through 15.

7. $(x,y) = (2,3) + (4,5)$
8. $(x,y) = (-2,1) + (3,-7)$
9. $(x,y) = (3,-1) - (4,-2)$
10. $(3,1) = (x,y) + (5,-1)$
11. $(x,y) = (3,1) \cdot (2,3)$
12. $(x,y) = (-1,2) \cdot (3,-5)$
13. $(x,y) = (2,-1)/(-1,3)$
14. $(x,y) = (23,11)/(5,-1)$
15. $(x,y) = \left(-\dfrac{1}{2}, \dfrac{\sqrt{3}}{2}\right)^3$

In Problems 16 through 19, give the expression as a single complex number.

16. (a) $(2 + 5i) + (4 - i)$, (b) $(2 + 5i) - (4 - i)$
17. (a) $(2 + 5i)(4 - i)$, (b) $(2 + 5i)(4 + i)$
18. (a) $\dfrac{2 + 5i}{4 - i}$, (b) $\dfrac{2 + 5i}{4 + i}$
19. (a) $i^3 = i \cdot i^2$ (b) $i^4 = i^2 \cdot i^2$ (c) i^5 (d) $1/i$
20. Prove, for any positive integer n,
   (a) $i^{4n} = 1$ (b) $i^{4n+1} = i$ (c) $i^{4n+2} = -1$ (d) $i^{4n+3} = -i$
   (e) $i^{n+4} = i^n$

In the following problems, use the ordinary properties of the real numbers, and the fact that $i^2 = -1$.

21. State Problems 7, 11 and 13, interpreting the ordered pairs as complex numbers, and solve.

22. If the complex number $x + yi = 0$, show algebraically that $x = 0$ and $y = 0$. *Hint:* Since $x = -yi$, $x^2 = -y^2$. What can be said about x and y?

23. If $x_1 + y_1 i$ and $x_2 + y_2 i$ are two complex numbers such that $x_1 + y_1 i = x_2 + y_2 i$, show by using algebra that $x_1 = x_2$ and $y_1 = y_2$. *Hint:* By transposing, $x_1 - x_2 + (y_1 - y_2)i = 0$. Then use Problem 22.

24. Prove that the sum and product of two conjugate complex numbers are both real.

25. Prove that if the sum and product of two imaginary numbers are real, the numbers are conjugate complex numbers.

15–3 Complex roots of an equation. In commenting on the complex roots of any equation, as at the end of the last article, we must state and prove a fundamental theorem.

THEOREM. *If the complex number $a + bi$, $b \neq 0$, is a root of the polynomial equation*

$$f(x) = a_0 x^n + a_1 x^{n-1} + a_2 x^{n-2} + \cdots + a_{n-1} x + a_n = 0, \quad (15\text{-}10)$$

where $a_0 \neq 0$, n is a positive integer, and a_i are real constants, then its conjugate $a - bi$ is also a root.

Proof. Since $a + bi$ is a root of $f(x) = 0$, $a + bi$ is a factor of $f(x)$. Let us divide $f(x)$ by the product $[x - (a + bi)][x - (a - bi)] \equiv x^2 - 2ax + a^2 + b^2$ until the remainder is of degree less than $x^2 - 2ax + a^2 + b^2$. The remainder will therefore be of the first degree at most. Symbolically,

$$\frac{f(x)}{x^2 - 2ax + a^2 + b^2} \equiv q(x) + \frac{Rx + S}{x^2 - 2ax + a^2 + b^2} \quad (15\text{-}11)$$

or

$$f(x) \equiv [x^2 - 2ax + a^2 + b^2] q(x) + Rx + S,$$

where R and S are real constants. Since this identity is true for all x, it is true for $x = a + bi$. Thus

$$f(a + bi) = 0 \cdot q(a + bi) + R(a + bi) + S = 0.$$

Consequently, by the property stated in Problem 22,

$$Ra + S = 0 \quad \text{and} \quad Rbi = 0.$$

Since $b \neq 0$, $R = 0$, and therefore $S = 0$, which shows that the division in (15–11) was exact, so that $a - bi$ is also a factor of $f(x)$, that is, $a - bi$ is a root of $f(x) = 0$.

As a result of this theorem, we have several remarks concerning the roots of Eq. (15–10).

1. The imaginary roots of such an equation occur in pairs.
2. Any such equation has an even number of imaginary roots.
3. If the degree of such an equation is odd, the equation has at least one real root.

Problems

Find the third degree equation with integral coefficients having the given numbers as roots.

1. $3, 2 - i$
2. $-4, 6i$
3. $1/2, -5 + i$
4. $3/2, i - 4$
5. $-1/3, 3 + \sqrt{2}i$
6. $2/3, (-3 + \sqrt{5}i)/2$

Solve the following equations given the one root in parentheses. *Hint:* Use division.

7. $x^3 - 4x^2 + 9x - 36 = 0$, $(3i)$
8. $2x^3 + 9x^2 + 14x + 5 = 0$, $(-2 + i)$
9. $x^3 - 8x^2 + 23x - 22 = 0$, $(3 - \sqrt{2}i)$
10. $2x^3 - 8x^2 + 11x - 5$, $\frac{1}{2}(3 + i)$

15–4 Graphical representation of complex numbers. The interpretation of the ordered pair (x,y) as the complex number $x + yi$ lends itself to a simple graphical representation of the complex numbers. Since (x,y) may be plotted as a point in the rectangular coordinate system, every complex number $x + yi$ may be associated with some point in the plane, and every point in the plane with some complex number. This plane is called the *complex plane*, and the figure on which the complex numbers are plotted is called the *Argand diagram*.* The real numbers lie on the x-axis, the pure imaginaries on the y-axis, and a number such as $5 - 2i$ is represented by the point $(5, -2)$.

Any point $P(x,y)$ in the plane, other than $(0,0)$, lies on some circle with center at $(0,0)$, whose radius is r, where $r = \sqrt{x^2 + y^2}$. Let θ be the angle in standard position having OP as its terminal side (see Fig. 15–1), so that

$$x = r \cos \theta, \qquad y = r \sin \theta. \qquad (15\text{–}12)$$

Thus the complex number, $x + yi$, may be written in *trigonometric form*,

$$x + yi = r(\cos \theta + i \sin \theta).† \qquad (15\text{–}13)$$

---

* The system of representing complex numbers graphically was discovered independently by Wessel (Norwegian), Argand (French), and Gauss (German) about 1800.

† This form, $r(\cos + i \sin \theta)$, is sometimes abbreviated $r \operatorname{cis} \theta$.

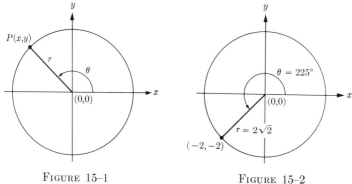

FIGURE 15-1   FIGURE 15-2

Since $\sin \theta$ and $\cos \theta$ are both periodic with period $2\pi = 360°$, for $r > 0$ and any integer k,

$$r[\cos(\theta + k360°) + i\sin(\theta + k360°)]$$

is also a trigonometric form for $x + yi$. We note that r must be greater than zero. This value of

$$r = \sqrt{x^2 + y^2} \qquad (15\text{-}14)$$

is called the *absolute value* or *modulus* of $x + yi$. The angle θ is called the *amplitude* or *argument* of $x + yi$.

EXAMPLE 1. Express the complex number $-2 - 2i$ in trigonometric form.

Solution. By locating the point $(-2, -2)$ which corresponds to $-2 - 2i$ (see Fig. 15-2), we have

$$r = \sqrt{(-2)^2 + (-2)^2} = 2\sqrt{2}$$

and $\tan \theta = 1$, with θ terminating in the third quadrant. Thus,

$$\theta = 225°,$$

and $$-2 - 2i = 2\sqrt{2}\,(\cos 225° + i \sin 225°).$$

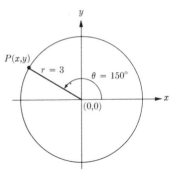

FIGURE 15-3

EXAMPLE 2. Express the complex number $3(\cos 150° + i \sin 150°)$ in the form $x + yi$.

Solution. On the terminal side of the angle of 150° in standard position, locate the point P, 3 units from the origin, as in Fig. 15-3. Since $P(x,y)$ represents the complex number, we have

$$x = 3 \cos 150° = -3\sqrt{3}/2,$$

and

$$y = 3 \sin 150° = 3/2.$$

Hence,

$$3(\cos 150° + i \sin 150°) = -\frac{3\sqrt{3}}{2} + \frac{3}{2}i.$$

One of the advantages of the trigonometric form for complex numbers is its usefulness in obtaining products. By letting $r_1(\cos \theta_1 + i \sin \theta_1)$ and $r_2(\cos \theta_2 + i \sin \theta_2)$ be two complex numbers, we find

$$r_1(\cos \theta_1 + i \sin \theta_1) \cdot r_2(\cos \theta_2 + i \sin \theta_2)$$
$$= r_1 r_2 (\cos \theta_1 \cos \theta_2 + i \cos \theta_1 \sin \theta_2 + i \sin \theta_1 \cos \theta_2 + i^2 \sin \theta_1 \sin \theta_2)$$
$$= r_1 r_2 [(\cos \theta_1 \cos \theta_2 - \sin \theta_1 \sin \theta_2) + i (\sin \theta_1 \cos \theta_2 + \cos \theta_1 \sin \theta_2)]$$
$$= r_1 r_2 [\cos (\theta_1 + \theta_2) + i \sin (\theta_1 + \theta_2)]. \qquad (15\text{-}15)$$

Thus, the modulus of the product of two complex numbers is the product of the moduli, and the amplitude is the sum of the amplitudes.

Problems

1. Locate the point representing graphically each of the following complex numbers. Give the trigonometric form for each, using the least positive or zero value of its amplitude.

(a) 2 (b) -2
(c) $3i$ (d) $-i$
(e) $2 - 2i$ (f) $-2 + 2i$
(g) $-\dfrac{1}{2} + \dfrac{\sqrt{3}}{2}i$ (h) $-\dfrac{1}{2} - \dfrac{\sqrt{3}}{2}i$

2. Express each of the following in the form $x + yi$.

(a) $3(\cos 0° + i \sin 0°)$ (b) $2(\cos 90° + i \sin 90°)$
(c) $\cos 180° + i \sin 180°$ (d) $2(\cos 225° + i \sin 225°)$
(e) $2(\cos 270° + i \sin 270°)$ (f) $8(\cos 135° + i \sin 135°)$
(g) $4(\cos 300° + i \sin 300°)$ (h) $6(\cos 150° + i \sin 150°)$

3. Perform the multiplications, expressing the final result in the form $x + yi$. Check your result by expressing each of the given numbers in the form $x + yi$, and then performing the multiplication algebraically.

(a) $3(\cos 60° + i \sin 60°) \cdot 2(\cos 30° + i \sin 30°)$
(b) $4(\cos 120° + i \sin 120°) \cdot 2(\cos 90° + i \sin 90°)$
(c) $3(\cos 135° + i \sin 135°) \cdot 4(\cos (-45°) + i \sin (-45°))$
(d) $[2(\cos 120° + i \sin 120°)]^3$

*4. Prove that the quotient of the two complex numbers $r_1(\cos \theta_1 + i \sin \theta_1)$ and $r_2(\cos \theta_2 + i \sin \theta_2)$ is given by

$$\frac{r_1}{r_2}[\cos (\theta_1 - \theta_2) + i \sin (\theta_1 - \theta_2)]. \qquad (15\text{-}16)$$

5. Use (15-16) to perform the following divisions. Check as in Problem 3.

(a) $4(\cos 60° + i \sin 60°) \div 2(\cos 30° + i \sin 30°)$
(b) $6(\cos 0° + i \sin 0°) \div 3(\cos 240° + i \sin 240°)$

6. Prove that the reciprocal of $r(\cos \theta + i \sin \theta)$ is

$$\frac{1}{r}(\cos \theta - i \sin \theta).$$

7. Prove that $[r(\cos \theta + i \sin \theta)]^2 = r^2(\cos 2\theta + i \sin 2\theta)$.

8. In considering formula (15-12), we recall a similar formula, (13-5), given in Chapter 13. These are indeed the same, if in the ordered pairs considered in this chapter the x and y are thought of as representing the x- and y-components of a vector from the origin to the point (x,y). More specifically, this algebra of ordered pairs may be interpreted as a study of vectors in

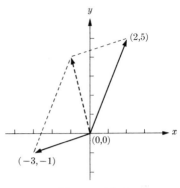

FIGURE 15-4

the plane. By plotting the following number pairs with their sum and drawing the appropriate vectors, show that the parallelogram law for the sum of two vectors is satisfied (see last paragraph in Article 13-3).

(a) $(2,5) + (-3,-1) =$ (b) $(-1,2) + (4,-5) =$

Hint: See Fig.15-4.

(c) $(3,-2) + (-2,3) =$ (d) $(-3,-2) + (6,4) =$

15-5 Powers and roots of complex numbers. In Article 15-4, we obtained an expression for the product of two complex numbers in trigonometric form. By using (15-15), we immediately have

$$[r(\cos\theta + i\sin\theta)]^2 = r^2(\cos 2\theta + i\sin 2\theta),$$

or

$$[r(\cos\theta + i\sin\theta)]^3 = r^3(\cos 3\theta + i\sin 3\theta).$$

In fact, we can prove by mathematical induction* a general theorem.

THEOREM. *If n is any positive integer,*

$$[r(\cos\theta + i\sin\theta)]^n = r^n(\cos n\theta + i\sin n\theta). \qquad (15\text{-}17)$$

This theorem holds for rational, irrational and even complex values of the exponent, but we shall make use of it only with integral values.

Proof. Part (a). *Verification.* This is done at the beginning of this article.

---

* This result, known as De Moivre's Theorem, was discovered by Abraham De Moivre (1667-1754).

15-5] POWERS AND ROOTS OF COMPLEX NUMBERS

Part (b). Assuming
$$[r(\cos\theta + i\sin\theta)]^k = r^k(\cos k\theta + i\sin k\theta), \quad (15\text{-}18)$$
we must show
$$[r(\cos\theta + i\sin\theta)]^{k+1} = r^{k+1}[\cos(k+1)\theta + i\sin(k+1)\theta]. \quad (15\text{-}19)$$
Multiplying each member of (15-18) by $r(\cos\theta + i\sin\theta)$, we have
$$[r(\cos\theta + i\sin\theta)]^{k+1} = r^k(\cos k\theta + i\sin k\theta)r(\cos\theta + i\sin\theta),$$
and by Eq. (15-15),
$$= r^{k+1}[\cos(k\theta + \theta) + i\sin(k\theta + \theta)]$$
$$= r^{k+1}[\cos(k+1)\theta + i\sin(k+1)\theta],$$
which is exactly (15-19).
Part (c) follows immediately.

EXAMPLE 1. Show that $z^3 = 1$ if $z = -1/2 + (\sqrt{3}/2)i$.

Solution. Putting z in trigonometric form, we have
$$z = \cos 120° + i\sin 120°.$$
Thus,
$$z^3 = (\cos 120° + i\sin 120°)^3$$
$$= \cos 3(120°) + i\sin 3(120°)$$
$$= \cos 360° + i\sin 360° = 1.$$

A more important use of (15-17) is made in finding the roots of complex numbers. In our discussion we should recall that for r positive, the notation $\sqrt[n]{r}$ represents the principal nth root of r, that is, the only nth root of r which is positive and real.

EXAMPLE 2. Find the three cube roots of $-2 - 2\sqrt{3}i$.

Solution. Expressing our answer in trigonometric form, $r(\cos\theta + i\sin\theta)$, we are seeking values of r and θ such that
$$[r(\cos\theta + i\sin\theta)]^3 = -2 - 2\sqrt{3}i,$$

or, also expressing $-2 - 2\sqrt{3}i$ in trigonometric form,

$$r^3 (\cos 3\theta + i \sin 3\theta) = 4 (\cos 240° + i \sin 240°).$$

When two complex numbers are equal, their moduli are equal, and their amplitudes are either equal or differ by integral multiples of 360°. Thus,

$$r^3 = 4, \quad \text{and} \quad 3\theta = 240° + k360°,$$

or

$$r = \sqrt[3]{4}, \qquad \theta = 80° + k120°,$$

where k is any positive or negative integer, or zero. Hence $r (\cos \theta + i \sin \theta)$ will be

$$\sqrt[3]{4} (\cos 80° + i \sin 80°), \quad \text{for } k = 0,$$

$$\sqrt[3]{4} (\cos 200° + i \sin 200°), \quad \text{for } k = 1,$$

$$\sqrt[3]{4} (\cos 320° + i \sin 320°), \quad \text{for } k = 2.$$

These three values are all distinct, and represent the three different cube roots of $-2 - 2\sqrt{3}i$. For any other integral value of k, the expression will reduce to one of these three, so that these three numbers are the only cube roots. A complex number has three and only three cube roots, four and only four fourth roots, and, in general, n and only n nth roots. Should it be required to reduce our answers to the form $x + yi$, we can use tables to find $\sqrt[3]{4}$ and the values of the functions of 80°, 200°, and 320°.

The general theorem concerning such roots should now be apparent.

THEOREM. *For any complex number* $r (\cos \theta + i \sin \theta)$ *and any positive integer* n.

$$\sqrt[n]{r} (\cos \theta_k + i \sin \theta_k), \qquad (15\text{--}20)$$

where

$$\theta_k = \frac{\theta + k360°}{n}, \quad k = 0, 1, 2, \cdots (n-1),$$

represents the n distinct nth roots of $r (\cos \theta + i \sin \theta)$.

POWERS AND ROOTS OF COMPLEX NUMBERS

Proof. To show that (15–20) is an nth root of $r(\cos\theta + i\sin\theta)$ for each k, we merely use (15–17) to raise it to the nth power:

$$[\sqrt[n]{r}(\cos\theta_k + i\sin\theta_k)]^n = r(\cos n\theta_k + i\sin n\theta_k)$$
$$= r[\cos(\theta + k360°) + i\sin(\theta + k360°)]$$
$$= r(\cos\theta + i\sin\theta).$$

Also we must notice that the n complex numbers given by (15–20) for the n different values of k are distinct, since no two of their amplitudes differ by a mutliple of 360°.

The use of formula (15–20) will enable us to find the n nth roots of any complex number directly by substitution.

PROBLEMS

Write each of the expressions in Problems 1–7 in the form $x + yi$.

1. $[2(\cos 15° + i\sin 15°)]^6$
2. $[3(\cos 120° + i\sin 120°)]^5$
3. $[2(\cos 315° + i\sin 315°)]^3$
4. $(\cos 36° + i\sin 36°)^{10}$
5. $\left(-\dfrac{\sqrt{3}}{2} + \dfrac{i}{2}\right)^5$
6. $(1-i)^8$
7. $\left(\dfrac{1}{\sqrt{2}} + \dfrac{i}{\sqrt{2}}\right)^{200}$

Find and represent graphically Problems 8–15.

8. The square roots of $4 + 4\sqrt{3}i$
9. The square roots of $-16i$
10. The cube roots of 1
11. The cube roots of -8
12. The fourth roots of $4 - 4\sqrt{3}i$
13. The fourth roots of $16(\cos 120° + i\sin 120°)$
14. The cube roots of $8(\cos 300° + i\sin 300°)$
15. The tenth roots of 1. Compare the figure with Fig. 4–10.

Solve the equations in Problems 16–19, expressing the roots in the form $x + yi$, and represent them graphically.

16. $z^6 = 64$
17. $z^4 = 1$
18. $z^3 + i = 0$
19. $z^5 + 32 = 0$

CHAPTER 16

PROGRESSIONS

If we consider the linear function $y = 2x - 1$ for positive integral values of x, beginning with 1, the function assumes the values 1, 3, 5, $\cdots$, $2n - 1$, $\cdots$. Similarly, if the function is $y = 2^x$, the corresponding values are 2, 4, 8, $\cdots$, 2^n, $\cdots$. Both of these sets of numbers are examples of sequences. In general, a *sequence* is a set of values of some function, the domain of whose independent variable is either all or a part of the set of positive integers. The integer 1 corresponds to the first *term* of the sequence, 2 to the second term, 3 to the third term, and so on. In this way, the first, second, third, etc., values in the sequence are specifically determined. If the entire set of integers is considered, the sequence is infinite; but if only the first n positive integers make up the domain of the independent variable, the sequence is finite. We shall be concerned in this chapter with a discussion of two special types of sequences.

16–1 Arithmetic progressions. An *arithmetic progression* is a sequence in which each term after the first is obtained by adding the same fixed number, called the *common difference*, to the preceding term.

ILLUSTRATIONS. (1) The finite sequence 2, 5, 8, 11, 14 is an arithmetic progression with the common difference of 3. The function defining this sequence is $f(x) = 3x - 1$.

(2) The infinite sequence 7, 2, -3, -8, -13, $\cdots$, is an arithmetic progression with the common difference -5. $f(x) = 12 - 5x$.

Let us use the following general notations for any arithmetic progression:

a, the first term,

d, the common difference,

n, the number of terms,

and

l, the last or nth term.

Thus, in Illustration (1), $a = 2$, $d = 3$, $n = 5$, and $l = 3n - 1$, while for Illustration (2), $a = 7$, $d = -5$, and $l = 12 - 5n$. Note that the nth term really represents the function which defines the sequence.

In general, the first n terms of an arithmetic progression may be represented by

$$a, a + d, a + 2d, a + 3d, \cdots, a + (n - 1)d.$$

The last value also gives us the expression for the nth term l, in terms of a, n, and d,

$$l = a + (n - 1)d. \tag{16-1}$$

EXAMPLE 1. Find the 25th term of the arithmetic progression, 2, 5, 8, 11, $\cdots$.

Solution. In this progression, since $a = 2$, and $d = 3$, we have

$$\begin{aligned} l &= a + (n - 1)d \\ &= 2 + (24)3 = 74. \end{aligned}$$

EXAMPLE 2. If the sixth term of an arithmetic progression is 27 and the twelfth term is 48, find the first term.

Solution. We have the two relations,

$$27 = a + 5d$$

and

$$48 = a + 11d.$$

Subtracting the respective members of the first equation from those of the second, $21 = 6d$ or

$$d = 3\tfrac{1}{2}.$$

Substituting this value in the first equation,

$$27 = a + 5(3\tfrac{1}{2}),$$

we find

$$a = 9\tfrac{1}{2}.$$

EXAMPLE 3. Find the arithmetic progression of 6 terms if the first is $\frac{2}{3}$ and the last is $7\frac{1}{3}$.

Solution. Using Eq. (16–1), we have

$$\frac{22}{3} = \frac{2}{3} + (5)d,$$

and solving,

$$d = \frac{4}{3}.$$

Therefore, the required progression is

$$\frac{2}{3},\ 2,\ \frac{10}{3},\ \frac{14}{3},\ 6,\ \frac{22}{3}.$$

We are often interested in the sum of the general finite arithmetic progression. Letting

$$S_n = a + (a + d) + (a + 2d) + \cdots + [a + (n - 1)d], \quad (16\text{–}2)$$

we may also write this expression in reverse order:

$$S_n = [a + (n - 1)d] + [a + (n - 2)d] + \cdots (a + d) + a.$$

Adding the respective members of these equations, and grouping the corresponding terms,

$$2S_n = [2a + (n - 1)d] + [2a + (n - 1)d] + [2a + (n - 1)d]$$
$$+ \cdots + [2a + (n - 1)d].$$

Since there are n terms, $2a + (n - 1)d$, on the right side of this equation, we have

$$2S_n = n[2a + (n - 1)d],$$

or

$$S_n = \frac{n[2a + (n - 1)d]}{2}.\ * \qquad (16\text{–}3)$$

Recalling that $l = a + (n - 1)d$, Eq. (16–3) may also be written

$$S_n = \frac{n(a + l)}{2} \qquad (16\text{–}4)$$

---

* This relation can also be proved by mathematical induction (see Problem 17, Article 11–1).

EXAMPLE 4. Find the sum of the first thirty terms of the arithmetic progression $-15, -13, -11, \cdots$.

Solution. Since $a = -15$, $d = 2$, and $n = 30$, we have

$$S_{30} = \frac{30[2(-15) + (30-1)2]}{2}$$

$$= \frac{30(-30 + 58)}{2} = 420.$$

EXAMPLE 5. The sum of the first 15 terms of an arithmetic progression is 270. Find the first term and the common difference if the fifteenth term is 39.

Solution. Using Eq. (16–4),

$$270 = \frac{15(a + 39)}{2}.$$

Solving for a, we have

$$15a + 585 = 540,$$

$$15a = -45,$$

$$a = -3.$$

Since $l = a + (n-1)d$,

$$39 = -3 + 14d,$$

or

$$d = 3.$$

Problems

Write the next three terms in each of the following arithmetic progressions, and find l and S_n.

1. $1, 4, 7, \cdots$ to 9 terms.
2. $27, 25, 23, \cdots$ to 30 terms.
3. $10, 7, 4, \cdots$ to 15 terms.
4. $-5/4, -1/4, 3/4, \cdots$ to 8 terms.

In Problems 5–11, three of the elements a, l, d, n and S_n, of the arithmetic progression are given. Find the missing elements in each case.

5. $a = 2, d = 4, n = 12$
6. $a = 3, n = 4, l = 12$
7. $a = -2, n = 14, S_n = 20$
8. $d = 3, n = 5, l = 14$
9. $d = \frac{1}{2}, n = 14, S_n = 30$
10. $a = 4, d = 4, S_n = 40$
11. $a = 6, d = 5, l = 36$

12. Find the value of k so that $8k + 4$, $6k - 2$, and $2k - 7$ will form an arithmetic progression.

13. What are the first three terms of an arithmetic progression whose 9th term is 16 and 40th term is 47?

14. The 18th and 52nd terms of an arithmetic progression are 3 and 173, respectively. Find the 25th term.

15. Find the sum of all the even integers from 12 to 864 inclusive.

16. Find the sum of all the odd integers from 27 to 495 inclusive.

*17. The terms between any two terms of an arithmetic progression are called the *arithmetic means* between these two terms. Insert four arithmetic means between -1 and 14.

18. Insert five arithmetic means between 14 and 86.

19. Insert three arithmetic means between -18 and 4.

*20. Insert one arithmetic mean between 24 and 68. Such a number is called *the arithmetic mean* of the two numbers.

21. Find the arithmetic mean of (a) 7 and -15, (b) $3/5$ and $5/3$.

*22. A *harmonic progression* is a sequence of numbers whose reciprocals form an arithmetic progression. Insert two harmonic means between 4 and 8.

23. For any sequence of numbers forming an arithmetic progression, show that the products formed by multiplying each term by any constant, also form an arithmetic progression.

24. If a^2, b^2, and c^2 form an arithmetic progression, show that $a + b$, $c + a$, and $b + c$ form a harmonic progression.

25. How many numbers between 10 and 200 are exactly divisible by 7? Find their sum.

26. How many numbers between 25 and 400 are exactly divisible by 11? Find their sum.

27. If a clock strikes the appropriate number of times on each hour, how many times will it strike in one week?

28. A man accepts a position at $3600 with the understanding that he will receive an increase of $250 every six months. What will his salary be after working 15 years? How much would his entire earnings have been?

29. Due to the force of gravity a body falls 16.1 ft during the first second, 48.3 the next second, 80.5 the third, and so on. How far will the body fall in 10 sec?

30. A man bought a house at the beginning of 1940 for $10,000. If it increased $500 in value each year, how much was it worth at the end of 1954?

31. A piece of equipment cost a certain factory $29,000. If it depreciates in value 15% the first year, 13.5% the second, 12% the third, and so on, what will its value be at the end of 10 years, all percentages applying to the original cost?

32. A certain antique originally worth $1600 in 80 years is evaluated at $5660. Find the value at the end of each ten-year period if the increase in value during each such period was $125 more than during the preceding ten years.

16–2 Geometric progressions. A *geometric progression* is a sequence in which each term after the first is obtained by multiplying the same fixed number, called the *common ratio*, by the preceding term.

ILLUSTRATIONS. (1) The sequence 2, 4, 8, $\cdots$, 2^n is a geometric progression with the common ratio 2. The function defining this sequence is 2^x.

(2) The sequence 3, 1, $\frac{1}{3}$, $\frac{1}{9}$, $\cdots$ is a geometric progression, with $\frac{1}{3}$ the common ratio, whose defining function is 3^{2-x}.

The general notation used for the geometric progression is similar to that of Article 16–1:

a, the first term,

r, the common ratio,

n, the number of terms,

l, the last or nth term.

Since the common ratio r is multiplied by each term to give the next following term, a geometric series may be represented by

$$a, ar, ar^2, ar^3, \cdots, ar^{n-1},$$

with the last value giving us the expression for the nth term,

$$l = ar^{n-1} \qquad (16\text{--}5)$$

EXAMPLE 1. Find the 10th term of the geometric progression, $-8, 4, -2, \cdots$.

Solution. With $a = -8$, $r = -\frac{1}{2}$, and $n = 10$, we have

$$l = (-8)(-\tfrac{1}{2})^9$$
$$= (-8)(-\tfrac{1}{512}) = \tfrac{1}{64}.$$

EXAMPLE 2. If the 8th term of a geometric progression is 243 and the 5th term is 9, write the first three terms.

Solution. Since we have for $n = 8$

$$243 = ar^7,$$

and for $n = 5$

$$9 = ar^4,$$

dividing the respective members,

$$r^3 = 27, \quad \text{or} \quad r = 3.$$

Substituting this value in the second equation,

$$9 = a(3)^4, \quad \text{or} \quad 81a = 9,$$

$$a = \tfrac{1}{9}.$$

Therefore the first three terms of the required series are

$$\tfrac{1}{9}, \tfrac{1}{3}, \text{ and } 1.$$

The expression for the value of the sum of n terms of any geometric progression is easily obtained. Writing out the sum, and multiplying this expression by r,

$$S_n = a + ar + ar^2 + \cdots + ar^{n-2} + ar^{n-1},$$

and

$$rS_n = ar + ar^2 + \cdots + ar^{n-1} + ar^n.$$

Subtracting the respective members, we have

$$S_n - rS_n = a - ar^n,$$

and solving for S_n,

$$S_n(1 - r) = a(1 - r^n)$$

$$S_n = \frac{a(1 - r^n)}{1 - r}, \quad (r \neq 1) \qquad (16\text{-}6)^*$$

Since $l = ar^{n-1}$, $rl = ar^n$, and we also have

$$S_n = \frac{a - rl}{1 - r}, \quad (r \neq 1) \qquad (16\text{-}7)$$

We notice in Eqs. (16-6) and (16-7) that $r \neq 1$. What can be said about any geometric progression when $r = 1$?

PROBLEMS

Write out the next three terms in each of the following geometric progressions, and find l and S_n:

1. $1, 4, 16, \cdots$ to 8 terms.
2. $27, 9, 3, \cdots$ to 9 terms.
3. $\frac{1}{125}, -\frac{1}{25}, \frac{1}{5}, \cdots$ to 7 terms.
4. $P(1 + r), P(1 + r)^2, P(1 + r)^3, \cdots$ to 10 terms.

In Problems 5–11, three of the elements a, l, r, n, and S_n, of the geometric progression are given. Find the missing elements in each case.

5. $a = 2, r = 3, n = 6$
6. $a = 2, n = 4, l = 16$
7. $a = 1, n = 3, S_n = 13$
8. $r = -3, n = 5, l = 162$
9. $r = \frac{1}{3}, n = 5, S_n = \frac{4}{9}$
10. $a = 3, r = \frac{2}{5}, S_n = \frac{609}{125}$
11. $a = -2, r = 2, l = -64$

12. Find the value of k so that $2k + 2$, $5k - 11$, and $7k - 13$ will form a geometric progression.

13. What are the first three terms of the geometric progression whose 3rd term is $\frac{25}{4}$ and whose 7th term is $\frac{4}{25}$?

14. The second term of a geometric progression is $\frac{5}{4}$ and its fourth term is $\frac{1}{5}$. Find its third term.

---

* This expression can also be proved by mathematical induction [see Problem 16, Article (11-1)].

15. In the geometric progression $18, -12, 8 \cdots$, which term is $\frac{512}{729}$?

*16. The terms between any two terms of a geometric progression are called the *geometric means* between these two terms. Insert four geometric means between $\frac{25}{4}$ and $\frac{8}{125}$.

17. Insert three geometric means between $\frac{27}{8}$ and $\frac{2}{3}$.

*18. Insert one geometric mean between $\frac{7}{8}$ and $\frac{175}{32}$. A single number, such as this, is called *the geometric mean* of two numbers.

19. What is the geometric mean of the two numbers a and b?

20. Find the geometric mean of (a) 12 and $\frac{4}{3}$, (b) 28 and 112.

21. Prove that for the two unequal positive numbers a and b, their arithmetic mean is greater than the geometric mean.

22. The population of a certain town is 5000. If it increases 5% every year, what will the approximate population be at the end of 10 yr?

23. A rubber ball is dropped from a height of 9 ft. If it rebounds one-third of the distance it has fallen after each fall, how far will it rebound the 6th time? Through what distance has it traveled when it strikes the ground the 7th time?

24. A man accepts a position at $3600 with the understanding that he will receive a 2% increase every year. What will his salary be after 10 years of service?

25. An automobile purchased for $3000 depreciates in value 12% every year. Find its value at the end of 5 yr.

26. Explain the statement that the compound interest discussion in Article 12-5 is an application of the concept of geometric series.

27. If a paper napkin .01 inch thick could be folded into a piece of paper half as large, but twice as thick, folded again in the same manner, and this process continued 30 times, approximately how thick would the resulting piece of paper be?

28. If you had the choice of a salary of $1000 a day for a month of 31 days, or $1 for the first day's work, $2 for the second, $4 for the third, and each day thereafter for the rest of the month your salary would be doubled, which choice would you make?

16–3 Geometric progressions with infinitely many terms. In all the discussion so far we have considered only the sum of a finite sequence. Because of the definition of the arithmetic progression, it should be clear that such a progression with an infinite number of terms would have no finite sum. Moreover, for a geometric progression with r greater than 1, each term would be larger than the preceding one, so that no definite value could exist representing an infinite sum. In fact, if $r = 1$, each term would be the same, and no infinite sum would exist.

If $|r| < 1$, we have a different situation. Consider the sum of the geometric series

$$S_n = \frac{1}{2} + \frac{1}{4} + \frac{1}{8} + \cdots + \frac{1}{2^n}, \tag{16-8}$$

where $r = \frac{1}{2}$.

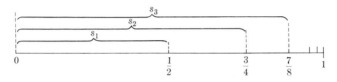

FIGURE 16-1

One interpretation of S_n may be seen in Fig. 16-1, which represents a line segment one unit in length. The brackets denote the terms, and the numbers below denote the sum of the progression at any stage. Each term that is added represents a length of half the total length remaining between the point representing the finite sum and the point 1. The more terms considered in Eq. (16-8) the closer we get to the point 1. In fact, we may get as close as we please to 1 by considering a sufficient number of terms, but no matter how many terms are considered, the sum will never exceed 1. Thus we say the limiting value of S_n is 1, and we write

$$S = \lim_{n \to \infty} S_n = 1. \tag{16-9}$$

The expression $\lim_{n \to \infty} S_n$ is read "the limit of S sub n, as n increases without limit."

This may also be shown algebraically. Equation (16-6) may be written

$$S_n = \frac{a}{1-r} - \frac{ar^n}{1-r}. \tag{16-10}$$

In the progression (16–8), $a = r = \frac{1}{2}$, so that

$$S_n = \frac{\frac{1}{2}}{1-\frac{1}{2}} - \frac{\frac{1}{2}(\frac{1}{2})^n}{1-\frac{1}{2}}$$
$$= 1 - (\tfrac{1}{2})^n.$$

As n increases without limit, $(\frac{1}{2})^n$ approaches nearer and nearer to zero. For example,

$$\left(\frac{1}{2}\right)^{10} = \frac{1}{1024}$$

and

$$\left(\frac{1}{2}\right)^{20} = \frac{1}{1,048,576}.$$

Thus, again,

$$\lim_{n \to \infty} S_n = 1.$$

This same process holds for any geometric progression where $|r| < 1$. Using Eq. (16–10), we have

$$S_n = \frac{a}{1-r} - \frac{ar^n}{1-r}.$$

If we let n increase without limit, since a, and r are fixed, and $|r| < 1$, $\lim_{n \to \infty} \frac{ar^n}{1-r} = 0.$ Therefore,

$$S = \lim_{n \to \infty} S_n = \frac{a}{1-r}, \quad |r| < 1. \qquad (16\text{–}11)$$

This limit, $S = a/(1-r)$, is called the sum of the geometric progression with infinitely many terms.

EXAMPLE 1. Find the sum of the infinite geometric progression $\frac{3}{2}$, $1, \frac{2}{3}, \frac{4}{9}, \cdots$.

Solution. With $a = \frac{3}{2}$, and $r = \frac{2}{3}$, we find

$$S = \frac{a}{1-r} = \frac{\frac{3}{2}}{1-\frac{2}{3}} = \frac{\frac{3}{2}}{\frac{1}{3}} = 4\tfrac{1}{2}.$$

16-3] PROGRESSIONS WITH INFINITELY MANY TERMS

EXAMPLE 2. Convert the repeating decimal $3.242424\cdots$ into an equivalent common fraction.

Solution. We can represent the number $3.242424\cdots$ as the number 3 plus a geometric progression with $a = .24$ and $r = .01$, since

$$3.242424\cdots = 3 + (.24 + .0024 + .000024 + \cdots).$$

The sum S in the parentheses may be found:

$$S = \frac{a}{1-r} = \frac{.24}{1-.01} = \frac{.24}{.99} = \frac{8}{33}.$$

Thus,

$$3.242424\cdots = 3 + \frac{8}{33} = \frac{107}{33}.$$

This result may be checked by actually dividing 107 by 33.

Any infinite repeating decimal can be converted into an equivalent common fraction. Moreover, any common fraction may be written as a periodic decimal (see Article 1–1).

PROBLEMS

Find the sum of each of the following infinite geometric progressions:

1. $1, \frac{1}{3}, \frac{1}{9}, \cdots$
2. $128, 48, 18, \cdots$
3. $16, -4, 1, \cdots$
4. $\frac{4}{3}, 1, \frac{3}{4}, \cdots$
5. $2, \sqrt{2}, 1, \cdots$
6. $\sqrt{2} + 1, 1, \sqrt{2} - 1, \cdots$

Convert each of the repeating decimals into equivalent common fractions:

7. $3.333\cdots$
8. $6.272727\cdots$
9. $.555\cdots$
10. $8.690909\cdots$
11. $5.818181\cdots$
12. $.142857142857\cdots$

13. The sum of an infinite geometric progression is $\frac{7}{2}$, and the first term is 3. What is the common ratio?

14. The length of the side of a square is 4 in. A second square is inscribed by connecting the mid-points of the sides of the first square, a third by connecting the mid-points of the sides of the second, and so on. Find the sum of the areas of the infinitely many squares thus formed, including the first.

15. Find the sum of the perimeters of all the squares described in Problem 14.

16. If the original figure in Problem 14 were an equilateral triangle, with sides 4 in., and new equilateral triangles were formed in the same way, what would the sum of their areas be, including the first?

17. A ball is dropped from a height of 48 ft, and rebounds two-thirds of the distance it falls. If it continues to fall and rebound in this way, how far will it travel before coming to rest?

18. From a line 1 ft long, the middle third is erased. From each of the two remaining thirds, the middle thirds are erased. From each of the four remaining ninths, the middle thirds are erased. If this process is continued indefinitely, what will the sum of the lengths of the remaining line segments be?

CHAPTER 17

PERMUTATIONS, COMBINATIONS, AND PROBABILITY

Although any extensive treatment is impossible, certain elementary topics will be discussed in connection with permutations and combinations because they are so useful in mathematics as well as other branches of the natural sciences. Their uses will be extended to simple probability.

17-1 The fundamental principle. We shall start our discussion by considering a simple example. Let us find how many numbers of two different digits can be formed from the four integers 1, 2, 3, and 4. Any one of the four may be chosen for the tens digit of the number. With each particular choice of this type, there will remain three integers from which to choose the units digit. Thus, for each of the four choices, there are three more choices, making a total of $4 \cdot 3 = 12$ numbers in all. Listing them, we have

$$\begin{aligned} &41, 42, 43, \text{ (4 for the tens digit)}, \\ &31, 32, 34, \text{ (3 for the tens digit)}, \\ &21, 23, 24, \text{ (2 for the tens digit)}, \\ &12, 13, 14, \text{ (1 for the tens digit)}, \end{aligned} \qquad (17\text{-}1)$$

as the possible two-digit numbers.

A slightly more complicated problem would be to find how many numbers of three different digits could be formed with the same integers 1, 2, 3, and 4. This is easily found, for each of the twelve numbers listed in (17–1) can be considered as representing the hundreds and tens digits. Since two of the four integers have already been selected in each case, the possibility for the new units digit must be one of the two remaining integers. For example, with the first number 41, we may have 2 or 3, forming the numbers 412 or 413. We therefore, have $(4 \cdot 3) \cdot 2$ or 24 numbers, each containing three digits, which can be formed from the given integers.

These examples illustrate the fundamental principle involved.

FUNDAMENTAL PRINCIPLE. *If one thing can be done in n_1 different ways, and if, after this is done, a second thing can be done in n_2 different ways, after which a third thing can be done in n_3 different ways, and so on (for any finite number of things), then the total number of ways in which all the things may be done in the stated order is $n_1 n_2 n_3 \cdots$.*

EXAMPLE 1. How many different committees consisting of one Democrat and one Republican can be formed from 7 Democrats and 4 Republicans?

Solution. The Democrat can be chosen in any one of seven ways, and independently, the Republican can be chosen in any one of four ways. Thus, by the fundamental principle, the total number of different possible committees is $7 \cdot 4 = 28$.

EXAMPLE 2. If two cubical dice are thrown, in how many ways can they fall?

Solution. Since each cube has six faces, each may fall independently any one of six different ways. Consequently, there are $6 \cdot 6 = 36$ possible ways for them to fall.

EXAMPLE 3. In a certain election, there are three candidates for president, four for vice-president, five for secretary, but only two for treasurer. In how many different possible ways may the election turn out?

Solution. With each of the three possibilities for president, there are four for vice-president. With each of these $4 \cdot 3 = 12$ possibilities, there are five possibilities for secretary, and so on. Thus the total number of different possible election results is

$$4(3)(5)(2) = 120.$$

PROBLEMS

1. A nickel and a dime are tossed on a table. In how many ways can they fall?

2. If all questions are answered in a true-false quiz of ten questions, how many possible ways are there of answering the entire quiz?

3. How many numbers of three different digits less than 500 can be formed from the integers 1, 2, 3, 4, 5, 6, 7?

PERMUTATIONS

4. If there are twelve milers entered in a race, in how many ways can first, second, and third place be awarded?

5. There are five main roads between the cities A and B, and four between B and C. In how many ways can a person drive from A to C and return, going through B on both trips, without driving on the same road twice?

6. How many numbers of at most three different digits can be formed from the integers 1, 2, 3, 4, 5, 6?

7. How many numbers of at least three different digits can be formed from the integers 1, 2, 3, 4, 5, 6?

8. A tennis club consists of 12 boys and 9 girls. How many mixed doubles teams (one boy and one girl) are possible? In how many ways can a mixed doubles match be arranged?

9. A freshman student must take a modern language, a natural science, a social science, and English. If there are four possible different modern languages, five natural sciences, three social sciences, but each student must take the same English course, in how many different ways can he select his course of study?

10. In choosing an ace, king, queen, and jack from an ordinary deck of fifty-two cards, how many ways may they be chosen if (a) they must be of different suits? (b) they may or may not be of different suits? (c) they must be of the same suit? (d) they must be in a particular suit?

11. If there are eight outside doors in a dormitory, in how many ways can a student enter one and (a) leave by a different door? (b) leave by any door?

12. A baseball stadium has four entrance gates, but nine exits. In how many ways may two men enter together, but leave by different exits?

17-2 Permutations. Each different arrangement or ordered set of things is called a *permutation* of those things. For example, the numbers 423 and 432 use the same numbers, but are different permutations of the three numbers. In general, if we have n things and arrange r of them in a definite order, such an arrangement is a permutation of n things taken r at a time.

A formula for the total number of permutations of n different things taken r at a time is easily established using the fundamental principle of the last article. This function of the integers n and r will be denoted by $P(n,r)$.* The possible number of arrangements

---

* The symbol $_nP_r$ is also frequently used.

of n things taken r at a time is equivalent to the number of ways of choosing from n different things to fill r positions. There are n choices for the first position, then $n - 1$ choices for the second position, $n - 2$ choices for the third, and so on. The rth position can be filled with any of the $n - (r - 1) = n - r + 1$ remaining things, so that the r positions can be filled in $n(n - 1)(n - 2) \cdots (n - r + 1)$ ways. Thus,

$$P(n,r) = n(n - 1)(n - 2) \cdots (n - r + 1). \qquad (17\text{--}2)$$

If we are interested in the permutation of n things taken n (or all) at a time, $r = n$ in (17–2), and we have

$$P(n,n) = n(n - 1)(n - 2) \cdots 3 \cdot 2 \cdot 1 = n!.* \qquad (17\text{--}3)$$

By multiplying the numerator and denominator of the right member of (17–2) by $(n - r)!$, we obtain an alternate formula for $P(n,r)$.

$$P(n,r) = \frac{n(n - 1)(n - 2) \cdots (n - r + 1)(n - r)!}{(n - r)!}$$

or

$$P(n,r) = \frac{n!}{(n - r)!}.^{\dagger} \qquad (17\text{--}4)$$

EXAMPLE 1. Find the number of permutations of the four integers 1, 2, 3, and 4 taken two at a time.

Solution. Since we wish to find the number of permutations of four things taken two at a time,

$$P(4,2) = 4 \cdot 3 = 12.$$

These twelve arrangements are listed in (17–1).

EXAMPLE 2. If four persons enter a bus in which there are ten vacant seats, how many possible ways are there for the four to be seated?

Solution. Since this represents the arrangement of ten seats,

---

* Recall the definition of factorial n, $n!$ (Article 7–2).
† Factorial zero, $0!$, is by definition equal to 1.

taken four at a time,

$$P(10,4) = 10 \cdot 9 \cdot 8 \cdot 7 = 5040.$$

EXAMPLE 3. In how many ways may five books be arranged on a shelf?

Solution. This represents the permutation of five things taken 5 at a time. Thus

$$P(5,5) = 5! = 120.$$

Although the number of permutations of the eight letters of the word "readings" is clearly 8!, the number of such permutations of the letters of "gargling" is somewhat less, since three of the letters are alike. An example will illustrate the next theorem.

EXAMPLE 4. Find the number of permutations of the eight letters of "gargling."

Solution. Considering all the letters as different, we would have P, the required number of permutations equal to 8!. This might be done by giving each of the g's a different subscript, such as g_1, g_2, and g_3. Actually, however, the three g's could be permuted among themselves in 3! ways. In fact, for each distinct arrangement of the other letters, this would be the case. Thus $3!P = 8!$ or

$$P = \frac{8!}{3!}.$$

A general theorem, whose proof follows the reasoning in Example 4 and is left as an exercise, is the following:

THEOREM 1. *If P represents the number of different permutations of n things taken all at a time, if p are of one kind, q are of another kind, r of a third kind, and so on, then*

$$P = \frac{n!}{p!q!r! \cdots}. \qquad (17\text{-}5)$$

EXAMPLE 5. How many permutations are there of the eleven letters in "Mississippi" taken all together?

Solution. Since there are four s's, four i's, and two p's, we have

$$P = \frac{11!}{4!4!2!} = 34{,}650.$$

Problems

1. How many numbers of three different digits each can be formed from the digits 1, 2, 3, 4, 5, 6, 7, 8, 9?

2. How many numbers of three different digits each less than 700 can be formed from the digits in Problem 1?

3. How many numbers of at most three different digits each can be formed from the digits in Problem 1?

4. In how many ways can a class elect a president, vice-president, secretary, and treasurer from a class of 100 students?

5. In how many ways can 4 boys and 3 girls be seated in a row containing 7 seats, (a) if they may sit anywhere? (b) if the boys and girls must alternate?

6. In how many ways can 4 boys and 4 girls be seated in a row containing 8 seats, (a) if they may sit anywhere? (b) if the boys and girls must alternate?

7. A baseball manager insists on having his best hitter bat fourth and the pitcher bat last. In such circumstances, how many batting orders are possible?

8. In how many ways can eight people be seated in a row of eight seats if two of them insist on sitting next to each other?

9. A language teacher insists on keeping books of the same language together on his shelf. If he has 12 spaces for 5 French, 4 Italian, and 3 German books, in how many ways can they be placed on his shelf?

10. In how many ways can eight people be seated around a table?

11. In how many ways can eight people be seated around a table, if two of them insist on sitting next to each other?

12. How many license plates can be made using any two letters for the first two places, and any of the numbers 0 through 9 for the last three?

13. Do Problem 12, if no letter or number is repeated.

14. How many permutations are there of the letters of the word (a) *algebra*? (b) *college*?

15. How many permutations are there of the letters of the word *Tennessee*?

16. In how many ways can 4 red beads, 5 white beads, and 3 blue beads be arranged in a row?

17. In how many ways can 7 different colored beads be made into a bracelet?

18. Show that $P(n + 1, r) \equiv (n + 1)P(n, r - 1)$.

19. Solve the equation $P(n,5) = 20P(n,3)$ for n.

20. Find the value of $P(5,1) + P(5,2) + P(5,3) + P(5,4) + P(5,5)$.

17–3 Combinations. In contrast to a permutation which is a certain ordered arrangement of different things, a *combination* is a set or collection of things without regard to order. Thus, by the combinations of n different things taken r at a time, we mean all the possible selections of r different things from the n things, with no regard to order or arrangement. This total number of combinations is written $C(n,r)$.*

The only difference between permutations and combinations is order or arrangement. For any one of the $C(n,r)$ combinations, consisting of r different things, these r elements may be rearranged by permutation in $r!$ different ways. Therefore, for all the possible combinations there will be $C(n,r)r!$ different permutations. Since these are all the possible permutations of the n things taken r at a time, we have the relation

$$C(n,r) \cdot r! = P(n,r)$$

or

$$C(n,r) = \frac{P(n,r)}{r!}. \quad (17\text{–}6)$$

This may be written in terms of factorials by using (17–4),

$$C(n,r) = \frac{n!}{(n-r)!r!}. \quad (17\text{–}7)$$

Using (17–7), it is clear that

$$C(n,r) = C(n, n-r) \quad (17\text{–}8)$$

(recall Problem 25, Article 11–2). The form given in (17–6) is most frequently used in problems.

EXAMPLE 1. In a class of 15 boys and 10 girls, in how many ways may a committee made up of 3 boys and 2 girls be selected?

Solution. Since order has nothing to do with membership on the committee, the boys may be selected in $C(15,3)$ ways, while the

---

* Recall Eq. (11–5) and accompanying footnote. The symbol $_nC_r$ is also frequently used.

girls may be selected in $C(10,2)$ ways. By the fundamental principle, we have

$$C(15,3) \cdot C(10,2) = \frac{15 \cdot 14 \cdot 13}{1 \cdot 2 \cdot 3} \cdot \frac{10 \cdot 9}{1 \cdot 2}$$
$$= 20{,}475.$$

EXAMPLE 2. On a certain examination, the student must answer 8 out of the 12 questions, including 5 of the first 6. In how many ways may he write the examination?

Solution. Since five of the first six must be answered, this may be done in $C(6,5)$ ways. Having now answered five questions, he must answer three of the remaining 7. This may be done in $C(7,3)$ ways. Therefore

$$C(6,5) \cdot C(7,3) = 210.$$

Frequently we wish to find the total number of combinations of n things taken 1, 2, 3, $\cdots$, or n at a time. In Problem 27, Article 11–2, we showed

$$2^n = 1 + C(n,1) + C(n,2) + C(n.3) + \cdots + C(n,n),$$

so that the total number of combinations taken 1, 2, 3, $\cdots$, or n at a time is

$$C(n,1) + C(n,2) + C(n,3) + \cdots + C(n,n) = 2^n - 1. \quad (17\text{--}9)$$

EXAMPLE 3. How many different sums of money can be made from a penny, a nickel, a dime, and a quarter?

Solution. Since there are four coins, the number of different amounts of money is

$$C(4,1) + C(4,2) + C(4,3) + C(4,4) = 2^4 - 1 = 15.$$

PROBLEMS

1. Find the value of (a) $C(7,4)$; (b) $C(10,2)$; (c) $C(21,19)$.
2. Find the value of

$$C(8,3) + C(8,4) + C(8,5) + C(8,6) + C(8,7) + C(8,8).$$

3. In how many ways may a committee of 4 be chosen from a group of 25?

4. From a group of 25 Democrats and 18 Republicans, how many committees consisting of 3 Democrats and 2 Republicans are possible?

5. From the group in Problem 4, if one of two specific Democrats is to be chairman, how many committees are possible with the same balance of Democrats and Republicans as previously stated?

6. How many football games are played if each of the nine football teams in the Midwestern Conference plays each of the other teams in that conference once?

7. On a college baseball squad, there are three catchers, five pitchers, seven infielders and seven outfielders. How many different baseball nines can be formed?

8. In how many ways can a bridge hand of 13 cards be chosen from a deck of 52 cards?

9. In how many ways can a person get a bridge hand consisting of only aces or face cards?

10. In how many ways can a person get a bridge hand which consists of cards seven or lower?

11. In how many ways can a person get a bridge hand consisting of 2 aces, 1 king, 1 queen, 3 jacks, and the 6 other cards ten or less?

12. From 4 red balls, 5 white balls, and 6 blue balls, how many selections consisting of 5 balls can be made if 2 are to be red, 1 white, and 2 blue?

13. Without considering special cases, (a) how many straight lines are determined by 9 points? (b) how many circles are determined by 9 points?

14. How many triangles are determined by 9 points, no three of which lie on the same straight line?

15. How many tetrahedrons are determined by 9 points, no four of which lie in the same plane?

16. From a group of 15 people, how many committees can be formed consisting of two, three, or four people?

17. In how many ways may a college president's wife invite (a) two, (b) three, (c) two or more, of eight faculty wives to a tea?

18. How many different sums of money can be formed from a penny, nickel, dime, quarter, and half dollar, if at least two coins are used?

19. Solve for n the equation $C(n + 2,4) = 6C(n,2)$.

20. Prove $C(n,r) + C(n,r - 1) \equiv C(n + 1,r)$.

17–4 Probability. The theory of probability, which had its origin in games of chance during the seventeenth century, and at that time was developed mathematically by Pascal (1623–1662) and Fermat

(1601–1665), has become one of the most important and useful branches of mathematics. It is extremely helpful in statistical studies of physics, biological and social phenomena, in life insurance, or industry, and, in fact, in all fields of study in which the future is under consideration. Anything more than an introduction to the subject is unfortunately beyond the scope of this book.

In considering any future event, an act which may result in the event occurring or failing to occur is called a *trial*. If the event occurs, it is called a *success*, but if it fails to occur, it is called a *failure*.

DEFINITION 1. If an event, may occur s ways, but fail to occur f ways and if each of the $s + f$ ways is equally likely,* the probability of obtaining a success in a trial of the event is

$$p = \frac{s}{s+f}, \qquad (17\text{--}10)$$

but the probability of obtaining a failure is

$$q = \frac{f}{s+f}. \qquad (17\text{--}11)$$

It is important to note that $0 \leq p \leq 1$, $0 \leq q \leq 1$, and $p + q = 1$. If a success is certain, $f = 0$, so that $p = 1$. Also, since $p + q = 1$, $q = 1 - p$. Thus, if the probability that an event will succeed is p, the probability that it will fail is $1 - p$.

In computing any probability, we must compute s, f, and $s + f$. This is often possible by making use of the formulas in the last two articles.

EXAMPLE 1. If a die, which is a cube that has spots one through six on its faces, is thrown, what is the probability that a four will show?

Solution. Since $s = 1$, and $s + f = 6$, $p = 1/6$.

EXAMPLE 2. If two cards are drawn from an ordinary deck of 52 cards, what is the probability that both are spades?

---

* By *equally likely*, we mean that there is no reason for expecting any one to happen rather than any other.

Solution. The number of ways of drawing 2 cards from a deck of 52 is $C(52,2)$, while the number of ways of drawing two spades is $C(13,2)$. Thus $s + f = C(52,2)$, and $s = C(13,2)$, so that

$$p = \frac{C(13,2)}{C(52,2)} = \frac{78}{1326} = \frac{1}{17}.$$

EXAMPLE 3. A committee of 5 is to be selected at random from 6 Democrats and 4 Republicans. What is the probability that the committee will consist of 3 Democrats and 2 Republicans?

Solution. The total number of possible five-man committees is $C(10,5)$, while the number of "successful" committees is $C(6,3) \cdot C(4,2)$. Thus

$$p = \frac{C(6,3)C(4,2)}{C(10,5)} = \frac{10}{21}.$$

It must be emphasized that a certain probability based on theory will guarantee nothing in any particular situation. Specifically, in Example 1, the probability of $\frac{1}{6}$ for a four showing does not mean that a four will show exactly once in six throws. Nor will it show exactly ten times in sixty throws. We merely expect, in any large number of throws, to have a four show approximately one-sixth of the time.

PROBLEMS

1. If a die is thrown, what is the probability that a three will turn up? (b) What is the probability that a one or two will turn up?

2. If a card is drawn from an ordinary deck of 52 cards, (a) what is the probability that it is a king, (b) a face card; (c) seven or below, (d) a diamond, (e) a black card?

3. If two dice are thrown, what is the probability of throwing a total of four? (The total may be four if each die shows a two, if one shows a one and the second a three, or vice versa.)

4. If two dice are thrown, what is the probability of throwing a total of seven or eleven?

5. If two dice are thrown, what is the probability of throwing a total of two, three, or twelve?

6. If two dice are thrown and the result is an eight, what is the probability it was thrown with two fours?

7. If two cards are drawn from an ordinary deck of 52 cards, what is the probability that (a) both will be aces, (b) both will be red, (c) both will be below a six?

8. A bag contains 8 red and 4 white balls. What is the probability that a ball drawn out will be red? white?

9. If three balls are drawn from the bag described in Problem 8, what is the probability (a) two will be red and one white, (b) all three will be red?

10. If four men and four women are seated at random in a row, what is the probability the men and women alternate?

11. If there are six different sections of an algebra course offered, what is the probability of three students, registering independently, having the same section? having Section One?

12. What is the probability of drawing a bridge hand from an ordinary pack of cards (a) all in the same suit, (b) all hearts?

13. Six couples are invited to play bridge. What is the probability that each man will have his wife for a partner?

14. If a family of four takes seats at random in a row, what is the probability the father and mother (a) sit together, (b) sit on the ends?

15. In a group of thirty, what is the probability that two people will have their birthday dates coincide? *Hint:* Find the probability that no one of the group will have coincident birthdays. Use logarithms.

17-5 Empirical probability. Often when the analysis cannot be exact, experimental or statistical data on past information may be beneficial in finding the probability of an event occurring. The larger the number of cases considered, the more accurate the results will be, and care should always be taken to secure as large a sample as possible. We use the following definition:

DEFINITION 2. If an event occurs h times out of n trials, the relative frequency of its occurrence is h/n.

It has been assumed, and with justification, that for large values of n, this relative frequency approximates the mathematical probability of an event occurring. In this way we define what is called *empirical* probability.

DEFINITION 3. If an event occurs h times out of n trials, where n is a large number, then the probability p that the event will occur in any one trial is defined as

$$p = \frac{h}{n}. \qquad (17\text{-}12)$$

This use of probability is used extensively in the social sciences, industry, actuarial work, and so on.

EXAMPLE 1. An investigator asked 1000 people in a large city which of two candidates they preferred in the coming election for mayor. If 679 preferred the first candidate, what is the probability that a person selected at random would prefer this candidate?

Solution. Since $h = 679$ and $n = 1000$, the desired probability is $p = 679/1000 = 0.679$.

Table V in the back of the book is the American Experience Table of Mortality which lists the number of persons surviving at different ages, based on an initial group of 100,000 at an age of ten years. It is useful in finding the life expectancy of a person at any age.

EXAMPLE 2. Using Table V, determine the probability that a person aged 21 will live to be 65.

Solution. From Table V, of the 100,000 people considered, 91,914 are living at age 21, and of these, 49,341 are still alive at age 65. The desired probability is thus

$$p = \frac{49{,}341}{91{,}914} = 0.54.$$

PROBLEMS

1. In a certain city there were 20,000 automobiles registered. If 625 traffic accidents were reported in that same year of registration, what was the probability of a specific car being involved in one of the accidents?

2. In the town of Oberlin in which there were 2000 bicycles registered, 25 bicycle thefts were reported in 1954. Assuming this ratio will continue, what is the probability that a certain individual will have his bicycle stolen in 1955?

308   PERMUTATIONS, COMBINATIONS, AND PROBABILITY   [CHAP. 17

3. A certain factory making small resistors finds that out of 100 resistors, three will be faulty. What is the probability that a given resistor from this company will not be accurate?

4. A baseball player had 192 safe hits out of 512 times at bat in the 1954 season. Assuming he will have the same batting average during the next season, what is the probability that he will get a hit any time he goes to bat?

5. A certain inland lake in northern Michigan has been completely frozen over by Christmas Day, 35 of the last 40 years. Based on this fact, what is the probability it will be frozen over by Christmas Day of this next year?

6. In a certain college, one out of every nine students left at the end of their first year due to scholastic difficulty. If next year's freshman class has 360 first year students, how many of these would be expected to continue for another year?

Use Table V in the following problems.

7. What is the probability that a person sixteen years old will live to be fifty?

8. What is the probability that a person fifty years old will live to be eighty?

9. What is the probability that a person eighty years old will die during the following year?

10. What is the probability that a person sixty-five years old will die during the next five years?

11. What is the probability that a person fifty years old will die between the ages of 70 and 71?

17–6 Probability of more than one event. Although the problems discussed have been concerned with only one event, it is also possible to consider probabilities of two or more events taking place. Two or more events are called *mutually exclusive* if not more than one of them can happen in a given trial. For example, a toss of a coin cannot result in both a head and a tail; on a single drawing from a pack of cards, the card drawn cannot be both a king and queen, and so on.

> THEOREM 2. *If two events are mutually exclusive with the probability for the first event happening p_1, and the probability for the second happening p_2, then the probability p that one or the other of the two events will happen is $p = p_1 + p_2$.*

Proof. If s is the number of ways in which one or the other event can happen, $s = s_1 + s_2$, where s_1 represents the number of ways the first event can happen, and s_2 the number of ways the second can happen. Letting f be the number of ways neither event can occur, $s + f$ will be the total number of possibilities. Thus

$$p = \frac{s_1 + s_2}{s + f} = \frac{s_1}{s + f} + \frac{s_2}{s + f},$$

and by definition we have

$$p = p_1 + p_2.$$

EXAMPLE 1. From a bag containing four red, five white, and three blue marbles, a marble is drawn. What is the probability that it is red or blue?

Solution. The probability of drawing a red ball is $4/12$, while the probability of drawing a blue ball is $3/12$. Since these two events or drawings are mutually exclusive,

$$p = \tfrac{4}{12} + \tfrac{3}{12} = \tfrac{7}{12}.$$

Two or more events are called *independent* if the happening of one event in no way affects the happening of any other of the events. For example, in Example 1, if a marble is drawn from the bag, after which it is replaced, and a second marble is drawn, these drawings or events are independent.

THEOREM 3. *If two events are independent with the probability of the first happening p_1, and the probability of the second happening p_2, then the probability p that both events will happen in a single trial is $p = p_1 p_2$.*

Proof. If s_1 and s_2 represent the number of ways the two events can happen, and f_1 and f_2 the number of ways the two can fail to happen, due to the fundamental principle, $s_1 s_2$ represents the number of ways in which the two events can happen, and $(s_1 + f_1)(s_2 + f_2)$ the total number of ways the two events can either happen or not happen, then,

$$p = \frac{s_1 s_2}{(s_1 + f_1)(s_2 + f_2)} = \frac{s_1}{s_1 + f_1} \cdot \frac{s_2}{s_2 + f_2},$$

which by definition is
$$p = p_1 p_2.$$

The proof of a fourth theorem is similar. Two or more events are called *dependent*, if the happening of any one affects the occurrence of the other events. If in Example 1 the second ball is drawn without replacing the first ball drawn, the probability of drawing a certain colored ball will depend on the result of the first drawing.

THEOREM 4. *If two events are dependent, the probability of the first event happening is p_1 and after it happens, the probability of the second happening is p_2, then the probability p, that both events will happen, is*
$$p = p_1 p_2.$$

The proof is left as an exercise.

EXAMPLE 2. From the bag described in Example 1, a marble is drawn and then replaced, and a second marble is drawn. What is the probability that both marbles drawn are red?

Solution. These two events are independent, for the probability involved in the second drawing is not affected by the result of the first. Since the probability of drawing a red marble is $\frac{4}{12}$ on either the first or second drawing,
$$p = \tfrac{1}{3}\left(\tfrac{1}{3}\right) = \tfrac{1}{9}.$$

EXAMPLE 3. From the bag described in Example 1, a marble is drawn, and *without* replacing it, a second marble is drawn. What is the probability that both are red?

Solution. Again the probability of drawing a red marble on the first drawing is $\tfrac{1}{3}$. The second drawing, however, is made from eleven marbles, of which three are red (assume the first marble drawn was red). Thus
$$p = \tfrac{1}{3} \cdot \tfrac{3}{11} = \tfrac{1}{11}.$$

Notice that the outcome of the second drawing depends on the result of the first.

In any problem involving two or more events extreme **care** should be taken to determine whether the events are independent, **dependent**, or mutually exclusive.

Although each of these theorems is proved for two events, the theorems may clearly be generalized to account for n events.

PROBLEMS

1. If two coins are tossed, what is the probability that both will (a) fall with the same side up? (b) turn up tails?

2. If the mathematics library has three copies of the same book, what is the probability that 4 students will use the same copy, if they borrow it at different times?

3. If a ball player is batting at .300 (3 hits out of ten times at bat), what is the probability he will get four hits in a row?

4. Find the probability of drawing (a) an ace or queen from an ordinary pack of playing cards if one card is drawn, (b) an ace and a queen if after the first drawing the card is returned and a second is drawn, (c) an ace and a queen if the cards are drawn in succession, without replacement.

5. If two dice are thrown three times, what is the probability of throwing (a) a seven once and only once? (b) a seven at least once?

6. If two cards are drawn from an ordinary pack of playing cards, what is the probability that (a) both are of the same denomination, (b) both are of the same suit, (c) both are of the same color, (d) each is of a different color?

7. A football team's probability of winning its first game is $\frac{3}{7}$ and its second $\frac{4}{7}$. What is the probability (a) of winning both games, (b) of winning the first but losing the second, (c) losing both games?

8. If the probability of passing the first test in an algebra course is $\frac{2}{3}$, and the second is $\frac{3}{4}$, what is the probability of (a) passing both tests, (b) failing both tests, (c) passing only one test?

9. The top seeded player in a tennis tournament has probabilities of $\frac{3}{4}$, $\frac{2}{3}$, and $\frac{1}{2}$ of winning his first, second, and third matches. Find the probability that (a) he will win all three, (b) he will win the first two but lose the third.

10. A young man twenty-four years old marries a young woman of twenty-two. Find the probability that (a) they will both be alive at the age of seventy, (b) he will live until seventy and she until seventy-five, (c) they both will die within twenty-five years of their marriage. *Hint:* Use Table V.

ANSWERS TO ODD-NUMBERED PROBLEMS

CHAPTER 1

ARTICLE 1-3

1. 36, 14, 275, 25/11
3. $-16, -32, -192, -3$
5. $-4, 4, 0, 0$
7. 5/6, 1/6, 1/6, 3/2
9. $-71/28, -1/28, 45/28, 36/35$
11. $-5/4, 5/4, 0, 0$

ARTICLE 1-4

1. $3a + b - 1, a + 5b - 7$
3. $6x + 6y - z, 2x + 3z$
5. $-8x - 11y, 12x - y$
7. $-2x - 4y$
9. $-6x - 9y - 4$
11. $16x - 3y - 9$
13. $13x - 10y$
15. $a^2 - (b^2 - 2bc + c^2)$
17. $4x^2 - (4y^2 + 4y + 1)$
19. $(c - a)x, -(a - c)x$
21. $(a + b)x, -(-a - b)x$
23. 7
25. 17
27. 5
29. 35

ARTICLE 1-5

1. a^9
3. $3x^{12}$
5. y^{24}
7. a^{20}
9. $125c^3$
11. $32a^{15}$
13. $a^{(r+s)t}$
15. $2^n x^{n^2}$
17. $2x^2 + x - 15$
19. $16x^2 - 4y^2$
21. $r^3 - r^2 s - rs^2 + s^3$
23. $x^3 + y^3$
25. $2x^4 - 3x^3 - 10x^2 + 6x + 8$
27. $x^3 - 2x^2 - 5x + 6$
29. $x^6 - 8y^6$
31. $a^{2n+1} - a^{2n} - 7a^{n+1} + 7a^n + 10a - 10$
33. $x^{4n} - 2x^{2n} y^{2n} + y^{4n}$
35. $x^4 - 4x^3 y + 6x^2 y^2 - 4xy^3 + y^4$

ARTICLE 1-6, (1)

1. $3y^2 - 2x^2$
3. $\dfrac{2x^2}{y} - 3x^3$
5. $3x - 4y + 6x^2 y^2$
7. $x^2 - 7x + 10 \equiv (x - 2)(x - 5)$
9. $3x^2 - 13x + 4 \equiv (3x - 1)(x - 4)$
11. $2x^3 - 7x^2 + 11x - 4 \equiv (x^2 - 3x + 4)(2x - 1)$
13. $x^2 y - 6x^3 - 12xy^2 - 6y^3 \equiv (-3x^2 - 4xy - 12y^2)(2x - 3y) - 42y^3$
15. $4x^3 + 5 + 4x^2 - 13x \equiv (2x^2 - 3x + 1)(2x + 5)$
17. $5x^3 - 2x^2 + 3x - 4 \equiv (5x + 8)(x^2 - 2x + 1) + (14x - 12)$
19. $x^6 - y^6 \equiv (x^5 + x^4 y + x^3 y^2 + x^2 y^3 + xy^4 + y^5)(x - y)$

ARTICLE 1-6, (2)

1. $Q = 3x + 7, R = 17$
3. $Q = x^2 - 4x + 8, R = -7$

5. (a) $Q = x^3 - 3x - 10, R = -28$
   (b) $Q = x^3 - 3x^2 - 4, R = -4$
7. (a) $Q = 3x^3 + 6x^2 + 12x + 17, R = 14$
   (b) $Q = 3x^3 - 6x^2 + 12x - 31, R = 42$
9. $x^3 - 2x^2 + 3x - 4 \equiv (x^2 + x + 6)(x - 3) + 14$
11. $x^4 - 5x^3 + x^2 - 6 \equiv (x^3 - 4x^2 - 3x - 3)(x - 1) - 9$

ARTICLE 1–7

1. $6ax - 8ay$ 3. $-21x^3y - 28xy^2$ 5. $4x^2 - 9y^2$
7. $x^4 - 16y^4$ 9. $4x^2 + 28xy + 49y^2$ 11. $x^2 - 7x + 10$
13. $x^2y^4 - 2xy^2z^2w + z^4w^2$ 15. $28x^2 - 9xy - 9y^2$
17. $4x^2 + 12xy + 9y^2 - 9$
19. $x^2 + 4y^2 + z^2 - 4xy - 2xz + 4yz$
21. $x^3 + 8$
23. $x^2 + 6xy + 9y^2 - 4z^2 + 16zw - 16w^2$
25. $a^2 + b^2 + c^2 + d^2 - 2ab + 2ac - 2ad - 2bc + 2bd - 2cd$
27. $4(x + 2y)^2 + 2(x + 2y) - 12$
29. $8x^3 + 36x^2y + 54xy^2 + 27y^3$

ARTICLE 1–8, (1)

1. $4(x - 5)$ 3. $3y(y - 3)$
5. $xyz^2(yz - 3x + 5y^2)$ 7. $(2x + 5)(3y - 4x)$
9. $2z(x + 3y)(z - 3x)$ 11. $(3 - a)(3 + a)$
13. $(15a^4 - 8b)(15a^4 + 8b)$ 15. $x(xy^2 - 5d^3)(xy^2 + 5d^3)$
17. $(x + 2y - z)(x + 2y + z)$
19. $(a + b + c + d)(a + b - c - d)$
21. $[9(4x - 3y) + 5(3z + w)][9(4x - 3y) - 5(3z + w)]$
23. $(x - 4)^2$ 25. $(3xy + 11)^2$
27. $5(z - 3w)^2$ 29. $(7 - x)^2$
31. $(a - 2)(a^2 + 2a + 4)$
33. $(2x^{2n} + 3y^m)(4x^{4n} - 6x^{2n}y^m + 9y^{2m})$
35. $(x - 5y)(19x^2 - 10xy + 7y^2)$
37. $(x - 4)(x - 3)$ 39. $(ab - 5)(ab + 4)$
41. $(7x - 2)(5x - 2)$ 43. $(3a - 4)(2a + 5)$
45. $(x + y - 5)(x + y - 2)$ 47. $(4x + 2y - 5)(2x + y + 2)$
49. $2(2a + 2b + c + d)(3a + 3b - 5c - 5d)$

ARTICLE 1–8, (2)

1. $(a + b)(x - y)$ 3. $(x - 2)(x^2 + 4)$
5. $(a - 3)(2 - b^2)$ 7. $(x - 1 + y)(x - 1 - y)$
9. $(2x + y - 2)(2x - y + 2)$

11. $(x + y + z - w)(x + y - z + w)$
13. $(x + 2y - 3)(x + 2y + 2)$
15. $(x^2 - xy - 3y^2)(x^2 + xy - 3y^2)$
17. $(a^2 - 2ab + 3b^2)(a^2 + 2ab + 3b^2)$
19. $(a^4 + b^4)(a^2 + b^2)(a + b)(a - b)$
21. $(x - z)(x + 2y + z)$ 23. $3(z - x)(x + 2y + z)$

CHAPTER 2

ARTICLE 2–1

1. $\dfrac{4}{9}$ 3. $\dfrac{a^2 x^2}{y^2}$ 5. $\dfrac{a}{x + y}$

7. $\dfrac{x + 1}{x}$ 9. $\dfrac{x + 4}{x - 4}$ 11. $\dfrac{y + 2}{y + 5}$

13. $\dfrac{3a + 1}{2a - 1}$ 15. $\dfrac{2(3 - x)}{x + 5}$ 17. $\dfrac{x + 6}{x^2 + 6x + 36}$

19. $\dfrac{x^2 + 2xy + y^2}{x^2 + y^2}$

ARTICLE 2–2

1. $\dfrac{6}{5}$ 3. $\dfrac{(3x - 4y)(3x + 4y)}{12xy}$ 5. $\dfrac{23 - 2x}{18}$

7. $\dfrac{2x^2 - y^2}{x - y}$ 9. $\dfrac{x - y}{5x - 3}$ 11. $\dfrac{5yz - 4xz + 3xy}{xyz}$

13. $\dfrac{5(1 - x)}{3(x - 4)}$ 15. $\dfrac{2x^2 - 9x - 9}{(2x - 3)(x - 5)(x - 6)}$

17. $\dfrac{4a^2 + 9a + 29}{a^3 - 27}$ 19. $-\dfrac{5}{(x + 2)(x + 3)}$

21. $\dfrac{8y^4 - 28y^3 + 21y^2 + 27y - 35}{(2y - 3)^2(y + 1)}$

23. $\dfrac{xz - x^2 + xy - y^2 + yz - z^2}{(x - y)(y - z)(z - x)}$

ARTICLE 2–3, (1)

1. 1/4   3. 7/17   5. $\dfrac{1}{a + b}$

7. $\dfrac{15x}{4y}$ 9. $\dfrac{40x^3}{81}$ 11. $\dfrac{x}{x^2 + xy + y^2}$

13. $\dfrac{1}{x + 3}$ 15. y 17. 1

19. $-\dfrac{2 + 3x}{x^2(x + 1)}$

ARTICLE 2-3, (2)

1. $-\dfrac{57}{5}$ 3. $\dfrac{x}{z}$ 5. $\dfrac{2y + 5x}{2y - 5x}$

7. $x - 1$ 9. $(3x + 2y)(y - 2x)$

ARTICLE 2-4

1. 8 3. $\dfrac{3y^4}{x^2}$ 5. $\dfrac{2x^5}{3y^5}$

7. $\dfrac{cd(a + b)}{ab}$ 9. $\dfrac{(a + b)^2}{ab}$ 11. $x^{nm}y^{2m}$

ARTICLE 2-5

1. 5 3. 4/7 5. 16/9
7. 1/64 9. 81 11. $\sqrt[20]{x}$
13. $\sqrt[20]{x}$ 15. $\sqrt[20]{x}$ 17. $\dfrac{1}{64xy^5}$
19. $\dfrac{5x^2}{3y}$ 21. $a + 2\sqrt{ab} + b$ 23. $x + y$

25. $x^2 + \dfrac{2x}{y} + \dfrac{1}{y^2}$

ARTICLE 2-6

1. $2\sqrt{2}$ 3. $5/2$ 5. $-5\sqrt[3]{5}$
7. $3xy^2\sqrt{3xy}$ 9. $3zx^2y\sqrt[3]{3zy^2}$ 11. $5\sqrt{35}/21$
13. $b\sqrt{a^2 + c^2}$ 15. $\sqrt{3xy}/y^2$ 17. $xy\sqrt[4]{27x^3y^2}/9$
19. $\sqrt{5}$ 21. $\sqrt[3]{45x}/3x$ 23. $x\sqrt{13xz}/y$

ARTICLE 2-7

1. $4\sqrt{3}$ 3. $11\sqrt{2}$ 5. $-\sqrt{x + y}$

7. $(ac^2 + b^3c + a^4b^2)\sqrt{abc}$ 9. $32\sqrt{3}/3$
11. 5, 2

ARTICLE 2–8
1. $\sqrt{65}$ 3. $2\sqrt[3]{13}$ 5. $(x-y)\sqrt{x+y}$
7. $(x+y)\sqrt{x^2 - xy + y^2}$ 9. $2\sqrt[3]{3}$
11. $2(\sqrt{3} + \sqrt{7})$ 13. -2 15. $-13 - \sqrt{15}$
17. $(3 - \sqrt{5})/2$ 19. $8/3$ 21. $\sqrt[3]{9}/3$
23. $\sqrt[4]{3a^2b^2}/b$ 25. $\sqrt{15}/10$ 27. $\sqrt[3]{4}$
29. $(5\sqrt{7} + 5\sqrt{3})/4$ 31. $\dfrac{x^2 - x\sqrt{y}}{x^2 - y}$
33. $-(57 + 13\sqrt{21})/12$ 35. $2(4 + 2\sqrt[3]{3} + \sqrt[3]{9})/5$
37. $(2\sqrt{3} + 3\sqrt{2} - \sqrt{30})/12$

CHAPTER 3

ARTICLE 3–1
1. $-6.5, -5, -1, 0, 0.333, 1/3, \sqrt{4}, 2.3, 2^3$
5. $(0), (-1), (1), \left(\dfrac{\sqrt{2} + \sqrt{3}}{2}\right), \left(\dfrac{x_1 + x_2}{2}\right)$
7. (a) ± 2 (b) $\pm\sqrt{5}$ (c) ± 3 (d) $\pm 1/4$
   (e) $7, -3$ (f) 4 (g) $9, -3$ (h) $6, -4$
   (i) $6, -2$ (j) none   (k) none   (l) $8, 2$
11. $|x - a| = r$, or $\sqrt{(x-a)^2} = r$

ARTICLE 3–2
3. (a) $(3,2)$ (b) $(-4,6)$ (c) $(5,0)$
5. II, IV, III, I, II, IV
7. (a) $(8,4), (4,-4), (-4,4)$ (b) $(-1,6), (3,-2), (-3,-4)$

ARTICLE 3–3
1. (a) $\sqrt{34}$ (b) $\sqrt{106}$ (c) $3\sqrt{2}/4$
   (d) 13 (e) 8 (f) $\sqrt{(x+1)^2 + (y-3)^2}$
7. Yes; No
9. $(2\sqrt{3}, -1 - 4\sqrt{3})$ or $(-2\sqrt{3}, -1 + 4\sqrt{3})$
11. $(\pm a\sqrt{2}/2, 0)$ and $(0, \pm a\sqrt{2}/2)$
13. $(1,0), (0,-1), (1/\sqrt{2}, 1/\sqrt{2}), (-1/2, \sqrt{3}/2)$

ANSWERS TO ODD-NUMBERED PROBLEMS 317

ARTICLE 3-4

1. (a) $(x - 3)^2 + (y - 1)^2 = 25$ (b) $(x - 4)^2 + (y + 2)^2 = 9$
   (c) $(x + 1)^2 + (y - 3)^2 = 9$ (d) $(x - 2)^2 + (y + 4)^2 = 25$
3. (a) All points inside or on the circle
   (b) All points outside or on the circle
   (c) The point $(3, -1)$
5. $4r\sqrt{3}$

ARTICLE 3-5

5. Problem 2: (a) $\frac{1}{8}$ (c) $-\frac{5}{8}$ (e) $\frac{2}{3}$ (g) 2
              (b) $\frac{3}{8}$ (d) $-\frac{5}{6}$ (f) $\frac{5}{4}$ (h) $-\frac{1}{3}$
   Problem 3: (a) $\frac{1}{12}$ (c) $\frac{1}{8}$ (e) $-\frac{3}{4}$ (g) $\frac{5}{24}$
              (b) $\frac{1}{3}$ (d) $\frac{2}{9}$ (f) $-\frac{5}{12}$ (h) $-\frac{5}{2}$
7. (a) .4712 (c) .8203 (e) 4.4186
   (b) 2.7285 (d) 3.3080 (f) -6.6116
9. (a) 47°41′ (b) 33°53′ (c) 51° (d) 151°
   (e) 81°21′ (f) 85°17′ (g) 164°47′ (h) 42°22′
   (i) 212°37′ (j) 206°12′ (k) 37°38′ (l) 4°26′
11. 3/2 radians, 85°57′ 13. 120°, 30°, 172°30′, 108°30′
15. 60 radians 17. 5, 1, θ

ARTICLE 3-6

1. $-5, -3, 1, -7$
3. $1, -1, -1/4.$ $f(3)$ does not exist
5. $0, \sqrt{2}, 2.$ All non-negative values
7. $2, 0, 2, 4.$ All real values, all non-negative real values
9. $P = 3x, A = x^2\sqrt{3}/4$

ARTICLE 3-7

1. $-5/2$ 3. 0 5. 2, 5
7. 1 9. 1 11. $0, \pm 1$

CHAPTER 4

ARTICLE 4-2

1. $+$ 3. $+$ 5. $+$ 7. $-$
9. $-$ 11. $+$ 13. $+$ 15. $-$
17. I or IV 19. III or IV 21. I or IV 23. IV
25. II
27. $\sin \theta = 4/5$ 29. $\sin \theta = 3/5$
    $\cos \theta = 3/5$ $\cos \theta = -4/5$

31. $\sin \theta = -15/17$
    $\cos \theta = -8/17$
33. $\sin \theta = -1$
    $\cos \theta = 0$
35. $\sin \theta = 0$
    $\cos \theta = 1$
37. $\sin \theta = -3/5$
    $\tan \theta = 3/4$
39. $\cos \theta = -\sqrt{5}/3$
    $\tan \theta = -2/\sqrt{5}$
41. $\sin \theta = -\sqrt{11}/6$
    $\tan \theta = -\sqrt{11}/5$
43. $\cos \theta = -4/5$
    $\tan \theta = 3/4$
45. $\sin \theta = \pm 2/\sqrt{13}$
    $\cos \theta = \pm 3/\sqrt{13}$
47. $\cos \theta = \pm \sqrt{95}/12$
    $\tan \theta = \pm 7/\sqrt{95}$
49. $\sin \theta = -15/17$
    $\cos \theta = -8/17$
51. $\sin \theta = -15/17$
    $\cos \theta = -8/17$

ARTICLE 4–3

1. $\sin \theta = 0$
   $\cos \theta = -1$
3. $\sin \theta = 0$
   $\cos \theta = 1$
5. (a) $\sin \theta = 0$
      $\cos \theta = -1$
   (b) $\sin \theta = -1$
      $\cos \theta = 0$
   (c) $\sin \theta = 1$
      $\cos \theta = 0$
   (d) $\sin \theta = 0$
      $\cos \theta = 1$
   (e) $\sin \theta = 0$
      $\cos \theta = -1$
   (f) $\sin \theta = 0$
      $\cos \theta = 1$
   (g) $\sin \theta = 1$
      $\cos \theta = 0$
   (h) $\sin \theta = 0$
      $\cos \theta = 1$
   (i) $\sin \theta = 0$
      $\cos \theta = -1$
25. (a) $\sqrt{2}$ (b) 2

ARTICLE 4–4

11. (a) 1 (b) 1
13. $-\sqrt{3}$
15. $\frac{5}{2} - \sqrt{3}$
17. -1
19. 2
21. $30° = \pi/6$
    $150° = 5\pi/6$
23. $30° = \pi/6$
    $210° = 7\pi/6$
25. $135° = 3\pi/4$
    $315° = 7\pi/4$
27. $45° = \pi/4$
    $135° = 3\pi/4$
29. $150° = 5\pi/6$
    $330° = 11\pi/6$

CHAPTER 5

ARTICLE 5–3

1. $\sin 75° = \dfrac{\sqrt{6} + \sqrt{2}}{4}$, $\cos 75° = \dfrac{\sqrt{6} - \sqrt{2}}{4}$, $\tan 75° = 2 + \sqrt{3}$

3. $\cos \dfrac{7\pi}{12} = \dfrac{\sqrt{2} - \sqrt{6}}{4}$, $\tan \dfrac{7\pi}{12} = -(2 + \sqrt{3})$

5. (a) $56/65$ (b) $-33/65$ (c) $-56/33$
   (d) $-16/65$ (e) $63/65$ (f) $-16/63$

ANSWERS TO ODD-NUMBERED PROBLEMS 319

7. (a) II (b) IV
9. $\sin(\alpha + \beta) = -304/425$, $\cos(\alpha + \beta) = 297/425$
11. $\dfrac{\sqrt{3}\tan\theta + 1}{\sqrt{3} - \tan\theta}$
13. $\dfrac{1 - \tan\theta}{1 + \tan\theta}$
15. $\dfrac{2}{\sqrt{3}\sin\theta - \cos\theta}$
17. $\dfrac{\sin\theta - \cos\theta}{\sqrt{2}}$
33. $13\sin(\theta + \theta_1)$, where $\sin\theta_1 = 12/13$ and $\cos\theta_1 = 5/13$
35. $5\sin(\theta + \theta_1)$, where $\sin\theta_1 = -3/5$ and $\cos\theta_1 = 4/5$
37. $\sqrt{2}\sin\left(\theta + \dfrac{\pi}{4}\right)$

ARTICLE 5–4

1. $-\sin 16°$
3. $-\sin 41°$
5. $-\cot 24°$
7. $\cos 5°$
9. $-\cot 14°$
11. $-\sin 16°18'$
13. $-\cos 24°46'$
15. $\sin 23°21'$
27. π

ARTICLE 5–5

3. $33° = 15° + 18°$, $39° = 75° - 36°$, $42° = 60° - 18°$
5. (a) .2476 (b) 1.3111
   (c) .8760 (d) .8949
   (e) .3872 (f) .3035
   (g) .9026 (h) $-.9465$
7. (a) $25°10'$ (b) $48°20'$
   (c) $27°40'$ (d) $57°40'$
   (e) $12°10'$ (f) $75°50'$
   (g) $20°20'$ (h) $38°20'$

CHAPTER 6

ARTICLE 6–1

1. -2
3. $5/6$
5. -3
7. -7
9. $11/9$
11. -5
13. $-16/9$
15. $\pi/3, 2\pi/3$
17. $0.3217, 3.4633$
19. $0.2195, 2.3985, 3.3611, 5.5401$
21. $\dfrac{c}{a - b}$
23. $2A/h$
25. $\dfrac{l - a}{n - 1}$
27. $\dfrac{Sr + a - S}{r}$
29. $\dfrac{2S - an}{n}$
31. $\dfrac{a\tan\theta_1}{\tan\theta_2 - \tan\theta_1}$
33. 18 years
35. \$45,000
37. $1\tfrac{5}{7}$ hrs

ARTICLE 6-2

1. Minimum -4, when $x = -3$.
3. Minimum $-121/8$, when $x = -5/4$
5. Minimum $-169/24$, when $x = 17/12$
7. Minimum 2, when $x = -3$
9. 8, 8   11. 40 ft, 40 ft

ARTICLE 6-3, (1)

1. $-1, -5$ 3. $3/2, -4$ 5. $1/3, 5/2$
7. none   9. $\pm 4/3$
11. $\pi/6, 5\pi/6, 7\pi/6, 11\pi/6$ 13. $10/3, -5/2$
15. $\pi/4, 3\pi/4, 5\pi/4, 7\pi/4$ 17. $-5/3, 2$
19. $\pi/3, \pi/2, 3\pi/2, 5\pi/3$ 21. $-a \pm b$
23. $\pi/2, 7\pi/6, 11\pi/6$ 25. $-1/2, -14/3$

ARTICLE 6-3, (2)

1. $3/2, -4$ 3. $\dfrac{-1 \pm \sqrt{5}}{2}$ 5. $0, \pi/3, 5\pi/3$

7. a, b 9. $1, \dfrac{c-a}{a-b}$

11. $3\pi/4, 7\pi/4, 0.5880, 3.7295$ 13. $1, -5/4$

15. $0.3142, 0.9425, 2.1991, 2.8275$
$3.4558, 4.0841, 5.3407, 5.9691$ 17. $\dfrac{v_0 \pm \sqrt{v_0^2 - 2gs}}{g}$

19. -3 21. $3/5$ or $5/3$ 23. 1

25. $\dfrac{-27 + \sqrt{909}}{2}$ yds

ARTICLE 6-4

5. $x < 6$ 7. $x > 3$ 9. $-\tfrac{7}{2} < x < 2$
11. $x > \tfrac{5}{2}$ or $x < -\tfrac{7}{3}$ 13. All real values
15. $x > 5$ or $x < 2$ 17. $x > 5$ or $x < 2$
19. $x > 5$ or $x < 1$

21. $\dfrac{3\pi}{4} < \theta < \dfrac{5\pi}{4}$ or $0 \leq \theta < \dfrac{\pi}{4}$ or $\dfrac{7\pi}{4} < \theta \leq 2\pi$

23. $\pi < \theta < 2\pi$ 25. $\dfrac{4\pi}{3} < \theta < 2\pi$

27. $0 \leq \theta \leq \dfrac{7\pi}{6}$ or $\dfrac{11\pi}{6} \leq \theta < 2\pi$ 29. All values

31. $\dfrac{\pi}{4} < \theta < \dfrac{5\pi}{4}$

ARTICLE 6-5

1. The sum of the zeros are -6, -1, $-5/2$, $11/2$, $17/6$, $5/2$, -6, and $5/3$. The product of the zeros are 5, -6, -6, $15/2$, $5/6$, -4, 11, $4/3$
3. $x^2 + x - 20 = 0$
5. $12x^2 + x - 6 = 0$
7. $x^2 - 4x + 1 = 0$
9. $8x^2 + 12x + 1 = 0$
11. $5x^2 - 8x - 4 = 0$
13. $2x^2 + 5x - 3 = 0$
15. $36x^2 + 24x - 5 = 0$
17. $6x^2 - 5x - 1 = 0$
19. $k = -1$
21. $k = 25$
23. $k = 49/12$
25. $k = -5$
27. $k = 0$
29. All values of k
31. $-8 < k < 8$
33. $k = 27/4$
35. $k > \frac{3}{2}$ or $k < -1$

ARTICLE 6-6

1. $\pm 2, \pm \sqrt{7}$
3. $\pm 1/3, \pm 1/2$
5. Real roots: $-2, 1$
7. $-64, 8$
9. $5, 2, \dfrac{7 \pm \sqrt{53}}{2}$
11. $0, 0$
13. $-\dfrac{23}{9}$
15. $5, -5/2, \dfrac{5 \pm \sqrt{57}}{4}$

ARTICLE 6-7

1. $11/2$
3. $7, 1$
5. 16
7. -5
9. 6
11. $2.09, 6.05$ (approx.)
13. $0, \pi/2$
15. $3\pi/4, 5.99$ (approx.)

ARTICLE 6-8

1. $z = \dfrac{kx}{y}$
3. $z = 6xy$
5. $C = kd^2$
7. $A = kx^2$
9. $\dfrac{1}{2}, \dfrac{\sqrt{3}}{2} 0.6157$
11. 90
13. 576π sq in.
15. 72 ergs
17. 2 ft
19. Multiplied by 16

ARTICLE 6-9

1. $x = 2, y = -1$
3. $x = 3, y = 5$
5. $x = 2/3, y = 3/2$
7. $\alpha = \pi/6, 5\pi/6$
   $\beta = 0$

9. $\alpha = \pi/6, 11\pi/6$
$\beta = \pi/6, 7\pi/6$

11. $x = 5, y = -2$

13. $x = \dfrac{a^3 + a^2b + 3ab^2 + b^3}{a^2 + b^2}$

$y = \dfrac{a^3 - a^2b + ab^2 + b^3}{a^2 + b^2}$

15. $x = \dfrac{a}{\tan k_2 - \tan k_1}$

$y = \dfrac{a \tan k_1}{\tan k_2 - \tan k_1}$

17. 5, 4

19. 62°, 28°

21. 150 mi/hr, 10 mi/hr

23. 12 hr, 15 hr

ARTICLE 6–10

1. $x = 1, y = 2, z = 3$
5. $x = \frac{1}{3}, y = -\frac{2}{5}, z = \frac{1}{2}$
9. $x = 3, y = 4, z = 6$
13. 90°, 15°, 75°.

3. $x = 3, y = -1, z = 4$
7. $x = 5, y = 6, z = 7$
11. 544
15. $x^2 + y^2 - x - 7y + 6 = 0$

ARTICLE 6–11

1. $x = 0, 2$
$y = 0, 16$

3. no real values

5. $x = \dfrac{3 \pm \sqrt{7}}{2}, y = \dfrac{5 \mp \sqrt{7}}{2}$

7. $x = 2, -\frac{8}{5}; y = 1, -\frac{1}{5}$

9. $x = \dfrac{-2 \pm 2\sqrt{5}}{3}$

11. $b = \pm a\sqrt{1 + m^2}$

$y = 1 \pm \sqrt{5}$

13. 23, 32

15. 7, 4

CHAPTER 7

ARTICLE 7–1 (1)

1. -2
11. $x = 1$

3. 3

5. 1
15. Real; equal; imaginary

ARTICLE 7–1 (2)

1. 5
17. $9x - 5y - 2 = 0$

3. 0

15. $\dfrac{4x + 3y - 17}{2}$
19. $x = -1, y = 4$.

ARTICLE 7–2

1. Rows and columns are interchanged

3. Two columns are identical
5. 1008 9. $x = a, b$ 11. $x = 3, 1$

ARTICLE 7–3

1. -110 3. -484 5. 0

ARTICLE 7–4

3. $x = -4, y = -3, z = 2, w = 1$

CHAPTER 8

ARTICLE 8–1 (1)

1. $R = 17$ 3. $R = -7$ 5. (a) -6, (b) 57
7. Yes 9. No 11. Yes

ARTICLE 8–1, (2)

1. 2, one; 3, two; -4, three
3. -7, single; $3/2$, three
5. $-5/3$, single; 3, four
7. upper limit 3, lower limit -4
9. upper limit 3, lower limit -2
11. upper limit 3, lower limit -3
13. upper limit 2, lower limit -1
15. upper limit 2, lower limit -3

ARTICLE 8–2

5. 2, double; single between 1 and 2; single between -5 and -6
7. single at -3; double between 0 and 1
9. single between -2 and -3

ARTICLE 8–4

1. $-2, 3, 1/2$ 3. $1, 3, 5, 7$ 5. $1, 2, 3, -5$
7. $-\frac{3}{2}, -\frac{3}{2}, \frac{1 \pm \sqrt{5}}{2}$ 9. $1, 1, 1, 1$
11. $\frac{1}{2}, \frac{1}{2}, \frac{1}{2}, \pm \sqrt{-1}$
13. $\pi/6, 5\pi/6, 7\pi/6, 11\pi/6, 3\pi/2$
15. $\pi/6, 5\pi/6, 7\pi/6, 11\pi/6, 3\pi/4, 7\pi/4$

ARTICLE 8–5

1. 0.75 3. 0.45 5. 1.36, 1.69
7. $-0.62, 1.62$ 9. 1.817 11. 1.189

CHAPTER 9

ARTICLE 9–1

1. $x = \dfrac{y + 6}{5}$

3. $x = 2 \pm \sqrt{y + 4}$

5. $x = \dfrac{1}{\pm \sqrt{1 - y}}$

7. $x = \sqrt[n]{y}$

9. $x = \dfrac{2y + 3 \pm \sqrt{4y - 7}}{2}$

ARTICLE 9–3

3. $\pm 2\pi/3 + 2n\pi$
5. $n\pi$
7. $\pm \pi/4 + 2n\pi$
9. $\pi/2$
11. $-\pi/2$
13. $(-1)^n 24° + n180°$
15. $\pm 51° + n360°$
17. $84°20'$
21. $\frac{1}{2}$ arc tan $y/3$
23. $\frac{1}{2}$ arc sec $y/2$
27. $\frac{1}{4}(4 + \cos 3y)$
29. $\tan(y + 2)$

31. $\dfrac{\cos\left(2y - \dfrac{\pi}{12}\right) - 1}{2}$

ARTICLE 9–4

1. $\pm 3/5$
3. $\pm 12/5$
5. $\sqrt{11}/6$
7. u
9. $\pm \sqrt{1 - u^2}/u$
11. $\pi/7$
13. $\pi/18$
15. $2\pi/5$
17. 0
19. $uv \pm \sqrt{(1 - u^2)(1 - v^2)}$
21. $\pm 3/5$
23. 1 or $-7/8$
25. $-16/63$
27. $(24\sqrt{5} - 14)/75$
37. $2n\pi \pm (\pi/2 - \theta)$
39. $n\pi + \theta$
41. 1
43. 0
45. 0

CHAPTER 10

ARTICLE 10–1

1. amp. 5, period 1
3. amp. 3, period 6
5. amp. 1.5, period $4\pi/3$
7. amp. 2, period 8π
9. amp. .5, period $2\pi/3$
11. amp. 100, period 200 (approx.)

ARTICLE 10–3

1. amp. 2, period 2π, phase displacement $\pi/6$
3. amp. 5, period π, phase displacement $\pi/16$

ANSWERS TO ODD-NUMBERED PROBLEMS 325

5. amp. 1, period 2π, phase displacement .25
7. amp. 1, period 2π, phase displacement 1.176
9. amp. 17, period 2π, phase displacement, .4900
11. amp. $\sqrt{29}$, period 2π, phase displacement, 1.1903
13. period 2π 15. period 4π 17. period 2π

ARTICLE 10–4

1. P_x has an x-coordinate, $x = a \cos \omega t$ 3. Yes; period π
5. $E = 8 \sin 120\pi t$ 7. $y = .001 \sin 800\pi t$

ARTICLE 10–5

1. amp., a
3. $A_0 = 80$, $A_1 = 9.5$, $A_2 = 1.44$, $\alpha_1 = .53$, $\alpha_2 = 2.16$

CHAPTER 11

ARTICLE 11–2

1. $a^7 + 7a^6b + 21a^5b^2 + 35a^4b^3 + 35a^3b^4 + 21a^2b^5 + 7ab^6 + b^7$
3. $32x^5 + 80x^4y^2 + 80x^3y^4 + 40x^2y^6 + 10xy^8 + y^{10}$
5. $625x^4 - 500x^3y^2 + 150x^2y^4 - 20xy^6 + y^8$
7. $x^2 + 6x^{5/3}y^{1/3} + 15x^{4/3}y^{2/3} + 20xy + 15x^{2/3}y^{4/3} + 6x^{1/3}y^{5/3} + y^2$
9. $x^2 - \dfrac{15x^{8/5}}{y^2} + \dfrac{90x^{6/5}}{y^4} - \dfrac{270x^{4/5}}{y^6} + \dfrac{405x^{2/5}}{y^8} - \dfrac{243}{y^{10}}$
11. $\dfrac{x^{24}}{4096} + \dfrac{3x^{22}}{256y^2} + \dfrac{33x^{20}}{128y^4} + \dfrac{55x^{18}}{16y^6} + \cdots$
13. $x^{11/3} - \dfrac{11x^{10/3}}{y^{1/3}} + \dfrac{55x^3}{y^{2/3}} - \dfrac{165x^{8/3}}{y} + \cdots$
15. $1 + kx + \dfrac{k(k-1)}{2}x^2 + \dfrac{k(k-1)(k-2)}{3!}x^3 + \cdots$
17. $59{,}136x^6y^6$ 19. $\dfrac{35y^8}{8}$ 21. $-414{,}720x^7$

ARTICLE 11–3

1. $1 - x + x^2 - x^3 + \cdots$ 3. $1 + \dfrac{x}{2} + \dfrac{3x^2}{8} + \dfrac{5x^3}{16} + \cdots$
5. 0.9803   7. 1.005   9. 5.745   11. 4.932

ANSWERS TO ODD-NUMBERED PROBLEMS

CHAPTER 12

ARTICLE 12–2

1. $\log_3 27 = 3$
3. $\log_4 1 = 0$
5. $\log_8 16 = \frac{4}{3}$
7. $\log_{10} .001 = -3$
9. $\log_4 4 = 1$
11. $36^{\frac{1}{2}} = 6$
13. $x = 2$
15. $x = -1$
17. $u = 25$
19. $x = 14$
21. $a = \frac{27}{8}$
27. $\log_b \dfrac{a+2}{a-3}$
29. $\log_a \dfrac{\sqrt{x}}{\sqrt[3]{y^2}}$
31. $\log_a x^5(x-1)^6$
33. $4 \log_{10} 60.3$
35. $3 \log_{10} 54.3 + \log_{10} 67 - \log_{10} 93.9 - 2 \log 32.5$

ARTICLE 12–3 (1)

1. 0.4099
3. 0.8401
5. 0.6730
7. 0.8373
9. 0.5153
11. 0.8858
13. 2.43
15. 4.12
17. 8.74
19. 1.851
21. 8.374
23. 3.213

ARTICLE 12–3 (2)

1. characteristic, 1; mantissa, .3782
3. characteristic, −3; mantissa, .5728
5. characteristic, 5; mantissa, .8723
7. characteristic, −4; mantissa, .2715
9. 2.5172
11. 3.6747
13. 8.8623 − 10
15. 6650
17. 267.2
19. .003653

ARTICLE 12–3 (3)

1. (a) 9.3629 − 10 (b) 9.8457 − 10
   (c) 9.9657 − 10 (d) 9.8944 − 10
   (e) .5089   (f) 9.9562 − 10
   (g) 9.5556 − 10 (h) 9.8954 − 10

ARTICLE 12–4

1. 32000
3. 34.6
5. 158,800
7. 0.5152
9. 2.642
11. 0.1313
13. 8.52
15. 0.923
17. 79.5
19. 396.1

ARTICLE 12–5

(All problems worked with four-place tables in back of book.)

1. (a) $295.80, (b) $297.10, (c) $297.10 (larger than (b) if more accurate tables were used), (d) $298.30

3. (a) 11.9 yrs, (b) 11.55 yrs
5. $r = 0.3467$ 7. (a) -0.2746; (b) .002

ARTICLE 12–6

1. 3.808   3. 2.493   5. 1.947
7. 0.5824   9. 7.052×10^{13} 11. 26.6
13. 2.33   17. 5   19. 3.155
21. 6.14   23. 0.402   25. $\log_e (1 \pm \sqrt{2})$
27. 2/11   29. 1/297

CHAPTER 13

ARTICLE 13–3

1. (a) $\beta = 52°40'$, $b = 319$, $c = 401$
   (b) $\beta = 27°20'$, $a = 1540$, $c = 1730$
   (c) $\alpha = 29°40'$, $\beta = 60°20'$, $c = 6.63$
   (d) $\alpha = 44°40'$, $\beta = 45°20'$, $a = 67.6$
   (e) $\alpha = 38°50'$, $a = .522$, $b = .648$
   (f) $\beta = 52°20'$, $b = 71.0$, $c = 89.7$
   (g) $\alpha = 33°30'$, $\beta = 56°30'$, $c = 68.7$
   (h) $\alpha = 48°35'$, $a = 2448$, $b = 2160$
   (i) $\alpha = 59°36'$, $\beta = 30°24'$, $b = 3185$
   (j) $\alpha = 23°10'$, $\beta = 66°50'$, $c = 8.320$
   (k) $\alpha = 27°3'$, $b = 1.619$, $c = 1.818$
   (l) $\alpha = 42°37'$, $a = 66.74$, $c = 98.58$
   (m) $\alpha = 42°30'$, $\beta = 47°30'$, $a = 3274$
   (n) $\beta = 65°13'$, $a = 147.0$, $c = 350.8$
3. 313 ft, 19°0'   5. 474 sq in   7. 55°
9. 7.075 in., 7.075 in., 10.62 in.
11. 172 ft   13. 78.8 ft   15. 2770 ft
17. 6.30 mi from A, 9.09 mi from B
19. 3195 ft south, 3485 ft west
21. (a) $v = 5$, $\theta = 53°$ (b) $v = 50$, $\theta = 63°$
    (c) $v = 32.9$, $\theta = 119°30'$ (d) $v = 858.2$, $\theta = 296°27'$
23. 56°, 22 ft/sec   25. N 47°0' W
27. (a) $f = 43$, $\theta = 122°$ (b) $f = 8250$, $\theta = 21°10'$ (c) $f = 3000$, $\theta = 255°$

ARTICLE 13–5

1. (a) $\gamma = 38°$, $b = 163$, $c = 102$
   (b) $\alpha = 40°26'$, $a = 36.27$, $b = 55.35$
   (c) $\beta = 13°52'$, $a = 270{,}400$, $c = 321{,}000$
   (d) $\alpha = 112°16'$, $a = 30.72$, $c = 26.56$

3. No
5. (a) $\alpha_1 = 91°50'$, $\gamma_1 = 49°30'$, $a_1 = 95.0$
   $\alpha_2 = 10°50'$, $\gamma_2 = 130°30'$, $a_2 = 17.9$
   (b) $\beta = 127°40'$, $\gamma = 7°10'$, $b = 55.0$
   (c) $\alpha = 17°49'$, $\gamma = 14°24'$, $c = 11.96$
   (d) $\alpha = 76°42'$, $\beta = 46°0'$, $a = 6.667$
7. (a) $\beta = 31°19'$, $\gamma = 96°23'$, $a = 28.35$
   (b) $\alpha = 45°50'$, $\beta = 76°30'$, $c = 545$
9. (a) $\alpha = 41°$, $\beta = 56°$, $\gamma = 83°$
   (b) $\alpha = 30°$, $\beta = 63°$, $\gamma = 87°$
15. 29.80 in., 48.52 in.   17. 100 mi, 60 mi   19. 25.5 ft
21. N $79°20'$ W   23. 148 mi/hr, S $63°$ E

CHAPTER 14

ARTICLE 14–1

1. $1/\sin \theta$
3. $\pm\sqrt{1 - \sin^2 \theta}$
5. $\pm \sin \theta/\sqrt{1 - \sin^2 \theta}$
7. $1/\cos \theta$
9. $\pm\sqrt{1 - \cos^2 \theta}$
11. $\pm\sqrt{1 - \cos^2 \theta}/\cos \theta$
13. $\sin \theta = \pm \tan \theta/\sqrt{1 + \tan^2 \theta}$, $\cos \theta = \pm 1/\sqrt{1 + \tan^2 \theta}$
    $\cot \theta = 1/\tan \theta$, $\sec \theta = \pm\sqrt{1 + \tan^2 \theta}$, $\csc \theta = \pm\sqrt{1 + \tan^2 \theta}/\tan \theta$
15. $1/(1 - \cos^2 \theta)$

ARTICLE 14–2

3. $\sin 7\pi/12 = \sqrt{2 + \sqrt{3}}/2$, $\cos 7\pi/12 = -\sqrt{2 - \sqrt{3}}/2$, $\tan 7\pi12 = -(2 + \sqrt{3})$
5. (a) $-120/169$ (b) $-119/169$ (c) $120/119$
   (d) $3/\sqrt{13}$ (e) $2/\sqrt{13}$ (f) $3/2$
7. $\dfrac{3 + 4 \cos 2\theta + \cos 4\theta}{8}$
33. $\pi[2n + \tfrac{1}{2}]$, $\pi\left[n + \dfrac{-1^n}{6}\right]$
35. $\dfrac{\pi}{3}\left[2n + \dfrac{1}{2}\right]$
37. $\pi[n + \tfrac{1}{12}]$
39. $36°52' + n360°$ or $0.6435 + 2n\pi$
41. $337°23' + n360°$ or $5.8885 + 2n\pi$
43. none
45. 0.940

ARTICLE 14–3

1. $(\sin 8\theta - \sin 2\theta)/2$
3. $(\sin 12\theta - \sin 2\theta)/2$
5. $(\sin 7\theta + \sin 3\theta)/2$
7. $[\sin (3\theta/2) - \sin (2\theta)]/2$
9. $2 \sin (\pi/6) \cos (\pi/18)$
11. $2 \cos (\pi/2) \cos (5\pi/18)$
13. $2 \sin 6\theta \cos 2\theta$
15. $2 \cos (5\theta/4) \sin (5\theta/12)$

ARTICLE 14-5

1. $.10414 < \sin \pi/30 < .10472$, $1 > \cos \pi/30 > .99452$,
$.10472 < \tan \pi/30 < .10530$

CHAPTER 15

ARTICLE 15-2

7. $(6,8)$
9. $(-1,1)$
11. $(3,11)$
13. $(-\frac{1}{2},-\frac{1}{2})$
15. $(1,0)$
17. (a) $13 + 18i$, (b) $3 + 22i$
19. (a) $-i$ (b) 1 (c) i (d) $-i$

ARTICLE 15-3

1. $x^3 - 7x^2 + 17x - 15 = 0$
3. $2x^3 + 19x^2 + 42x - 26 = 0$
5. $3x^3 - 17x^2 + 27x + 11 = 0$
7. $4, \pm 3i$
9. $2, 3 \pm \sqrt{2}i$

ARTICLE 15-4

1. (a) $2(\cos 0° + i \sin 0°)$ (e) $2\sqrt{2}(\cos 315° + i \sin 315°)$
   (b) $2(\cos 180° + i \sin 180°)$ (f) $2\sqrt{2}(\cos 135° + i \sin 135°)$
   (c) $3(\cos 90° + i \sin 90°)$ (g) $\cos 120° + i \sin 120°$
   (d) $\cos 270° + i \sin 270°$ (h) $\cos 240° + i \sin 240°$
3. (a) $6i$ (b) $-4\sqrt{3} - 4i$ (c) $12i$
   (d) 8
5. (a) $\sqrt{3} + i$ (b) $-1 + \sqrt{3}i$

ARTICLE 15-5

1. $64i$
3. $-4\sqrt{2} - 4\sqrt{2}i$
5. $\dfrac{\sqrt{3}}{2} + \dfrac{i}{2}$
7. 1
9. $-2\sqrt{2} + 2\sqrt{2}i, 2\sqrt{2} - 2\sqrt{2}i$
11. $1 + \sqrt{3}i, -2, 1 - \sqrt{3}i$
13. $\sqrt{3} + i, -1 + \sqrt{3}i, -\sqrt{3} - i, 1 - \sqrt{3}i$
15. $(\cos 36° + i \sin 36°), (\cos 72° + i \sin 72°), \cdots$
17. $1, i, -1, -i$
19. $\dfrac{\sqrt{5}+1}{2} + \sqrt{\dfrac{5-\sqrt{5}}{2}}\,i,$

$$\frac{1-\sqrt{5}}{2}+\sqrt{\frac{5+\sqrt{5}}{2}}\,i,$$

$$-2,$$

$$\frac{1-\sqrt{5}}{2}-\sqrt{\frac{5+\sqrt{5}}{2}}\,i,$$

$$\frac{\sqrt{5}+1}{2}-\sqrt{\frac{5-\sqrt{5}}{2}}\,i$$

CHAPTER 16

ARTICLE 16-1

1. 10, 13, 16; $l = 25$, $S_9 = 117$
3. 1, -2, -5; $l = -32$, $S_{15} = -165$
5. $l = 46$, $S_{12} = 288$
7. $d = 48/91$, $l = 34/7$
9. $a = -31/28$, $l = 151/28$
11. $n = 7$, $S_7 = 147$
13. 8, 9, 10
15. 187,026
17. 2, 5, 8, 11
19. $-25/2$, -7, $-3/2$
21. (a) -4; (b) 17/15
25. 27 numbers; $S = 2835$
27. 1092 times
29. 1610 ft
31. $5075

ARTICLE 16-2

1. 64, 256, 1024; $l = 16{,}384$, $S_8 = 21{,}845$
3. $-1, 5, -25$; $l = 125$, $S_7 = 13{,}021/125$
5. $l = 486$, $S_6 = 728$
7. $r = 1, 3,$ or -4; $l = 1, 9,$ or 16
9. $a = 36/121$; $l = 4/1089$
11. $n = 6$; $S = -126$
13. $a = 625/16, 125/8, 25/4$
15. $n = 9$
17. $\pm 9/4, 3/2, \pm 1$
19. $\pm\sqrt{ab}$
23. 2/81 ft; $26\frac{79}{81}$ ft
25. $1583
27. 10,000,000 in.

ARTICLE 16-3

1. 3/2
3. 64/5
5. $4 + 2\sqrt{2}$
7. 10/3
9. 5/9
11. 64/11
13. $r = 1/7$
15. $16[2 + \sqrt{2}]$
17. 240 ft

CHAPTER 17

ARTICLE 17-1

1. 4
3. 120
5. 240
7. 1920
9. 60
11. (a) 56, (b) 64

ARTICLE 17-2

1. 504
5. (a) 5040, (b) 144
9. 103,680   11. 1440
17. 360

3. 585
7. 5040
13. 468,000   15. 3780
19. $n = 8$

ARTICLE 17-3

1. (a) 35, (b) 45, (c) 210
5. 42,228   7. 18,375
11. $384C(36,6)$
15. 126
19. $n = 7$

3. 12,650
9. 560
13. (a) 36, (b) 84
17. (a) 28, (b) 56, (c) 247

ARTICLE 17-4

1. (a) 1/6, (b) 1/3   3. 1/12   5. 1/9
7. (a) 1/221, (b) 25/102, (c) 20/221
9. (a) 28/165, (b) 14/55   11. 1/36, 1/216
13. 1/66 if anyone may play with anyone else
    1/36 if men must have women for partners
15. .7 (approximate)

ARTICLE 17-5

1. 1/32   3. 3/100   5. 7/8
7. 0.731   9. 0.146   11. 0.034

ARTICLE 17-6

1. (a) 1/2, (b) 1/4
5. (a) 25/72, (b) 91/216
9. (a) 1/4, (b) 1/4

3. .0081
7. (a) 12/49, (b) 9/49, (c) 12/49

APPENDIX

TABLE I
Values of Trigonometric Functions

Degrees	Radians	Sine	Tangent	Cotangent	Cosine		
0° 00′	.0000	.0000	.0000		1.0000	1.5708	90° 00′
10′	.0029	.0029	.0029	343.77	1.0000	1.5679	50′
20′	.0058	.0058	.0058	171.89	1.0000	1.5650	40′
30′	.0087	.0087	.0087	114.59	1.0000	1.5621	30′
40′	.0116	.0116	.0116	85.940	.9999	1.5592	20′
50′	.0145	.0145	.0145	68.750	.9999	1.5563	10′
1° 00′	.0175	.0175	.0175	57.290	.9998	1.5533	89° 00′
10′	.0204	.0204	.0204	49.104	.9998	1.5504	50′
20′	.0233	.0233	.0233	42.964	.9997	1.5475	40′
30′	.0262	.0262	.0262	38.188	.9997	1.5446	30′
40′	.0291	.0291	.0291	34.368	.9996	1.5417	20′
50′	.0320	.0320	.0320	31.242	.9995	1.5388	10′
2° 00′	.0349	.0349	.0349	28.636	.9994	1.5359	88° 00′
10′	.0378	.0378	.0378	26.432	.9993	1.5330	50′
20′	.0407	.0407	.0407	24.542	.9992	1.5301	40′
30′	.0436	.0436	.0437	22.904	.9990	1.5272	30′
40′	.0465	.0465	.0466	21.470	.9989	1.5243	20′
50′	.0495	.0494	.0495	20.206	.9988	1.5213	10′
3° 00′	.0524	.0523	.0524	19.081	.9986	1.5184	87° 00′
10′	.0553	.0552	.0553	18.075	.9985	1.5155	50′
20′	.0582	.0581	.0582	17.169	.9983	1.5126	40′
30′	.0611	.0610	.0612	16.350	.9981	1.5097	30′
40′	.0640	.0640	.0641	15.605	.9980	1.5068	20′
50′	.0669	.0669	.0670	14.924	.9978	1.5039	10′
4° 00′	.0698	.0698	.0699	14.301	.9976	1.5010	86° 00′
10′	.0727	.0727	.0729	13.727	.9974	1.4981	50′
20′	.0756	.0756	.0758	13.197	.9971	1.4952	40′
30′	.0785	.0785	.0787	12.706	.9969	1.4923	30′
40′	.0814	.0814	.0816	12.251	.9967	1.4893	20′
50′	.0844	.0843	.0846	11.826	.9964	1.4864	10′
5° 00′	.0873	.0872	.0875	11.430	.9962	1.4835	85° 00′
10′	.0902	.0901	.0904	11.059	.9959	1.4806	50′
20′	.0931	.0929	.0934	10.712	.9957	1.4777	40′
30′	.0960	.0958	.0963	10.385	.9954	1.4748	30′
40′	.0989	.0987	.0992	10.078	.9951	1.4719	20′
50′	.1018	.1016	.1022	9.7882	.9948	1.4690	10′
6° 00′	.1047	.1045	.1051	9.5144	.9945	1.4661	84° 00′
10′	.1076	.1074	.1080	9.2553	.9942	1.4632	50′
20′	.1105	.1103	.1110	9.0098	.9939	1.4603	40′
30′	.1134	.1132	.1139	8.7769	.9936	1.4573	30′
40′	.1164	.1161	.1169	8.5555	.9932	1.4544	20′
50′	.1193	.1190	.1198	8.3450	.9929	1.4515	10′
7° 00′	.1222	.1219	.1228	8.1443	.9925	1.4486	83° 00′
10′	.1251	.1248	.1257	7.9530	.9922	1.4457	50′
20′	.1280	.1276	.1287	7.7704	.9918	1.4428	40′
30′	.1309	.1305	.1317	7.5958	.9914	1.4399	30′
40′	.1338	.1334	.1346	7.4287	.9911	1.4370	20′
50′	.1367	.1363	.1376	7.2687	.9907	1.4341	10′
8° 00′	.1396	.1392	.1405	7.1154	.9903	1.4312	82° 00′
10′	.1425	.1421	.1435	6.9682	.9899	1.4283	50′
20′	.1454	.1449	.1465	6.8269	.9894	1.4254	40′
30′	.1484	.1478	.1495	6.6912	.9890	1.4224	30′
40′	.1513	.1507	.1524	6.5606	.9886	1.4195	20′
50′	.1542	.1536	.1554	6.4348	.9881	1.4166	10′
9° 00′	.1571	.1564	.1584	6.3138	.9877	1.4137	81° 00′
		Cosine	Cotangent	Tangent	Sine	Radians	Degrees

TABLE I (Continued)
Values of Trigonometric Functions

Degrees	Radians	Sine	Tangent	Cotangent	Cosine		
9° 00'	.1571	.1564	.1584	6.3138	.9877	1.4137	81° 00'
10'	.1600	.1593	.1614	6.1970	.9872	1.4108	50'
20'	.1629	.1622	.1644	6.0844	.9868	1.4079	40'
30'	.1658	.1650	.1673	5.9758	.9863	1.4050	30'
40'	.1687	.1679	.1703	5.8708	.9858	1.4021	20'
50'	.1716	.1708	.1733	5.7694	.9853	1.3992	10'
10° 00'	.1745	.1736	.1763	5.6713	.9848	1.3963	80° 00'
10'	.1774	.1765	.1793	5.5764	.9843	1.3934	50'
20'	.1804	.1794	.1823	5.4845	.9838	1.3904	40'
30'	.1833	.1822	.1853	5.3955	.9833	1.3875	30'
40'	.1862	.1851	.1883	5.3093	.9827	1.3846	20'
50'	.1891	.1880	.1914	5.2257	.9822	1.3817	10'
11° 00'	.1920	.1908	.1944	5.1446	.9816	1.3788	79° 00'
10'	.1949	.1937	.1974	5.0658	.9811	1.3759	50'
20'	.1978	.1965	.2004	4.9894	.9805	1.3730	40'
30'	.2007	.1994	.2035	4.9152	.9799	1.3701	30'
40'	.2036	.2022	.2065	4.8430	.9793	1.3672	20'
50'	.2065	.2051	.2095	4.7729	.9787	1.3643	10'
12° 00'	.2094	.2079	.2126	4.7046	.9781	1.3614	78° 00'
10'	.2123	.2108	.2156	4.6382	.9775	1.3584	50'
20'	.2153	.2136	.2186	4.5736	.9769	1.3555	40'
30'	.2182	.2164	.2217	4.5107	.9763	1.3526	30'
40'	.2211	.2193	.2247	4.4494	.9757	1.3497	20'
50'	.2240	.2221	.2278	4.3897	.9750	1.3468	10'
13° 00'	.2269	.2250	.2309	4.3315	.9744	1.3439	77° 00'
10'	.2298	.2278	.2339	4.2747	.9737	1.3410	50'
20'	.2327	.2306	.2370	4.2193	.9730	1.3381	40'
30'	.2356	.2334	.2401	4.1653	.9724	1.3352	30'
40'	.2385	.2363	.2432	4.1126	.9717	1.3323	20'
50'	.2414	.2391	.2462	4.0611	.9710	1.3294	10'
14° 00'	.2443	.2419	.2493	4.0108	.9703	1.3265	76° 00'
10'	.2473	.2447	.2524	3.9617	.9696	1.3235	50'
20'	.2502	.2476	.2555	3.9136	.9689	1.3206	40'
30'	.2531	.2504	.2586	3.8667	.9681	1.3177	30'
40'	.2560	.2532	.2617	3.8208	.9674	1.3148	20'
50'	.2589	.2560	.2648	3.7760	.9667	1.3119	10'
15° 00'	.2618	.2588	.2679	3.7321	.9659	1.3090	75° 00'
10'	.2647	.2616	.2711	3.6891	.9652	1.3061	50'
20'	.2676	.2644	.2742	3.6470	.9644	1.3032	40'
30'	.2705	.2672	.2773	3.6059	.9636	1.3003	30'
40'	.2734	.2700	.2805	3.5656	.9628	1.2974	20'
50'	.2763	.2728	.2836	3.5261	.9621	1.2945	10'
16° 00'	.2793	.2756	.2867	3.4874	.9613	1.2915	74° 00'
10'	.2822	.2784	.2899	3.4495	.9605	1.2886	50'
20'	.2851	.2812	.2931	3.4124	.9596	1.2857	40'
30'	.2880	.2840	.2962	3.3759	.9588	1.2828	30'
40'	.2909	.2868	.2994	3.3402	.9580	1.2799	20'
50'	.2938	.2896	.3026	3.3052	.9572	1.2770	10'
17° 00'	.2967	.2924	.3057	3.2709	.9563	1.2741	73° 00'
10'	.2996	.2952	.3089	3.2371	.9555	1.2712	50'
20'	.3025	.2979	.3121	3.2041	.9546	1.2683	40'
30'	.3054	.3007	.3153	3.1716	.9537	1.2654	30'
40'	.3083	.3035	.3185	3.1397	.9528	1.2625	20'
50'	.3113	.3062	.3217	3.1084	.9520	1.2595	10'
18° 00'	.3142	.3090	.3249	3.0777	.9511	1.2566	72° 00'
		Cosine	Cotangent	Tangent	Sine	Radians	Degrees

TABLE I (Continued)
Values of Trigonometric Functions 337

Degrees	Radians	Sine	Tangent	Cotangent	Cosine		
18° 00′	.3142	.3090	.3249	3.0777	.9511	1.2566	72° 00′
10′	.3171	.3118	.3281	3.0475	.9502	1.2537	50′
20′	.3200	.3145	.3314	3.0178	.9492	1.2508	40′
30′	.3229	.3173	.3346	2.9887	.9483	1.2479	30′
40′	.3258	.3201	.3378	2.9600	.9474	1.2450	20′
50′	.3287	.3228	.3411	2.9319	.9465	1.2421	10′
19° 00′	.3316	.3256	.3443	2.9042	.9455	1.2392	71° 00′
10′	.3345	.3283	.3476	2.8770	.9446	1.2363	50′
20′	.3374	.3311	.3508	2.8502	.9436	1.2334	40′
30′	.3403	.3338	.3541	2.8239	.9426	1.2305	30′
40′	.3432	.3365	.3574	2.7980	.9417	1.2275	20′
50′	.3462	.3393	.3607	2.7725	.9407	1.2246	10′
20° 00′	.3491	.3420	.3640	2.7475	.9397	1.2217	70° 00′
10′	.3520	.3448	.3673	2.7228	.9387	1.2188	50′
20′	.3549	.3475	.3706	2.6985	.9377	1.2159	40′
30′	.3578	.3502	.3739	2.6746	.9367	1.2130	30′
40′	.3607	.3529	.3772	2.6511	.9356	1.2101	20′
50′	.3636	.3557	.3805	2.6279	.9346	1.2072	10′
21° 00′	.3665	.3584	.3839	2.6051	.9336	1.2043	69° 00′
10′	.3694	.3611	.3872	2.5826	.9325	1.2014	50′
20′	.3723	.3638	.3906	2.5605	.9315	1.1985	40′
30′	.3752	.3665	.3939	2.5386	.9304	1.1956	30′
40′	.3782	.3692	.3973	2.5172	.9293	1.1926	20′
50′	.3811	.3719	.4006	2.4960	.9283	1.1897	10′
22° 00′	.3840	.3746	.4040	2.4751	.9272	1.1868	68° 00′
10′	.3869	.3773	.4074	2.4545	.9261	1.1839	50′
20′	.3898	.3800	.4108	2.4342	.9250	1.1810	40′
30′	.3927	.3827	.4142	2.4142	.9239	1.1781	30′
40′	.3956	.3854	.4176	2.3945	.9228	1.1752	20′
50′	.3985	.3881	.4210	2.3750	.9216	1.1723	10′
23° 00′	.4014	.3907	.4245	2.3559	.9205	1.1694	67° 00′
10′	.4043	.3934	.4279	2.3369	.9194	1.1665	50′
20′	.4072	.3961	.4314	2.3183	.9182	1.1636	40′
30′	.4102	.3987	.4348	2.2998	.9171	1.1606	30′
40′	.4131	.4014	.4383	2.2817	.9159	1.1577	20′
50′	.4160	.4041	.4417	2.2637	.9147	1.1548	10′
24° 00′	.4189	.4067	.4452	2.2460	.9135	1.1519	66° 00′
10′	.4218	.4094	.4487	2.2286	.9124	1.1490	50′
20′	.4247	.4120	.4522	2.2113	.9112	1.1461	40′
30′	.4276	.4147	.4557	2.1943	.9100	1.1432	30′
40′	.4305	.4173	.4592	2.1775	.9088	1.1403	20′
50′	.4334	.4200	.4628	2.1609	.9075	1.1374	10′
25° 00′	.4363	.4226	.4663	2.1445	.9063	1.1345	65° 00′
10′	.4392	.4253	.4699	2.1283	.9051	1.1316	50′
20′	.4422	.4279	.4734	2.1123	.9038	1.1286	40′
30′	.4451	.4305	.4770	2.0965	.9026	1.1257	30′
40′	.4480	.4331	.4806	2.0809	.9013	1.1228	20′
50′	.4509	.4358	.4841	2.0655	.9001	1.1199	10′
26° 00′	.4538	.4384	.4877	2.0503	.8988	1.1170	64° 00′
10′	.4567	.4410	.4913	2.0353	.8975	1.1141	50′
20′	.4596	.4436	.4950	2.0204	.8962	1.1112	40′
30′	4625	.4462	.4986	2.0057	.8949	1.1083	30′
40′	.4654	.4488	.5022	1.9912	.8936	1.1054	20′
50′	.4683	.4514	.5059	1.9768	.8923	1.1025	10′
27° 00′	.4712	.4540	.5095	1.9626	.8910	1.0996	63° 00′
		Cosine	Cotangent	Tangent	Sine	Radians	Degrees

TABLE I (Continued)
Values of Trigonometric Functions

Degrees	Radians	Sine	Tangent	Cotangent	Cosine	
27° 00′	.4712	.4540	.5095	1.9626	.8910	1.0996	63° 00′
10′	.4741	.4566	.5132	1.9486	.8897	1.0966	50′
20′	.4771	.4592	.5169	1.9347	.8884	1.0937	40′
30′	.4800	.4617	.5206	1.9210	.8870	1.0908	30′
40′	.4829	.4643	.5243	1.9074	.8857	1.0879	20′
50′	.4858	.4669	.5280	1.8940	.8843	1.0850	10′
28° 00′	.4887	.4695	.5317	1.8807	.8829	1.0821	62° 00′
10′	.4916	.4720	.5354	1.8676	.8816	1.0792	50′
20′	.4945	.4746	.5392	1.8546	.8802	1.0763	40′
30′	.4974	.4772	.5430	1.8418	.8788	1.0734	30′
40′	.5003	.4797	.5467	1.8291	.8774	1.0705	20′
50′	.5032	.4823	.5505	1.8165	.8760	1.0676	10′
29° 00′	.5061	.4848	.5543	1.8040	.8746	1.0647	61° 00′
10′	.5091	.4874	.5581	1.7917	.8732	1.0617	50′
20′	.5120	.4899	.5619	1.7796	.8718	1.0588	40′
30′	.5149	.4924	.5658	1.7675	.8704	1.0559	30′
40′	.5178	.4950	.5696	1.7556	.8689	1.0530	20′
50′	.5207	.4975	.5735	1.7437	.8675	1.0501	10′
30° 00′	.5236	.5000	.5774	1.7321	.8660	1.0472	60° 00′
10′	.5265	.5025	.5812	1.7205	.8646	1.0443	50′
20′	.5294	.5050	.5851	1.7090	.8631	1.0414	40′
30′	.5323	.5075	.5890	1.6977	.8616	1.0385	30′
40′	.5352	.5100	.5930	1.6864	.8601	1.0356	20′
50′	.5381	.5125	.5969	1.6753	.8587	1.0327	10′
31° 00′	.5411	.5150	.6009	1.6643	.8572	1.0297	59° 00′
10′	.5440	.5175	.6048	1.6534	.8557	1.0268	50′
20′	.5469	.5200	.6088	1.6426	.8542	1.0239	40′
30′	.5498	.5225	.6128	1.6319	.8526	1.0210	30′
40′	.5527	.5250	.6168	1.6212	.8511	1.0181	20′
50′	.5556	.5275	.6208	1.6107	.8496	1.0152	10′
32° 00′	.5585	.5299	.6249	1.6003	.8480	1.0123	58° 00′
10′	.5614	.5324	.6289	1.5900	.8465	1.0094	50′
20′	.5643	.5348	.6330	1.5798	.8450	1.0065	40′
30′	.5672	.5373	.6371	1.5697	.8434	1.0036	30′
40′	.5701	.5398	.6412	1.5597	.8418	1.0007	20′
50′	.5730	.5422	.6453	1.5497	.8403	.9977	10′
33° 00′	.5760	.5446	.6494	1.5399	.8387	.9948	57° 00′
10′	.5789	.5471	.6536	1.5301	.8371	.9919	50′
20′	.5818	.5495	.6577	1.5204	.8355	.9890	40′
30′	.5847	.5519	.6619	1.5108	.8339	.9861	30′
40′	.5876	.5544	.6661	1.5013	.8323	.9832	20′
50′	.5905	.5568	.6703	1.4919	.8307	.9803	10′
34° 00′	.5934	.5592	.6745	1.4826	.8290	.9774	56° 00′
10′	.5963	.5616	.6787	1.4733	.8274	.9745	50′
20′	.5992	.5640	.6830	1.4641	.8258	.9716	40′
30′	.6021	.5664	.6873	1.4550	.8241	.9687	30′
40′	.6050	.5688	.6916	1.4460	.8225	.9657	20′
50′	.6080	.5712	.6959	1.4370	.8208	.9628	10′
35° 00′	.6109	.5736	.7002	1.4281	.8192	.9599	55° 00′
10′	.6138	.5760	.7046	1.4193	.8175	.9570	50′
20′	.6167	.5783	.7089	1.4106	.8158	.9541	40′
30′	.6196	.5807	.7133	1.4019	.8141	.9512	30′
40′	.6225	.5831	.7177	1.3934	.8124	.9483	20′
50′	.6254	.5854	.7221	1.3848	.8107	.9454	10′
36° 00′	.6283	.5878	.7265	1.3764	.8090	.9425	54° 00′
		Cosine	Cotangent	Tangent	Sine	Radians	Degrees

TABLE I (Continued)
Values of Trigonometric Functions

Degrees	Radians	Sine	Tangent	Cotangent	Cosine		
36° 00'	.6283	.5878	.7265	1.3764	.8090	.9425	54° 00'
10'	.6312	.5901	.7310	1.3680	.8073	.9396	50'
20'	.6341	.5925	.7355	1.3597	.8056	.9367	40'
30'	.6370	.5948	.7400	1.3514	.8039	.9338	30'
40'	.6400	.5972	.7445	1.3432	.8021	.9308	20'
50'	.6429	.5995	.7490	1.3351	.8004	.9279	10'
37° 00'	.6458	.6018	.7536	1.3270	.7986	.9250	53° 00'
10'	.6487	.6041	.7581	1.3190	.7969	.9221	50'
20'	.6516	.6065	.7627	1.3111	.7951	.9192	40'
30'	.6545	.6088	.7673	1.3032	.7934	.9163	30'
40'	.6574	.6111	.7720	1.2954	.7916	.9134	20'
50'	.6603	.6134	.7766	1.2876	.7898	.9105	10'
38° 00'	.6632	.6157	.7813	1.2799	.7880	.9076	52° 00'
10'	.6661	.6180	.7860	1.2723	.7862	.9047	50'
20'	.6690	.6202	.7907	1.2647	.7844	.9018	40'
30'	.6720	.6225	.7954	1.2572	.7826	.8988	30'
40'	.6749	.6248	.8002	1.2497	.7808	.8959	20'
50'	.6778	.6271	.8050	1.2423	.7790	.8930	10'
39° 00'	.6807	.6293	.8098	1.2349	.7771	.8901	51° 00'
10'	.6836	.6316	.8146	1.2276	.7753	.8872	50'
20'	.6865	.6338	.8195	1.2203	.7735	.8843	40'
30'	.6894	.6361	.8243	1.2131	.7716	.8814	30'
40'	.6923	.6383	.8292	1.2059	.7698	.8785	20'
50'	.6952	.6406	.8342	1.1988	.7679	.8756	10'
40° 00'	.6981	.6428	.8391	1.1918	.7660	.8727	50° 00'
10'	.7010	.6450	.8441	1.1847	.7642	.8698	50'
20'	.7039	.6472	.8491	1.1778	.7623	.8668	40'
30'	.7069	.6494	.8541	1.1708	.7604	.8639	30'
40'	.7098	.6517	.8591	1.1640	.7585	.8610	20'
50'	.7127	.6539	.8642	1.1571	.7566	.8581	10'
41° 00'	.7156	.6561	.8693	1.1504	.7547	.8552	49° 00'
10'	.7185	.6583	.8744	1.1436	.7528	.8523	50'
20'	.7214	.6604	.8796	1.1369	.7509	.8494	40'
30'	.7243	.6626	.8847	1.1303	.7490	.8465	30'
40'	.7272	.6648	.8899	1.1237	.7470	.8436	20'
50'	.7301	.6670	.8952	1.1171	.7451	.8407	10'
42° 00'	.7330	.6691	.9004	1.1106	.7431	.8378	48° 00'
10'	.7359	.6713	.9057	1.1041	.7412	.8348	50'
20'	.7389	.6734	.9110	1.0977	.7392	.8319	40'
30'	.7418	.6756	.9163	1.0913	.7373	.8290	30'
40'	.7447	.6777	.9217	1.0850	.7353	.8261	20'
50'	.7476	.6799	.9271	1.0786	.7333	.8232	10'
43° 00'	.7505	.6820	.9325	1.0724	.7314	.8203	47° 00'
10'	.7534	.6841	.9380	1.0661	.7294	.8174	50'
20'	.7563	.6862	.9435	1.0599	.7274	.8145	40'
30'	.7592	.6884	.9490	1.0538	.7254	.8116	30'
40'	.7621	.6905	.9545	1.0477	.7234	.8087	20'
50'	.7650	.6926	.9601	1.0416	.7214	.8058	10'
44° 00'	.7679	.6947	.9657	1.0355	.7193	.8029	46° 00'
10'	.7709	.6967	.9713	1.0295	.7173	.7999	50'
20'	.7738	.6988	.9770	1.0235	.7153	.7970	40'
30'	.7767	.7009	.9827	1.0176	.7133	.7941	30'
40'	.7796	.7030	.9884	1.0117	.7112	.7912	20'
50'	.7825	.7050	.9942	1.0058	.7092	.7883	10'
45° 00'	.7854	.7071	1.0000	1.0000	.7071	.7854	45° 00'
		Cosine	Cotangent	Tangent	Sine	Radians	Degrees

TABLE II
Logarithms of Numbers

N	0	1	2	3	4	5	6	7	8	9
1.0	.0000	.0043	.0086	.0128	.0170	.0212	.0253	.0294	.0334	.0374
1.1	.0414	.0453	.0492	.0531	.0569	.0607	.0645	.0682	.0719	.0755
1.2	.0792	.0828	.0864	.0899	.0934	.0969	.1004	.1038	.1072	.1106
1.3	.1139	.1173	.1206	.1239	.1271	.1303	.1335	.1367	.1399	.1430
1.4	.1461	.1492	.1523	.1553	.1584	.1614	.1644	.1673	.1703	.1732
1.5	.1761	.1790	.1818	.1847	.1875	.1903	.1931	.1959	.1987	.2014
1.6	.2041	.2068	.2095	.2122	.2148	.2175	.2201	.2227	.2253	.2279
1.7	.2304	.2330	.2355	.2380	.2405	.2430	.2455	.2480	.2504	.2529
1.8	.2553	.2577	.2601	.2625	.2648	.2672	.2695	.2718	.2742	.2765
1.9	.2788	.2810	.2833	.2856	.2878	.2900	.2923	.2945	.2967	.2989
2.0	.3010	.3032	.3054	.3075	.3096	.3118	.3139	.3160	.3181	.3201
2.1	.3222	.3243	.3263	.3284	.3304	.3324	.3345	.3365	.3385	.3404
2.2	.3424	.3444	.3464	.3483	.3502	.3522	.3541	.3560	.3579	.3598
2.3	.3617	.3636	.3655	.3674	.3692	.3711	.3729	.3747	.3766	.3784
2.4	.3802	.3820	.3838	.3856	.3874	.3892	.3909	.3927	.3945	.3962
2.5	.3979	.3997	.4014	.4031	.4048	.4065	.4082	.4099	.4116	.4133
2.6	.4150	.4166	.4183	.4200	.4216	.4232	.4249	.4265	.4281	.4298
2.7	.4314	.4330	.4346	.4362	.4378	.4393	.4409	.4425	.4440	.4456
2.8	.4472	.4487	.4502	.4518	.4533	.4548	.4564	.4579	.4594	.4609
2.9	.4624	.4639	.4654	.4669	.4683	.4698	.4713	.4728	.4742	.4757
3.0	.4771	.4786	.4800	.4814	.4829	.4843	.4857	.4871	.4886	.4900
3.1	.4914	.4928	.4942	.4955	.4969	.4983	.4997	.5011	.5024	.5038
3.2	.5051	.5065	.5079	.5092	.5105	.5119	.5132	.5145	.5159	.5172
3.3	.5185	.5198	.5211	.5224	.5237	.5250	.5263	.5276	.5289	.5302
3.4	.5315	.5328	.5340	.5353	.5366	.5378	.5391	.5403	.5416	.5428
3.5	.5441	.5453	.5465	.5478	.5490	.5502	.5514	.5527	.5539	.5551
3.6	.5563	.5575	.5587	.5599	.5611	.5623	.5635	.5647	.5658	.5670
3.7	.5682	.5694	.5705	.5717	.5729	.5740	.5752	.5763	.5775	.5786
3.8	.5798	.5809	.5821	.5832	.5843	.5855	.5866	.5877	.5888	.5899
3.9	.5911	.5922	.5933	.5944	.5955	.5966	.5977	.5988	.5999	.6010
4.0	.6021	.6031	.6042	.6053	.6064	.6075	.6085	.6096	.6107	.6117
4.1	.6128	.6138	.6149	.6160	.6170	.6180	.6191	.6201	.6212	.6222
4.2	.6232	.6243	.6253	.6263	.6274	.6284	.6294	.6304	.6314	.6325
4.3	.6335	.6345	.6355	.6365	.6375	.6385	.6395	.6405	.6415	.6425
4.4	.6435	.6444	.6454	.6464	.6474	.6484	.6493	.6503	.6513	.6522
4.5	.6532	.6542	.6551	.6561	.6571	.6580	.6590	.6599	.6609	.6618
4.6	.6628	.6637	.6646	.6656	.6665	.6675	.6684	.6693	.6702	.6712
4.7	.6721	.6730	.6739	.6749	.6758	.6767	.6776	.6785	.6794	.6803
4.8	.6812	.6821	.6830	.6839	.6848	.6857	.6866	.6875	.6884	.6893
4.9	.6902	.6911	.6920	.6928	.6937	.6946	.6955	.6964	.6972	.6981
5.0	.6990	.6998	.7007	.7016	.7024	.7033	.7042	.7050	.7059	.7067
5.1	.7076	.7084	.7093	.7101	.7110	.7118	.7126	.7135	.7143	.7152
5.2	.7160	.7168	.7177	.7185	.7193	.7202	.7210	.7218	.7226	.7235
5.3	.7243	.7251	.7259	.7267	.7275	.7284	.7292	.7300	.7308	.7316
5.4	.7324	.7332	.7340	.7348	.7356	.7364	.7372	.7380	.7388	.7396
N	0	1	2	3	4	5	6	7	8	9

TABLE II (Continued)
Logarithms of Numbers

N	0	1	2	3	4	5	6	7	8	9
5.5	.7404	.7412	.7419	.7427	.7435	.7443	.7451	.7459	.7466	.7474
5.6	.7482	.7490	.7497	.7505	.7513	.7520	.7528	.7536	.7543	.7551
5.7	.7559	.7566	.7574	.7582	.7589	.7597	.7604	.7612	.7619	.7627
5.8	.7634	.7642	.7649	.7657	.7664	.7672	.7679	.7686	.7694	.7701
5.9	.7709	.7716	.7723	.7731	.7738	.7745	.7752	.7760	.7767	.7774
6.0	.7782	.7789	.7796	.7803	.7810	.7818	.7825	.7832	.7839	.7846
6.1	.7853	.7860	.7868	.7875	.7882	.7889	.7896	.7903	.7910	.7917
6.2	.7924	.7931	.7938	.7945	.7952	.7959	.7966	.7973	.7980	.7987
6.3	.7993	.8000	.8007	.8014	.8021	.8028	.8035	.8041	.8048	.8055
6.4	.8062	.8069	.8075	.8082	.8089	.8096	.8102	.8109	.8116	.8122
6.5	.8129	.8136	.8142	.8149	.8156	.8162	.8169	.8176	.8182	.8189
6.6	.8195	.8202	.8209	.8215	.8222	.8228	.8235	.8241	.8248	.8254
6.7	.8261	.8267	.8274	.8280	.8287	.8293	.8299	.8306	.8312	.8319
6.8	.8325	.8331	.8338	.8344	.8351	.8357	.8363	.8370	.8376	.8382
6.9	.8388	.8395	.8401	.8407	.8414	.8420	.8426	.8432	.8439	.8445
7.0	.8451	.8457	.8463	.8470	.8476	.8482	.8488	.8494	.8500	.8506
7.1	.8513	.8519	.8525	.8531	.8537	.8543	.8549	.8555	.8561	.8567
7.2	.8573	.8579	.8585	.8591	.8597	.8603	.8609	.8615	.8621	.8627
7.3	.8633	.8639	.8645	.8651	.8657	.8663	.8669	.8675	.8681	.8686
7.4	.8692	.8698	.8704	.8710	.8716	.8722	.8727	.8733	.8739	.8745
7.5	.8751	.8756	.8762	.8768	.8774	.8779	.8785	.8791	.8797	.8802
7.6	.8808	.8814	.8820	.8825	.8831	.8837	.8842	.8848	.8854	.8859
7.7	.8865	.8871	.8876	.8882	.8887	.8893	.8899	.8904	.8910	.8915
7.8	.8921	.8927	.8932	.8938	.8943	.8949	.8954	.8960	.8965	.8971
7.9	.8976	.8982	.8987	.8993	.8998	.9004	.9009	.9015	.9020	.9025
8.0	.9031	.9036	.9042	.9047	.9053	.9058	.9063	.9069	.9074	.9079
8.1	.9085	.9090	.9096	.9101	.9106	.9112	.9117	.9122	.9128	.9133
8.2	.9138	.9143	.9149	.9154	.9159	.9165	.9170	.9175	.9180	.9186
8.3	.9191	.9196	.9201	.9206	.9212	.9217	.9222	.9227	.9232	.9238
8.4	.9243	.9248	.9253	.9258	.9263	.9269	.9274	.9279	.9284	.9289
8.5	.9294	.9299	.9304	.9309	.9315	.9320	.9325	.9330	.9335	.9340
8.6	.9345	.9350	.9355	.9360	.9365	.9370	.9375	.9380	.9385	.9390
8.7	.9395	.9400	.9405	.9410	.9415	.9420	.9425	.9430	.9435	.9440
8.8	.9445	.9450	.9455	.9460	.9465	.9469	.9474	.9479	.9484	.9489
8.9	.9494	.9499	.9504	.9509	.9513	.9518	.9523	.9528	.9533	.9538
9.0	.9542	.9547	.9552	.9557	.9562	.9566	.9571	.9576	.9581	.9586
9.1	.9590	.9595	.9600	.9605	.9609	.9614	.9619	.9624	.9628	.9633
9.2	.9638	.9643	.9647	.9652	.9657	.9661	.9666	.9671	.9675	.9680
9.3	.9685	.9689	.9694	.9699	.9703	.9708	.9713	.9717	.9722	.9727
9.4	.9731	.9736	.9741	.9745	.9750	.9754	.9759	.9763	.9768	.9773
9.5	.9777	.9782	.9786	.9791	.9795	.9800	.9805	.9809	.9814	.9818
9.6	.9823	.9827	.9832	.9836	.9841	.9845	.9850	.9854	.9859	.9863
9.7	.9868	.9872	.9877	.9881	.9886	.9890	.9894	.9899	.9903	.9908
9.8	.9912	.9917	.9921	.9926	.9930	.9934	.9939	.9943	.9948	.9952
9.9	.9956	.9961	.9965	.9969	.9974	.9978	.9983	.9987	.9991	.9996
N	0	1	2	3	4	5	6	7	8	9

TABLE III

Logarithms of Trigonometric Functions

342

Degrees	Log₁₀ Sine	Log₁₀ Tangent	Log₁₀ Cotangent	Log₁₀ Cosine	
0° 00′					90° 00′
10′	.4637 −3	.4637 −3	2.5363	.0000	50′
20′	.7648 −3	.7648 −3	2.2352	.0000	40′
30′	9408 −3	.9409 −3	2.0591	.0000	30′
40′	.0658 −2	.0658 −2	1.9342	.0000	20′
50′	.1627 −2	.1627 −2	1.8373	.0000	10′
1° 00′	.2419 −2	.2419 −2	1.7581	.9999 −1	89° 00′
10′	.3088 −2	.3089 −2	1.6911	.9999 −1	50′
20′	.3668 −2	.3669 −2	1.6331	.9999 −1	40′
30′	.4179 −2	.4181 −2	1.5819	.9999 −1	30′
40′	.4637 −2	.4638 −2	1.5362	.9998 −1	20′
50′	.5050 −2	.5053 −2	1.4947	.9998 −1	10′
2° 00′	.5428 −2	.5431 −2	1.4569	.9997 −1	88° 00′
10′	.5776 −2	.5779 −2	1.4221	.9997 −1	50′
20′	.6097 −2	.6101 −2	1.3899	.9996 −1	40′
30′	.6397 −2	.6401 −2	1.3599	.9996 −1	30′
40′	.6677 −2	.6682 −2	1.3318	.9995 −1	20′
50′	.6940 −2	.6945 −2	1.3055	.9995 −1	10′
3° 00′	.7188 −2	.7194 −2	1.2806	.9994 −1	87° 00′
10′	.7423 −2	.7429 −2	1.2571	.9993 −1	50′
20′	.7645 −2	.7652 −2	1.2348	.9993 −1	40′
30′	.7857 −2	.7865 −2	1.2135	.9992 −1	30′
40′	.8059 −2	.8067 −2	1.1933	.9991 −1	20′
50′	.8251 −2	.8261 −2	1.1739	.9990 −1	10′
4° 00′	.8436 −2	.8446 −2	1.1554	.9989 −1	86° 00′
10′	.8613 −2	.8624 −2	1.1376	.9989 −1	50′
20′	.8783 −2	.8795 −2	1.1205	.9988 −1	40′
30′	.8946 −2	.8960 −2	1.1040	.9987 −1	30′
40′	.9104 −2	.9118 −2	1.0882	.9986 −1	20′
50′	.9256 −2	.9272 −2	1.0728	.9985 −1	10′
5° 00′	.9403 −2	.9420 −2	1.0580	.9983 −1	85° 00′
10′	.9545 −2	.9563 −2	1.0437	.9982 −1	50′
20′	.9682 −2	.9701 −2	1.0299	.9981 −1	40′
30′	.9816 −2	.9836 −2	1.0164	.9980 −1	30′
40′	.9945 −2	.9966 −2	1.0034	.9979 −1	20′
50′	.0070 −1	.0093 −1	.9907	.9977 −1	10′
6° 00′	.0192 −1	.0216 −1	.9784	.9976 −1	84° 00′
10′	.0311 −1	.0336 −1	.9664	.9975 −1	50′
20′	.0426 −1	.0453 −1	.9547	.9973 −1	40′
30′	.0539 −1	.0567 −1	.9433	.9972 −1	30′
40′	.0648 −1	.0678 −1	.9322	.9971 −1	20′
50′	.0755 −1	.0786 −1	.9214	.9969 −1	10′
7° 00′	.0859 −1	.0891 −1	.9109	.9968 −1	83° 00′
10′	.0961 −1	.0995 −1	.9005	.9966 −1	50′
20′	.1060 −1	.1096 −1	.8904	.9964 −1	40′
30′	.1157 −1	.1194 −1	.8806	.9963 −1	30′
40′	.1252 −1	.1291 −1	.8709	.9961 −1	20′
50′	.1345 −1	.1385 −1	.8615	.9959 −1	10′
8° 00′	.1436 −1	.1478 −1	.8522	.9958 −1	82° 00′
10′	.1525 −1	.1569 −1	.8431	.9956 −1	50′
20′	.1612 −1	.1658 −1	.8342	.9954 −1	40′
30′	.1697 −1	.1745 −1	.8255	.9952 −1	30′
40′	.1781 −1	.1831 −1	.8169	.9950 −1	20′
50′	.1863 −1	.1915 −1	.8085	.9948 −1	10′
9° 00′	.1943 −1	.1997 −1	.8003	.9946 −1	81° 00′
	Log₁₀ Cosine	Log₁₀ Cotangent	Log₁₀ Tangent	Log₁₀ Sine	Degrees

TABLE III (Continued)
Logarithms of Trigonometric Functions 343

Degrees	Log₁₀ Sine	Log₁₀ Tangent	Log₁₀ Cotangent	Log₁₀ Cosine	
9° 00'	.1943 − 1	.1997 − 1	.8003	.9946 − 1	81° 00'
10'	.2022 − 1	.2078 − 1	.7922	.9944 − 1	50'
20'	.2100 − 1	.2158 − 1	.7842	.9942 − 1	40'
30'	.2176 − 1	.2236 − 1	.7764	.9940 − 1	30'
40'	.2251 − 1	.2313 − 1	.7687	.9938 − 1	20'
50'	.2324 − 1	.2389 − 1	.7611	.9936 − 1	10'
10° 00'	.2397 − 1	.2463 − 1	.7537	.9934 − 1	80° 00'
10'	.2468 − 1	.2536 − 1	.7464	.9931 − 1	50'
20'	.2538 − 1	.2609 − 1	.7391	.9929 − 1	40'
30'	.2606 − 1	.2680 − 1	.7320	.9927 − 1	30'
40'	.2674 − 1	.2750 − 1	.7250	.9924 − 1	20'
50'	.2740 − 1	.2819 − 1	.7181	.9922 − 1	10'
11° 00'	.2806 − 1	.2887 − 1	.7113	.9919 − 1	79° 00'
10'	.2870 − 1	.2953 − 1	.7047	.9917 − 1	50'
20'	.2934 − 1	.3020 − 1	.6980	.9914 − 1	40'
30'	.2997 − 1	.3085 − 1	.6915	.9912 − 1	30'
40'	.3058 − 1	.3149 − 1	.6851	.9909 − 1	20'
50'	.3119 − 1	.3212 − 1	.6788	.9907 − 1	10'
12° 00'	.3179 − 1	.3275 − 1	.6725	.9904 − 1	78° 00'
10'	.3238 − 1	.3336 − 1	.6664	.9901 − 1	50'
20'	.3296 − 1	.3397 − 1	.6603	.9899 − 1	40'
30'	.3353 − 1	.3458 − 1	.6542	.9896 − 1	30'
40'	.3410 − 1	.3517 − 1	.6483	.9893 − 1	20'
50'	.3466 − 1	.3576 − 1	.6424	.9890 − 1	10'
13° 00'	.3521 − 1	.3634 − 1	.6366	.9887 − 1	77° 00'
10'	.3575 − 1	.3691 − 1	.6309	.9884 − 1	50'
20'	.3629 − 1	.3748 − 1	.6252	.9881 − 1	40'
30'	.3682 − 1	.3804 − 1	.6196	.9878 − 1	30'
40'	.3734 − 1	.3859 − 1	.6141	.9875 − 1	20'
50'	.3786 − 1	.3914 − 1	.6086	.9872 − 1	10'
14° 00'	.3837 − 1	.3968 − 1	.6032	.9869 − 1	76° 00'
10'	.3887 − 1	.4021 − 1	.5979	.9866 − 1	50'
20'	.3937 − 1	.4074 − 1	.5926	.9863 − 1	40'
30'	.3986 − 1	.4127 − 1	.5873	.9859 − 1	30'
40'	.4035 − 1	.4178 − 1	.5822	.9856 − 1	20'
50'	.4083 − 1	.4230 − 1	.5770	.9853 − 1	10'
15° 00'	.4130 − 1	.4281 − 1	.5719	.9849 − 1	75° 00'
10'	.4177 − 1	.4331 − 1	.5669	.9846 − 1	50'
20'	.4223 − 1	.4381 − 1	.5619	.9843 − 1	40'
30'	.4269 − 1	.4430 − 1	.5570	.9839 − 1	30'
40'	.4314 − 1	.4479 − 1	.5521	.9836 − 1	20'
50'	.4359 − 1	.4527 − 1	.5473	.9832 − 1	10'
16° 00'	.4403 − 1	.4575 − 1	.5425	.9828 − 1	74° 00'
10'	.4447 − 1	.4622 − 1	.5378	.9825 − 1	50'
20'	.4491 − 1	.4669 − 1	.5331	.9821 − 1	40'
30'	.4533 − 1	.4716 − 1	.5284	.9817 − 1	30'
40'	.4576 − 1	.4762 − 1	.5238	.9814 − 1	20'
50'	.4618 − 1	.4808 − 1	.5192	.9810 − 1	10'
17° 00'	.4659 − 1	.4853 − 1	.5147	.9806 − 1	73° 00'
10'	.4700 − 1	.4898 − 1	.5102	.9802 − 1	50'
20'	.4741 − 1	.4943 − 1	.5057	.9798 − 1	40'
30'	.4781 − 1	.4987 − 1	.5013	.9794 − 1	30'
40'	.4821 − 1	.5031 − 1	.4969	.9790 − 1	20'
50'	.4861 − 1	.5075 − 1	.4925	.9786 − 1	10'
18° 00'	.4900 − 1	.5118 − 1	.4882	.9782 − 1	72° 00'
	Log₁₀ Cosine	Log₁₀ Cotangent	Log₁₀ Tangent	Log₁₀ Sine	Degrees

TABLE III (Continued)
Logarithms of Trigonometric Functions

Degrees	Log₁₀ Sine	Log₁₀ Tangent	Log₁₀ Cotangent	Log₁₀ Cosine	
18° 00′	.4900 −1	.5118 −1	.4882	.9782 −1	72° 00′
10′	.4939 −1	.5161 −1	.4839	.9778 −1	50′
20′	.4977 −1	.5203 −1	.4797	.9774 −1	40′
30′	.5015 −1	.5245 −1	.4755	.9770 −1	30′
40′	.5052 −1	.5287 −1	.4713	.9765 −1	20′
50′	.5090 −1	.5329 −1	.4671	.9761 −1	10′
19° 00′	.5126 −1	.5370 −1	.4630	.9757 −1	71° 00′
10′	.5163 −1	.5411 −1	.4589	.9752 −1	50′
20′	.5199 −1	.5451 −1	.4549	.9748 −1	40′
30′	.5235 −1	.5491 −1	.4509	.9743 −1	30′
40′	.5270 −1	.5531 −1	.4469	.9739 −1	20′
50′	.5306 −1	.5571 −1	.4429	.9734 −1	10′
20° 00′	.5341 −1	.5611 −1	.4389	.9730 −1	70° 00′
10′	.5375 −1	.5650 −1	.4350	.9725 −1	50′
20′	.5409 −1	.5689 −1	.4311	.9721 −1	40′
30′	.5443 −1	.5727 −1	.4273	.9716 −1	30′
40′	.5477 −1	.5766 −1	.4234	.9711 −1	20′
50′	.5510 −1	.5804 −1	.4196	.9706 −1	10′
21° 00′	.5543 −1	.5842 −1	.4158	.9702 −1	69° 00′
10′	.5576 −1	.5879 −1	.4121	.9697 −1	50′
20′	.5609 −1	.5917 −1	.4083	.9692 −1	40′
30′	.5641 −1	.5954 −1	.4046	.9687 −1	30′
40′	.5673 −1	.5991 −1	.4009	.9682 −1	20′
50′	.5704 −1	.6028 −1	.3972	.9677 −1	10′
22° 00′	.5736 −1	.6064 −1	.3936	.9672 −1	68° 00′
10′	.5767 −1	.6100 −1	.3900	.9667 −1	50′
20′	.5798 −1	.6136 −1	.3864	.9661 −1	40′
30′	.5828 −1	.6172 −1	.3828	.9656 −1	30′
40′	.5859 −1	.6208 −1	.3792	.9651 −1	20′
50′	.5889 −1	.6243 −1	.3757	.9646 −1	10′
23° 00′	.5919 −1	.6279 −1	.3721	.9640 −1	67° 00′
10′	.5948 −1	.6314 −1	.3686	.9635 −1	50′
20′	.5978 −1	.6348 −1	.3652	.9629 −1	40′
30′	.6007 −1	.6383 −1	.3617	.9624 −1	30′
40′	.6036 −1	.6417 −1	.3583	.9618 −1	20′
50′	.6065 −1	.6452 −1	.3548	.9613 −1	10′
24° 00′	.6093 −1	.6486 −1	.3514	.9607 −1	66° 00′
10′	.6121 −1	.6520 −1	.3480	.9602 −1	50′
20′	.6149 −1	.6553 −1	.3447	.9596 −1	40′
30′	.6177 −1	.6587 −1	.3413	.9590 −1	30′
40′	.6205 −1	.6620 −1	.3380	.9584 −1	20′
50′	.6232 −1	.6654 −1	.3346	.9579 −1	10′
25° 00′	.6259 −1	.6687 −1	.3313	.9573 −1	65° 00′
10′	.6286 −1	.6720 −1	.3280	.9567 −1	50′
20′	.6313 −1	.6752 −1	.3248	.9561 −1	40′
30′	.6340 −1	.6785 −1	.3215	.9555 −1	30′
40′	.6366 −1	.6817 −1	.3183	.9549 −1	20′
50′	.6392 −1	.6850 −1	.3150	.9543 −1	10′
26° 00′	.6418 −1	.6882 −1	.3118	.9537 −1	64° 00′
10′	.6444 −1	.6914 −1	.3086	.9530 −1	50′
20′	.6470 −1	.6946 −1	.3054	.9524 −1	40′
30′	.6495 −1	.6977 −1	.3023	.9518 −1	30′
40′	.6521 −1	.7009 −1	.2991	.9512 −1	20′
50′	.6546 −1	.7040 −1	.2960	.9505 −1	10′
27° 00′	.6570 −1	.7072 −1	.2928	.9499 −1	63° 00′
	Log₁₀ Cosine	Log₁₀ Cotangent	Log₁₀ Tangent	Log₁₀ Sine	Degrees

TABLE III (Continued)
Logarithms of Trigonometric Functions

Degrees	Log₁₀ Sine	Log₁₀ Tangent	Log₁₀ Cotangent	Log₁₀ Cosine	
27° 00′	.6570 −1	.7072 −1	.2928	.9499 −1	63° 00′
10′	.6595 −1	.7103 −1	.2897	.9492 −1	50′
20′	.6620 −1	.7134 −1	.2866	.9486 −1	40′
30′	.6644 −1	.7165 −1	.2835	.9479 −1	30′
40′	.6668 −1	.7196 −1	.2804	.9473 −1	20′
50′	.6692 −1	.7226 −1	.2774	.9466 −1	10′
28° 00′	.6716 −1	.7257 −1	.2743	.9459 −1	62° 00′
10′	.6740 −1	.7287 −1	.2713	.9453 −1	50′
20′	.6763 −1	.7317 −1	.2683	.9446 −1	40′
30′	.6787 −1	.7348 −1	.2652	.9439 −1	30′
40′	.6810 −1	.7378 −1	.2622	.9432 −1	20′
50′	.6833 −1	.7408 −1	.2592	.9425 −1	10′
29° 00′	.6856 −1	.7438 −1	.2562	.9418 −1	61° 00′
10′	.6878 −1	.7467 −1	.2533	.9411 −1	50′
20′	.6901 −1	.7497 −1	.2503	.9404 −1	40′
30′	.6923 −1	.7526 −1	.2474	.9397 −1	30′
40′	.6946 −1	.7556 −1	.2444	.9390 −1	20′
50′	.6968 −1	.7585 −1	.2415	.9383 −1	10′
30° 00′	.6990 −1	.7614 −1	.2386	.9375 −1	60° 00′
10′	.7012 −1	.7644 −1	.2356	.9368 −1	50′
20′	.7033 −1	.7673 −1	.2327	.9361 −1	40′
30′	.7055 −1	.7701 −1	.2299	.9353 −1	30′
40′	.7076 −1	.7730 −1	.2270	.9346 −1	20′
50′	.7097 −1	.7759 −1	.2241	.9338 −1	10′
31° 00′	.7118 −1	.7788 −1	.2212	.9331 −1	59° 00′
10′	.7139 −1	.7816 −1	.2184	.9323 −1	50′
20′	.7160 −1	.7845 −1	.2155	.9315 −1	40′
30′	.7181 −1	.7873 −1	.2127	.9308 −1	30′
40′	.7201 −1	.7902 −1	.2098	.9300 −1	20′
50′	.7222 −1	.7930 −1	.2070	.9292 −1	10′
32° 00′	.7242 −1	.7958 −1	.2042	.9284 −1	58° 00′
10′	.7262 −1	.7986 −1	.2014	.9276 −1	50′
20′	.7282 −1	.8014 −1	.1986	.9268 −1	40′
30′	.7302 −1	.8042 −1	.1958	.9260 −1	30′
40′	.7322 −1	.8070 −1	.1930	.9252 −1	20′
50′	.7342 −1	.8097 −1	.1903	.9244 −1	10′
33° 00′	.7361 −1	.8125 −1	.1875	.9236 −1	57° 00′
10′	.7380 −1	.8153 −1	.1847	.9228 −1	50′
20′	.7400 −1	.8180 −1	.1820	.9219 −1	40′
30′	.7419 −1	.8208 −1	.1792	.9211 −1	30′
40′	.7438 −1	.8235 −1	.1765	.9203 −1	20′
50′	.7457 −1	.8263 −1	.1737	.9194 −1	10′
34° 00′	.7476 −1	.8290 −1	.1710	.9186 −1	56° 00′
10′	.7494 −1	.8317 −1	.1683	.9177 −1	50′
20′	.7513 −1	.8344 −1	.1656	.9169 −1	40′
30′	.7531 −1	.8371 −1	.1629	.9160 −1	30′
40′	.7550 −1	.8398 −1	.1602	.9151 −1	20′
50′	.7568 −1	.8425 −1	.1575	.9142 −1	10′
35° 00′	.7586 −1	.8452 −1	.1548	.9134 −1	55° 00′
10′	.7604 −1	.8479 −1	.1521	.9125 −1	50′
20′	.7622 −1	.8506 −1	.1494	.9116 −1	40′
30′	.7640 −1	.8533 −1	.1467	.9107 −1	30′
40′	.7657 −1	.8559 −1	.1441	.9098 −1	20′
50′	.7675 −1	.8586 −1	.1414	.9089 −1	10′
36° 00′	.7692 −1	.8613 −1	.1387	.9080 −1	54° 00′
	Log₁₀ Cosine	Log₁₀ Cotangent	Log₁₀ Tangent	Log₁₀ Sine	Degrees

TABLE III (Continued)
Logarithms of Trigonometric Functions

Degrees	Log₁₀ Sine	Log₁₀ Tangent	Log₁₀ Cotangent	Log₁₀ Cosine	
36° 00′	.7692 −1	.8613 −1	.1387	.9080 −1	**54° 00′**
10′	.7710 −1	.8639 −1	.1361	.9070 −1	50′
20′	.7727 −1	.8666 −1	.1334	.9061 −1	40′
30′	.7744 −1	.8692 −1	.1308	.9052 −1	30′
40′	.7761 −1	.8718 −1	.1282	.9042 −1	20′
50′	.7778 −1	.8745 −1	.1255	.9033 −1	10′
37° 00′	.7795 −1	.8771 −1	.1229	.9023 −1	**53° 00′**
10′	.7811 −1	.8797 −1	.1203	.9014 −1	50′
20′	.7828 −1	.8824 −1	.1176	.9004 −1	40′
30′	.7844 −1	.8850 −1	.1150	.8995 −1	30′
40′	.7861 −1	.8876 −1	.1124	.8985 −1	20′
50′	.7877 −1	.8902 −1	.1098	.8975 −1	10′
38° 00′	.7893 −1	.8928 −1	.1072	.8965 −1	**52° 00′**
10′	.7910 −1	.8954 −1	.1046	.8955 −1	50′
20′	.7926 −1	.8980 −1	.1020	.8945 −1	40′
30′	.7941 −1	.9006 −1	.0994	.8935 −1	30′
40′	.7957 −1	.9032 −1	.0968	.8925 −1	20′
50′	.7973 −1	.9058 −1	.0942	.8915 −1	10′
39° 00′	.7989 −1	.9084 −1	.0916	.8905 −1	**51° 00′**
10′	.8004 −1	.9110 −1	.0890	.8895 −1	50′
20′	.8020 −1	.9135 −1	.0865	.8884 −1	40′
30′	.8035 −1	.9161 −1	.0839	.8874 −1	30′
40′	.8050 −1	.9187 −1	.0813	.8864 −1	20′
50′	.8066 −1	.9212 −1	.0788	.8853 −1	10′
40° 00′	.8081 −1	.9238 −1	.0762	.8843 −1	**50° 00′**
10′	.8096 −1	.9264 −1	.0736	.8832 −1	50′
20′	.8111 −1	.9289 −1	.0711	.8821 −1	40′
30′	.8125 −1	.9315 −1	.0685	.8810 −1	30′
40′	.8140 −1	.9341 −1	.0659	.8800 −1	20′
50′	.8155 −1	.9366 −1	.0634	.8789 −1	10′
41° 00′	.8169 −1	.9392 −1	.0608	.8778 −1	**49° 00′**
10′	.8184 −1	.9417 −1	.0583	.8767 −1	50′
20′	.8198 −1	.9443 −1	.0557	.8756 −1	40′
30′	.8213 −1	.9468 −1	.0532	.8745 −1	30′
40′	.8227 −1	.9494 −1	.0506	.8733 −1	20′
50′	.8241 −1	.9519 −1	.0481	.8722 −1	10′
42° 00′	.8255 −1	.9544 −1	.0456	.8711 −1	**48° 00′**
10′	.8269 −1	.9570 −1	.0430	.8699 −1	50′
20′	.8283 −1	.9595 −1	.0405	.8688 −1	40′
30′	.8297 −1	.9621 −1	.0379	.8676 −1	30′
40′	.8311 −1	.9646 −1	.0354	.8665 −1	20′
50′	.8324 −1	.9671 −1	.0329	.8653 −1	10′
43° 00′	.8338 −1	.9697 −1	.0303	.8641 −1	**47° 00′**
10′	.8351 −1	.9724 −1	.0278	.8629 −1	50′
20′	.8365 −1	.9747 −1	.0253	.8618 −1	40′
30′	.8378 −1	.9772 −1	.0228	.8606 −1	30′
40′	.8391 −1	.9798 −1	.0202	.8594 −1	20′
50′	.8405 −1	.9823 −1	.0177	.8582 −1	10′
44° 00′	.8418 −1	.9848 −1	.0152	.8569 −1	**46° 00′**
10′	.8431 −1	.9874 −1	.0126	.8557 −1	50′
20′	.8444 −1	.9899 −1	.0101	.8545 −1	40′
30′	.8457 −1	.9924 −1	.0076	.8532 −1	30′
40′	.8469 −1	.9949 −1	.0051	.8520 −1	20′
50′	.8482 −1	.9975 −1	.0025	.8507 −1	10′
45° 00′	.8495 −1	.0000	.0000	.8495 −1	**45° 00′**
	Log₁₀ Cosine	Log₁₀ Cotangent	Log₁₀ Tangent	Log₁₀ Sine	Degrees

TABLE IV
Powers and Roots

No.	Sq.	Sq. Root	Cube	Cube Root	No.	Sq.	Sq. Root	Cube	Cube Root
1	1	1.000	1	1.000	51	2,601	7.141	132,651	3.708
2	4	1.414	8	1.260	52	2,704	7.211	140,608	3.733
3	9	1.732	27	1.442	53	2,809	7.280	148,877	3.756
4	16	2.000	64	1.587	54	2,916	7.348	157,464	3.780
5	25	2.236	125	1.710	55	3,025	7.416	166,375	3.803
6	36	2.449	216	1.817	56	3,136	7.483	175,616	3.826
7	49	2.646	343	1.913	57	3,249	7.550	185,193	3.849
8	64	2.828	512	2.000	58	3,364	7.616	195,112	3.871
9	81	3.000	729	2.080	59	3,481	7.681	205,379	3.893
10	100	3.162	1,000	2.154	60	3,600	7.746	216,000	3.915
11	121	3.317	1,331	2.224	61	3,721	7.810	226,981	3.936
12	144	3.464	1,728	2.289	62	3,844	7.874	238,328	3.958
13	169	3.606	2,197	2.351	63	3,969	7.937	250,047	3.979
14	196	3.742	2,744	2.410	64	4,096	8.000	262,144	4.000
15	225	3.873	3,375	2.466	65	4,225	8.062	274,625	4.021
16	256	4.000	4,096	2.520	66	4,356	8.124	287,496	4.041
17	289	4.123	4,913	2.571	67	4,489	8.185	300,763	4.062
18	324	4.243	5,832	2.621	68	4,624	8.246	314,432	4.082
19	361	4.359	6,859	2.668	69	4,761	8.307	328,509	4.102
20	400	4.472	8,000	2.714	70	4,900	8.367	343,000	4.121
21	441	4.583	9,261	2.759	71	5,041	8.426	357,911	4.141
22	484	4.690	10,648	2.802	72	5,184	8.485	373,248	4.160
23	529	4.796	12,167	2.844	73	5,329	8.544	389,017	4.179
24	576	4.899	13,824	2.884	74	5,476	8.602	405,224	4.198
25	625	5.000	15,625	2.924	75	5,625	8.660	421,875	4.217
26	676	5.099	17,576	2.962	76	5,776	8.718	438,976	4.236
27	729	5.196	19,683	3.000	77	5,929	8.775	456,533	4.254
28	784	5.292	21,952	3.037	78	6,084	8.832	474,552	4.273
29	841	5.385	24,389	3.072	79	6,241	8.888	493,039	4.291
30	900	5.477	27,000	3.107	80	6,400	8.944	512,000	4.309
31	961	5.568	29,791	3.141	81	6,561	9.000	531,441	4.327
32	1,024	5.657	32,768	3.175	82	6,724	9.055	551,368	4.344
33	1,089	5.745	35,937	3.208	83	6,889	9.110	571,787	4.362
34	1,156	5.831	39,304	3.240	84	7,056	9.165	592,704	4.380
35	1,225	5.916	42,875	3.271	85	7,225	9.220	614,125	4.397
36	1,296	6.000	46,656	3.302	86	7,396	9.274	636,056	4.414
37	1,369	6.083	50,653	3.332	87	7,569	9.327	658,503	4.431
38	1,444	6.164	54,872	3.362	88	7,744	9.381	681,472	4.448
39	1,521	6.245	59,319	3.391	89	7,921	9.434	704,969	4.465
40	1,600	6.325	64,000	3.420	90	8,100	9.487	729,000	4.481
41	1,681	6.403	68,921	3.448	91	8,281	9.539	753,571	4.498
42	1,764	6.481	74,088	3.476	92	8,464	9.592	778,688	4.514
43	1,849	6.557	79,507	3.503	93	8,649	9.644	804,357	4.531
44	1,936	6.633	85,184	3.530	94	8,836	9.695	830,584	4.547
45	2,025	6.708	91,125	3.557	95	9,025	9.747	857,375	4.563
46	2,116	6.782	97,336	3.583	96	9,216	9.798	884,736	4.579
47	2,209	6.856	103,823	3.609	97	9,409	9.849	912,673	4.595
48	2,304	6.928	110,592	3.634	98	9,604	9.899	941,192	4.610
49	2,401	7.000	117,649	3.659	99	9,801	9.950	970,299	4.626
50	2,500	7.071	125,000	3.684	100	10,000	10.000	1,000,000	4.642

TABLE V

American Experience Table of Mortality

Age	Number living	Number dying	Age	Number living	Number dying	Age	Number living	Number dying
10	100,000	749	40	78,106	765	70	38,569	2,391
11	99,251	746	41	77,341	774	71	36,178	2,448
12	98,505	743	42	76,567	785	72	33,730	2,487
13	97,762	740	43	75,782	797	73	31,243	2,505
14	97,022	737	44	74,985	812	74	28,738	2,501
15	96,285	735	45	74,173	828	75	26,237	2,476
16	95,550	732	46	73,345	848	76	23,761	2,431
17	94,818	729	47	72,497	870	77	21,330	2,369
18	94,089	727	48	71,627	896	78	18,961	2,291
19	93,362	725	49	70,731	927	79	16,670	2,196
20	92,637	723	50	69,804	962	80	14,474	2,091
21	91,914	722	51	68,842	1,001	81	12,383	1,964
22	91,192	721	52	67,841	1,044	82	10,419	1,816
23	90,471	720	53	66,797	1,091	83	8,603	1,648
24	89,751	719	54	65,706	1,143	84	6,955	1,470
25	89,032	718	55	64,563	1,199	85	5,485	1,292
26	88,314	718	56	63,364	1,260	86	4,193	1,114
27	87,596	718	57	62,104	1,325	87	3,079	933
28	86,878	718	58	60,779	1,394	88	2,146	744
29	86,160	719	59	59,385	1,468	89	1,402	555
30	85,441	720	60	57,917	1,546	90	847	385
31	84,721	721	61	56,371	1,628	91	462	246
32	84,000	723	62	54,743	1,713	92	216	137
33	83,277	726	63	53,030	1,800	93	79	58
34	82,551	729	64	51,230	1,889	94	21	18
35	81,822	732	65	49,341	1,980	95	3	3
36	81,090	737	66	47,361	2,070			
37	80,353	742	67	45,291	2,158			
38	79,611	749	68	43,133	2,243			
39	78,862	756	69	40,890	2,321			

INDEX

Abscissa, 45
Absolute inequality, 113, 262
Absolute value, 42
 of complex numbers, 275
Addition formulas, 88
Addition,
 of fractions, 26
 of ordinates, 189
 of radicals, 37
 of two sine functions of different frequencies, 196
Algebraic expressions, 6
Ambiguous case of triangle, 241
Amplitude, 190
 of complex numbers, 275
Angle, 53
 complementary, 87
 of depression, 237
 of elevation, 237
 negative, 53
 positive, 53
 quadrantal, 74
Angular velocity, 191
Approximations of circular function values for small angles, 264
Arc,
 functions, 177
 length, 50
Area,
 of a circle, 60
 of a triangle, 145, 247
Argand diagram, 274
Argument of complex numbers, 275
Arithmetic mean, 286
Arithmetic progressions, 282
Associative law,
 for addition, 4
 for multiplication, 4
Atmospheric waves, 196

Bearing, 238
Beats, 195

Binomial, 7
 coefficients, 204
 theorem, 203
Braces, 4
Brackets, 4

Cancellation, 25
Characteristic of logarithm, 219
Chord, 78
Circle, 44, 50
 unit, 73
Circular functions, 67
 inverse, 177
Circumference of circle, 52
Coefficient, 6
 binomial, 204
Cofunction, 87
Columns, 141
Combinations, 301
Common difference, 282
Common logarithms, 217
Commutative law,
 for addition, 4
 for multiplication, 4
Complementary angle, 87
Completing the square, 110
Complex numbers, 3, 268
Complex plane, 274
Complex roots of an equation, 273
Components of vectors, 236
Composite numbers, 19
Composition of ordinates, 189, 197
Compound interest, 223
Computation by use of logarithms, 221
Conditional equation, 3
Conditional inequalities, 112
Conjugate complex numbers, 271
Constant, 60
Continuous curve, 197
Converse of Factor Theorem, 161

Conversion of sums and products, 260
Coordinate, 41
 axes, 45
 one-dimensional system, 41
 three-dimensional system, 135
 two-dimensional system, 44
Cosecant function, 71
Cosine function, 68
Cotangent function, 71
Cramer's rule, 157
Curve, 62
Cycle, 190

Decomposition into prime factors, 19
Defective equation, 101
Degree, 12, 55
 of a polynomial, 12
 of a term, 12
De Moivre's Theorem, 278
Depressed equation, 169
Depression, angle of, 237
Determinants, 141
 nth order, 141
 second order, 141
 third order, 141
Difference of ordered pairs, 269
Direct variation, 126
Discriminant, 119
Distance between two points,
 in one dimension, 43
 in two dimensions, 47
Distributive law, 4
Division,
 of algebraic expressions, 11
 of fractions, 28
 of radicals, 37
 synthetic, 15, 162
Domain, 60
Double-angle identities, 254

e (the base for natural logarithms), 2
Earthquakes, 199
Electric current, 192
Electromotive force, 192, 196
Elevation, angle of, 237

Empirical probability, 306
Equality, 3
 of ordered pairs, 268
Equation,
 conditional, 3
 of a curve, 50
 depressed, 169
 exponential, 229
 identity, 3
 involving radicals, 123
 logarithmic, 229
 in quadratic form, 122
 of a straight line, 145
Equivalent equation, 100
Even function, 62
Expansion of a determinant, 152
Explicit functions, 174
Exponent, 9
 integral, 31
 negative, 31
 rational, 31
 zero, 31
Exponential functions, 211

Factorial $n!$, 148
Factoring, 19
Factor, 4, 19
Factor Theorem, 161
 Converse of, 161
First Law of Tangents, 243
Formula for cos $(\alpha-\beta)$, 85
Fourier's Theorem, 198
Frequency, 190
Function, 59
 arc, 177
 circular, 67
 cosecant, 71
 cosine, 68
 cotangent, 71
 explicit, 174
 exponential, 211
 implicit, 174
 increasing, 75
 inverse, 174
 logarithmic, 213
 periodic, 71
 reciprocal, 72

INDEX 351

Function, (cont'd)
  secant, 71
  sine, 68
  single-valued, 60
  tangent, 68
  trigonometric, 67
Functional values of any angle, 95
Functional values of 3°, 96
Fundamental identities, 249
Fundamental Theorem for algebra, 163

Geometric mean, 290
Geometric progression, 287
Golden Section, 82
Graph,
  of circular functions, 185
  of a function, 62
  of a quadric function, 104
  of $y = a \sin kx$, 185
  of $y = a \sin (kx + b)$, 187
  of $y = \arccos x$, 178
  of $y = \arcsin x$, 178
  of $y = \arctan x$, 179
  of $y = \cos \theta$, 76
  of $y = \sin \theta$, 74
  of $y = \tan \theta$, 79
Graphical representation,
  of complex numbers, 274
  of empirical data, 64
  of a function, 62
  of polynomial functions, 165
Graphical solution of one linear and one quadratic equation, 176

Half-angle identities, 254
Harmonic analysis, 197
Harmonic synthesis, 197

Identity, 3
Identities, 249
  double-angle, 254
  general addition, 253
  half-angle, 254
Imaginary numbers, 270
Imaginary roots, 166
Implicit function, 174

Increasing function, 75
Index, 33
Induction, 200
Inductive property, 200
Inequalities, 112, 262, 42
  absolute, 113
  conditional, 112
  limiting, 265
Infinite geometric progression, 290
Initial side of an angle, 54
Integers,
  negative, 1
  positive, 1
  set of, 1
Integral rational expression, 12
Integral rational term, 12
Interpolation,
  for angles in radian measure, 56
  for circular function values, 97
Inverse circular functions, 177
Inverse functions, 174
Inverse variation, 126
Inversions, 147
Irrational numbers, 2
Irrational roots, 171

Law,
  of cosines, 240
  of natural growth, 224
  of signs, 5
  of sines, 231
  of tangents (first), 243
  of tangents (second), 244
Laws of algebra, 4
Least common denominator, 26
Length,
  of circular arc, 57
  of circular chord, 78
Limit of $\dfrac{\sin \theta}{\theta}$, 265
Limiting inequalities, 265
Linear equation in three variables, 134
Linear function, 100, 129
Linear interpolation, 97, 172, 217
Line values of circular functions, 79
Literal equation, 102

Logarithm, 213
  common, 217
  natural, 228
Logarithmic functions, 213
Lower limit for zeros, 162

Mantissa, 219
Mathematical induction, 200
Matrix, 141
Maximum, 106
Mean,
  arithmetic, 286
  geometric, 290
Measurement of an angle, 54
Mid-point, 44
Minimum, 106
Minor, 141, 152
Minutes, 55
Modulus of complex number, 275
Monomial, 7
Multinomial, 7
Multiplication,
  of algebraic expression, 9
  of fractions, 28
  of radicals, 37

Natural growth, 224
Natural logarithms, 228
Nature of roots, 119
nth root of complex number, 280
Numbers, 1

Oblique triangles, 240
Odd function, 62
One-dimensional coordinate system, 41
Ordered pairs, 268
Ordinate, 46
Origin, 41, 45
Oscillations, 195

Parabola, 106
Parentheses, 4
Pentagon, 82
Period, 71
Periodic decimal, 2
Periodic functions, 71, 93

Periodic piecewise continuous curve, 197
Permissible values, 3
Permutation, 297
Phase,
  difference, 190
  displacement, 190
Pi, 2
Polynomial, 12, 160
Power, 9, 193
  of electrical current, 193
Powers of complex numbers, 278
Prime numbers, 19
Principal nth roots, 33
Principal values, 180
Principle of superposition, 197
Probability, 303
  of dependent events, 310
  of independent events, 309
  of more than one event, 308
Product,
  of ordered pairs, 269
  of zeros of quadratic function, 118
Progressions, 282
  arithmetic, 282
  geometric, 287
Properties,
  of determinants, 148
  of logarithms, 214
Proportion, 125
Protractor, 95
Pure imaginary numbers, 270
Pythagorean Theorem, 240

Quadrantal angles, 74
Quadrants, 46, 74
Quadratic formula, 110
Quadratic function, 104

Radian, 55
Radian system, 55
Radical, 33, 34
Radical equation, 123
Radicand, 33
Range, 60
Rational integral function, 160
Rationalizing the denominator, 39

INDEX

Rational numbers, 1
Rational roots, 168
Real numbers, 2
Reciprocal functions, 72
Reduction formulas,
　general, 92
　special, 86
Redundant equation, 101
Relation between zeros and coefficients of the quadratic function, 117
Relatively prime numbers, 19
Remainder Theorem, 160
Resultant, 236, 248
Revolutions, 54
Roots,
　of an equation, 100
　of complex numbers, 278
Rows, 141

Secant function, 71
Second, 55
Second Law of Tangents, 244
Seismology, 199
Sexagesimal system, 55
Significant digits, 233
Simple harmonic motion, 190
Simple sound, 193
Sine function, 68
Single-valued function, 60
Solution,
　of an equation, 100
　of inequalities, 114
　of oblique triangles, 240
　of one linear and one quadratic equation (algebraically), 138
　of one linear and one quadratic equation (graphically), 139
　of quadratic equation, 107
　　by completing the square, 110
　　by quadratic formula, 109
　　factoring method, 107
　of right triangles, 232
　of systems of linear equations, 157
　of three linear equations, 134
　of triangles, 231
　of two linear equations, 129

Solution, (cont'd)
　algebraically, 132
　graphically, 130
Sound waves, 198
Sound wave (travelling), 193
Standard position,
　of an angle, 54
　of decimal, 219
Standing waves, 197
Straight line, 129
Subtraction,
　of radicals, 37
Subtrahend, 5
Sum,
　of arithmetic progression, 284
　of geometric progression, 287
　of ordered pairs, 268
　of vectors, 236
　of zeros of quadratic function, 117
Synthetic division, 15, 162

Tangent function, 68
Term, 6
Terminal side of an angle, 54
Three-dimensional coordinate system, 135
Tide Tables, 199
Transverse wave (travelling), 193
Trigonometry, 67
Two-dimensional coordinate system, 44

Unit circle, 73
Upper limit for zeros, 162
Use of Table 1, 96

Value,
　of functions of multiples of $30°$, 80
　of functions of multiples of $45°$, 80
　of functions of $\pi/5$, 82
　of functions of $3°$, 96
Values of special angle functions, 79
Variable,
　dependent, 59
　independent, 59
Variation, 125
　direct, 126

Variation, (cont'd)
  inverse, 126
  joint, 126
Vectors, 235
Vertex of parabola, 106

Waves,
  atmospheric, 196

Waves, (cont'd)
  sound, 193
  standing, 197
  transverse, 193

Zero,
  of a function, 63, 100
  of polynomial function, 162